FINITE OR INFINITE DIMENSIONAL COMPLEX ANALYSIS AND APPLICATIONS

Advances in Complex Analysis and Its Applications

Volume 2

FINITE OR INFINITE DIMENSIONAL COMPLEX ANALYSIS AND APPLICATIONS

Edited by
LE HUNG SON
Hanoi University of Technology
Faculty of Applied Mathematics
Hanoi, Vietnam

W. TUTSCHKE
Graz University of Technology
Department of Mathematics D
Graz, Austria

C.C. YANG
Hong Kong University of Sciences and Technology
Department of Mathematics
Hong Kong, China

Kluwer Academic Publishers
Boston/Dordrecht/London

Distributors for North, Central and South America:
Kluwer Academic Publishers
101 Philip Drive
Assinippi Park
Norwell, Massachusetts 02061 USA
Telephone (781) 871-6600
Fax (781) 871-6528
E-Mail <kluwer@wkap.com>

Distributors for all other countries:
Kluwer Academic Publishers Group
Post Office Box 322
3300 AH Dordrecht, THE NETHERLANDS
Telephone 31 78 6576 000
Fax 31 78 6576 474
E-Mail <orderdept@wkap.nl>

Electronic Services <http://www.wkap.nl>

Library of Congress Cataloging-in-Publication

Son, Le Hung/ Tutschke, W./ Yang, C.C.
Finite or Infinite Dimensional Complex Analysis and Applications
ISBN-13: 978-1-4613-7948-5 e-ISBN-13: 978-1-4613-0221-6
DOI: 10.1007/978-1-4613-0221-6

Softcover reprint of the hardcover 1st edition 2004

Printed on acid-free paper.

Contents

Preface ix

I. PLENARY TALKS AND SURVEYS

1

Nevanlinna theory and Diophantine approximations 1
Pei-Chu Hu , Chung-Chun Yang

2

In Ramanujan's Shadow 21
Irwin Kra

3

Holomorphic functions on the Lie ball and related topics 35
Mitsuo Morimoto, Keiko Fujita

4

Invariance of Plurigenera and Torsion-Freeness of Direct Image Sheaves of Pluricanonial Bundles 45
Yum-Tong Siu

5

Maxwell's equations in Clifford calculus framework – an overview on the development 85
Wolfgang Sprößig

6

Generalized Analytic Functions and their contributions to the development of Mathematical Analysis 101
W. Tutschke

II. COMPLEX METHODS IN THE PLANE

7

On q_p sequences 117
Rauno Aulaskari, Shamil Makhmutov, Hasi Wulan

8

Reconstruction of an Analytic Function from a Sequence of Values: Existence and Regularization 127
Dang Dinh Ang, Vy Khoi Le, Dang Duc Trong

9

p-adic Interpolation and Applications 143
Ha Huy Khoai, Le Thi Hoai Thu

10

Refined Order and the Paley Problem in Vector Spaces 153
Chuan-Gan Hu, Chao Liu

11

Integral representation of p-adic functions 169
Le Thi Hoai Thu

12

Malmquist-Yosida type theorems for algebraic differential equations 181
Liangwen Liao, Chung-Chun Yang

13

Transformation techniques for partial differential equations 193
Jürgen Püngel

14

Besov-type Characterizations for Quaternionic Bloch Functions 207
L. F. Reséndis O., L. M. Tovar S.

15

Bergman and Bauer Operators for Elliptic Equations in two Independent Variables 221
K.W. Tomantschger

16

Some results on Q_K-type spaces 245
Hasi Wulan

III. ANALYSIS IN HIGHER DIMENSIONS

17

Representations in Polydomains 255
H. Begehr

18

On B^q spaces of hyperholomorphic functions and the Bloch space in R^3 269
K. Gürlebeck, A. El-Sayed Ahmed

19

Regeneration in complex, quaternion and Clifford analysis 287
Joji Kajiwara, Xiao Dong Li, Kwang Ho Shon

20

A note on sesquilinear forms and the generalized uncertainty relations in *-algebras 299
Ken Kuriyama, Shigeru Furuichi, Kenjiro Yanagi

21

The Hartogs Extension Theorem for the Multi-regular functions Taking Values in a Clifford algebra 309
Le Hung Son , Nguyen Thanh Van

22

The extension of entire functions of nuclear type on locally convex spaces 323
Masaru Nishihara

23

Some Properties of the Inhomogeneous Generalized Cauchy-Riemann Systems in Quaternionic Algebra and its Applications 331
Nguyen Thanh Van

24

The Fundamental Solution of the Hyperbolic Dirac-Equation 345
D. Eelbode, F. Sommen

25

Inhomogeneous ordinary differential equations and local cohomologies and residues 361
S. Tajima

26

Infinite dimensional CR category 371
Shigeru Takeuchi

Preface

There is almost no field in Mathematics which does not use Mathematical Analysis. Computer methods in Applied Mathematics, too, are often based on statements and procedures of Mathematical Analysis.

An important part of Mathematical Analysis is Complex Analysis because it has many applications in various branches of Mathematics.

Since the field of Complex Analysis and its applications is a focal point in the Vietnamese research programme, the Hanoi University of Technology organized an International Conference on

Finite or Infinite Dimensional Complex Analysis and Applications

which took place in Hanoi from August 8 - 12, 2001. This conference was the 9th one in a series of conferences which take place alternately in China, Japan, Korea and Vietnam each year. The first one took place at Pusan University in Korea in 1993. The preceding 8th conference was held in Shandong in China in August 2000. The 9th conference of the above mentioned series of conferences was the first one which took place in Vietnam.

Present trends in Complex Analysis reflected in the present volume are mainly concentrated in the following four research directions:

1 Value distribution theory (including meromorphic funtions, meromorphic mappings, as well as p-adic functions over fields of finite or zero characteristic) and its applications,

2 Holomorphic functions in several (finitely or infinitely many) complex variables,

3 Clifford Analysis, i.e., complex methods in higher-dimensional real Euclidian spaces,

4 Generalized analytic functions.

Nevanlinna's value distribution theory (of one complex variable) is a generalization of the theory of entire and meromorphic functions, which enables one to give a quantitative estimate of a value that is taken by a certain function. Since then the theory itself has been generalized to holomorphic or meromorphic mappings as well as to p-adic functions over fields of finite or zero characteristic. Recently the theory has been used by many contemporary analysts in the studies of factorization and unicity theory of complex or p-adic meromorphic functions (mappings), complex dynamics (in one or higher dimensional spaces), and the growth estimate of meromorphic solutions of differential equations. More recently, the theory has been utilized to desire analogs of the *abc*-conjecture (in number theory) and to obtain some results of Hilbert's Tenth Problem type for (complex or p-adic) meromorphic functions in one and several complex variables.

Holomorphic functions in several complex variables are functions depending on several complex variables which are holomorphic in each separate complex variable. Such functions could also be denoted as multi-holomorphic functions. One of the goals of this theory is to reveal both common and different properties of holomorphic functions in one and in several complex variables. The branch of holomorphic functions in infinitely many variables, developing rapidly at present, is aimed at creating a coordinate-free version of the theory of holomorphic functions in several complex variables.

Classical Complex Analysis dealt mainly with holomorphic functions, and, therefore, its interactions to Partial Differential Equations were restricted to the Cauchy-Riemann system and the Laplace equation. Present trends in Complex Analysis try to apply complex methods to more general partial differential equations in the plane and in higher dimensions. This can be done in the frameworks of Clifford Analysis and the theory of generalized analytic functions.

Clifford Analysis resulted from generalizing the Cauchy-Riemann operator in the plane to higher-dimensional real Euclidian spaces. Denoting this operator by D, a function $u = u(x)$ where x runs in $\mathbf{R}^{n+1}$ is called (left-)monogenic if $Du = 0$. E.g., at present Clifford Analysis investigates multi-monogenic functions. These are functions depending on several variables $x, y, \ldots$ in a monogenic matter where each of the variables $x, y, \ldots$ runs in a respective real Euclidian space. Obviously, methods of multi-holomorphic functions can be applied to multi-monogenic functions. Therefore it makes sense to include both Clifford Analysis and the theory of multi-holomorphic functions in one conference. So it happened in the Hanoi Conference whose Proceedings are contained in the present volume.

The theory of generalized analytic functions, finally, replaces the Cauchy-Riemann system by more general uniformly elliptic linear or nonlinear systems in the plane. In the beginning only linear systems were under consideration. Uniformly elliptic linear systems for two desired real valued functions in the plane can be reduced to complex equations of the form $\partial w/\partial \bar{z} = a(z)w + b(z)\bar{w}$. Later on fully non-linear systems were reduced to equations of form $\partial w/\partial \bar{z} = f(z, w, \partial w/\partial z)$ which also can be solved with methods of Complex Analysis. Further results can be expected by combining the ideas of the theory of generalized analytic functions with methods of Clifford Analysis. This means, for instance, to solve equations of the form $Du = f(x, u)$ where D is again the Cauchy-Riemann operator of Clifford Analysis and x runs in a domain in a Euclidian space.

The editors of the present volume would like to thank the Hanoi University of Technology and its Faculty of Applied Mathematics, the Hanoi Institute of Mathematics, and the Hanoi University of Natural Sciences. Without the support of those institutions, it would not have been possible to organize the conference and to prepare the present Proceedings as one of its results. The chairman of the conference was Prof. Le Hung Son from the Faculty of Applied Mathematics of the Hanoi University of Technology. He was endowed also with grants from

- the Vietnamese National Research Programme on Natural Sciences,
- the International Mathematical Union,
- and the Vietnamese National Research Programme on Complex Analysis and its Applications.

Many referees helped to choose suitable contributions for the present book, and their comments improved the papers submitted for publication. The editors thank all of the referees. The editors thank also Dr. Nguyen Huu Dien for his cooperation by the preparation of the camera-ready manuscript.

The editors.

I. PLENARY TALKS AND SURVEYS

1

NEVANLINNA THEORY AND DIOPHANTINE APPROXIMATIONS

Pei-Chu Hu*

Department of Mathematics
Shandong University
Jinan250100, Shandong, China
E-mail: pchu@sdu.edu.cn

Chung-Chun Yang
Department of Mathematics
The Hong Kong University of Science and Technology
Clear Water Bay, Kowloon, Hong Kong
E-mail: mayang@ust.hk

Abstract: In this note, we will introduce some basic problems and progresses in Nevanlinna theory and Diophantine approximations, say, discuss the *abc*-conjecture and Hall's conjecture for integers, and prove their analogue for polynomials or entire functions by dint of Nevanlinna's value distribution theory.

Keywords: Nevanlinna theory, Mason theorem, *abc*-conjecture, integer, polynomial, meromorphic function

AMS-classification: Primary 32A22, 32H30. Secondary 11A41, 11A51

*The work of first author was partially supported by Natural Science Foundation of China and second author was partially supported by a UGC grant of Hong Kong, project no. HKUST6180/99p.

1. Lang's conjecture

We list three main open problems and introduce their progresses. The standard terminologies and notations in value distribution theory will be used. We start with the classic Picard theorem which states that a nonconstant meromorphic function f on $\mathbb{C}$ omits at most two distinct values. Further refinement of the Picard theorem is the defect relation

$$\sum_{a\in\mathbb{P}^1} \delta_f(a) \leq 2$$

due to Nevanlinna. In fact, Nevanlinna proved a more stronger result:

Theorem 1. *Let f be a nonconstant meromorphic function on $\mathbb{C}$ and let*

$$\mathcal{A} = \{a_0, a_1, ..., a_q\}$$

be a family of distinct points $a_j \in \mathbb{P}^1$. Then

$$\begin{aligned}(q+1-2)T(r,f) &\leq \sum_{j=0}^{q} N\left(r, \frac{1}{f-a_j}\right) - N_{\mathrm{Ram}}(r,f)\\ &\quad + \log\left\{\left(\frac{\rho}{r}\right)^{2m-1} \frac{T(R,f)}{\rho - r}\right\} + O(1)\end{aligned}$$

holds for any $r_0 < r < \rho < R$, and hence for any $\varepsilon > 0$,

$$\begin{aligned}(q+1-2)T(r,f) \leq \sum_{j=0}^{q} N\left(r, \frac{1}{f-a_j}\right) - N_{\mathrm{Ram}}(r,f)\\ + \{\log T(r,f) + (1+\varepsilon)\log\log T(r,f) - \log r\} + O(1)\end{aligned}$$

holds for all large r outside a set E with $\int_E d\log r < \infty$. In particular, if f is of finite order λ, then for any $\varepsilon > 0$, one has

$$\begin{aligned}(q+1-2)T(r,f) &\leq \sum_{j=0}^{q} N\left(r, \frac{1}{f-a_j}\right) - N_{\mathrm{Ram}}(r,f)\\ &\quad + (\lambda + \varepsilon - 1)\log r + O(1).\end{aligned}$$

Generally, P. Griffiths and J. King [9] proved the following result:

Theorem 2. *Let M be a connected complex manifold of dimension m. Let $\pi : M \longrightarrow \mathbb{C}^m$ be a proper surjective holomorphic mapping. Assume that N is a compact complex manifold of dimension n. Let L be a positive line bundle over N. Let K be the canonical divisor of N. Take $0 \neq s_j \in \Gamma(N,L), j = 1, \cdots, q$, such that the divisor $D = D_1 + \cdots + D_q$ has normal* *crossings in N, where $D_j = (s_j)$. Let $f : M \longrightarrow N$ be a meromorphic mapping of rank n. Then*

$$\| \quad qT_f(r,L) + T_f(r,[K]) \leq \sum_{j=1}^{q} N_f(r,D_j) + N_\beta(r) + O(\log^+ T_f(r,L)) + o(\log r). \qquad (1)$$

In Theorem 2, the condition $\operatorname{rank}(f) = n \leq m$ is crucial. One of major unsolved open problems in value distribution is the following Griffiths' conjecture:

Conjecture 1 (cf. [8], [30]). *Let M be a connected complex manifold of dimension m. Let $\pi : M \longrightarrow \mathbb{C}^m$ be a proper surjective holomorphic mapping. Assume that N is a compact complex manifold of dimension n. Let L be a positive line bundle over N. Let K be the canonical divisor of N. Take $0 \neq s_j \in \Gamma(N, L), j = 1, \cdots, q$, such that the divisor $D = D_1 + \cdots + D_q$ has normal crossings in N, where $D_j = (s_j)$. Let $f : M \longrightarrow N$ be a meromorphic mapping which is algebraically non-degenerate. Then*

$$\| \quad qT_f(r,L) + T_f(r,[K]) \leq \sum_{j=1}^{q} N_f(r,D_j) + N_\beta(r) + O(\log^+ T_f(r,L)) + o(\log r). \qquad (2)$$

If $N = \mathbb{P}^n$, L is the hyperplane section bundle over $\mathbb{P}^n$, Conjecture 1 is the classic second main theorems due to Nevanlinna ($M = \mathbb{C}, n = 1$), Cartan ($M = \mathbb{C}$), Weyl-Ahfors ($m = 1$) and Stoll, respectively. Under an additional condition, Hu [10] confirmed the conjecture, that is, for the case $n > m$, if there exists a holomorphic field φ on f over M of degree $n - m$ which is effective for f such that for some positive number c, α, β and for any $\varepsilon > 0$,

$$\| \quad \int_{r_0}^{r} \left(t^{2-2m} \int_{\|\pi\| \leq t} |\varphi|^{2/\alpha} (dd^c \|\pi\|^2)^m \right) \frac{dt}{t} \leq c \left(T_f(r,L)^\beta + r^\varepsilon \right),$$

then Conjecture 1 holds.

Conjecture 2 (cf. [20]). *Let M be a connected complex manifold of dimension m. Let $\pi : M \longrightarrow \mathbb{C}^m$ be a proper surjective holomorphic mapping. Let D be a divisor with normal crossings on an algebraic manifold N. Let K be the canonical divisor of N. There exists a proper algebraic subset Z_D of N having the following property. Let $f : M \longrightarrow N$*

be a meromorphic mapping such that $f(M) \not\subset Z_D$. Let E be an ample divisor. Then

$$\| \quad T_f(r,[D]) + T_f(r,[K]) \leq N_f(r,D) + N_\beta(r) + O(\log^+ T_f(r,[E])) + o(\log r). \qquad (3)$$

Obviously, Conjecture 2 implies Conjecture 1. A special case of Conjecture 2 is confirmed by Siu and Yeung [33] as follows:

Theorem 3. *Let N be an abelian variety and D be an ample divisor in N. Let $f : \mathbb{C} \longrightarrow N$ be a nonconstant holomorphic mapping. Then the image of f must intersect D.*

The above theorem has been open for a quite long time, named as Lang's conjecture (cf. [21]). The proof uses essentially only the techniques devised by Bloch [2] for the original proof of Bloch's theorem (that is, the Zariski closure of the image of a holomorphic map from $\mathbb{C}$ to an abelian variety is precisely the translate of an abelian subvariety), which explicitly constructs a suitable higher-order jet differential in terms of theta functions.

Conjecture 3. *Let $\mathcal{G}$ be a finite admissible family of meromorphic mappings from $\mathbb{C}^m$ into $\mathbb{P}(\amalg_p V^*)$, where V is a complex vector space of dimension $n+1$. Let $f : \mathbb{C}^m \longrightarrow \mathbb{P}(V)$ be an algebraically non-degenerate meromorphic mapping. Assume that for all $g \in \mathcal{G}$, g grows slower than f and that (f^p, g) is free for the interior product $\angle$. Then*

$$\| \quad \sum_{g\in\mathcal{G}} \{pT_f(r) - N_{f^p}(r,g)\} \leq \{n+1+o(1)\}pT_f(r). \qquad (4)$$

Shiffman [31] made the above conjecture for a family $\mathcal{G}$ of constants. For some discussion related to Conjecture 3, we refer the reader to [11] and [13].

Mason (see [22],[23],[24]) and Stothers [34] started one recent trend of thoughts by discovering an entirely new relation among polynomials as follows. Let $f(z)$ be a polynomial with coefficients in an algebraically closed field κ of characteristic 0 and let $r(f)$ be the number of distinct zeros of f.

Theorem 4. *Let $a(z)$, $b(z)$, $c(z)$ be relatively prime polynomials in κ and not all constants such that $a+b=c$. Then*

$$\max\{\deg(a), \deg(b), \deg(c)\} \leq r(abc) - 1.$$

Proof. Generally, if f is a meromorphic function of order ρ ($< +\infty$) in $\mathbb{C}$, then for any $\varepsilon > 0$, the classic second main theorem of Nevanlinna implies

$$T(r,f) \leq \overline{N}(r,f) + \overline{N}\left(r,\frac{1}{f}\right) + \overline{N}\left(r,\frac{1}{f-1}\right) + (\rho+\varepsilon-1)\log r + O(1).$$

Next we prove Theorem 4 for the case $\kappa = \mathbb{C}$. Since $a(z)$, $b(z)$ and $c(z)$ have no common zeros and not all constants such that $a + b = c$, then for any $\varepsilon > 0$, applying the above inequality to $f = a/c$ which has order 0 one has

$$\begin{aligned} T\left(r,\frac{a}{c}\right) &\leq \overline{N}\left(r,\frac{1}{a}\right) + \overline{N}\left(r,\frac{1}{b}\right) + \overline{N}\left(r,\frac{1}{c}\right) + (\varepsilon-1)\log r + O(1) \\ &= \overline{N}\left(r,\frac{1}{abc}\right) + (\varepsilon-1)\log r + O(1). \end{aligned}$$

Dividing the inequality by $\log r$ and letting $r \to +\infty$, we have

$$\max\{\deg(a), \deg(c)\} \leq r\,(abc) - 1.$$

Similarly, we can obtain

$$\max\{\deg(b), \deg(c)\} \leq r\,(abc) - 1.$$

So the claim is proved. □

2. Non-Archimedean analysis

Let κ be an algebraically closed field of characteristic zero, complete for a non-trivial non-Archimedean absolute value. Hu-Yang [13] translated Conjecture 2 into a problem over non-Archimedean fields:

Conjecture 4. *Let M be a nonsingular projective variety over κ. Let K be the canonical divisor of M, and let D be a simple normal crossings divisor on M. Let E be an almost ample divisor. Let $f : \kappa \longrightarrow M$ be an holomorphic curves. Then there exists a proper algebraic subset Z_D having the following property: when $f(\kappa) \not\subset Z_D$,*

$$T_f(r,[D]) + T_f(r,[K]) \leq N_f(r,D) + o(T_f(r,[E])).$$

The analogue of Conjecture 3 over non-Archimedean fields is confirmed by Hu-Yang [13] and Ru [27] for a family $\mathcal{G}$ of constants.

Theorem 5 (Hu-Yang [13], [12]). *Let $\mathcal{A} = \{a_j\}_{j=0}^{q}$ be a finite admissible family of points $a_j \in \mathbb{P}\left(\amalg_{m_j} V^*\right)$ with $n < q$. Let $f : \kappa \longrightarrow \mathbb{P}(V)$*

be a holomorphic curve which is algebraically non-degenerate. Set

$$d = \max_{\lambda\in J_n^q}\left\{\frac{1}{n+1}\sum_{i=0}^{n} m_{\lambda(i)}\right\}.$$

Then

$$\left(\sum_{j=0}^{q} m_j - d(n+1)\right)T(r,f) \le \sum_{j=0}^{q} N_{f^{\amalg m_j}}(r,a_j) - N_{\mathrm{Ram}}(r,f) - \frac{n(n+1)}{2}\log r + O(1),$$

where

$$N_{\mathrm{Ram}}(r,f) = \min_{\lambda\in J_n^q} N\left(r,\frac{1}{\mathbf{W}_\lambda}\right)$$

in which $\mathbf{W}_\lambda$ *is the Wronskian of* $\tilde{\alpha}_{\lambda(0)}\circ\tilde{f},\ldots,\tilde{\alpha}_{\lambda(n)}\circ\tilde{f}$.

If the ramification term is not involved in the second main theorem, Min Ru [27] observed that the first main theorem yields directly the following fact:

Theorem 6. *Let* $\mathcal{A} = \{a_j\}_{j=0}^q$ *be a finite admissible family of points* $a_j \in \mathbb{P}(\amalg_m V^*)$ *with* $n < q$. *Let* $f : \kappa \longrightarrow \mathbb{P}(V)$ *be a holomorphic curve satisfying* $f(\kappa) \not\subset \ddot{E}[a_j]$ *for* $j = 0,\ldots,q$. *Then*

$$(q+1-n)\,mT(r,f) \le \sum_{j=0}^{q} N_{f^{\amalg m}}(r,a_j) + O(1).$$

In fact, under the conditions of Theorem 6, we have the following more stronger formula (cf. [13]):

$$mT(r,f) = \max_{0\le i\le n} N_{f^{\amalg m}}(r,a_{\lambda(i)}) + O(1), \quad \lambda\in J_n^q. \tag{5}$$

Theorem 6 can then be deduced immediately by (5). W. l. o. g., we may assume that

$$N_{f^{\amalg m}}(r,a_0) \le N_{f^{\amalg m}}(r,a_1) \le \cdots \le N_{f^{\amalg m}}(r,a_q).$$

By using (5), we have

$$mT(r,f) = N_{f^{\amalg m}}(r,a_j) + O(1), \quad j \ge n,$$

which means

$$(q+1-n)mT(r,f) = \sum_{j=n}^{q} N_{f^{\amalg m}}(r,a_j) + O(1) \le \sum_{j=0}^{q} N_{f^{\amalg m}}(r,a_j) + O(1).$$

Over the field of non-Archimedean meromorphic functions, we proved the analogue of Theorem 4 as follows:

Theorem 7 ([14]). *Let $a(z)$, $b(z)$, $c(z)$ be entire functions in κ without common zeros and not all constants such that $a+b=c$. Then*

$$\max\{T(r,a),T(r,b),T(r,c)\} \leq \overline{N}\left(r,\frac{1}{abc}\right) - \log r + O(1).$$

Proof. Write

$$f=\frac{a}{c}, \quad g=\frac{b}{c}.$$

Then f and g all are not constants by our assumptions, and satisfy $f+g=1$. By the second main theorem, and noting that

$$\overline{N}\left(r,\frac{1}{f-1}\right)=\overline{N}\left(r,\frac{1}{g}\right)=\overline{N}\left(r,\frac{1}{b}\right),$$

we obtain

$$\begin{aligned} T(r,f) &\leq \overline{N}(r,f)+\overline{N}\left(r,\frac{1}{f}\right)+\overline{N}\left(r,\frac{1}{f-1}\right)-\log r+O(1) \\ &= \overline{N}\left(r,\frac{1}{c}\right)+\overline{N}\left(r,\frac{1}{a}\right)+\overline{N}\left(r,\frac{1}{b}\right)-\log r+O(1) \\ &= \overline{N}\left(r,\frac{1}{abc}\right)-\log r+O(1). \end{aligned}$$

Similarly, we have

$$T(r,g) \leq \overline{N}\left(r,\frac{1}{abc}\right)-\log r+O(1).$$

By a basic formula in [13], we have

$$\begin{aligned} T(r,f) &= \max\left\{N(r,f),N\left(r,\frac{1}{f}\right)\right\}+O(1) \\ &= \max\left\{N\left(r,\frac{1}{c}\right),N\left(r,\frac{1}{a}\right)\right\}+O(1) \\ &= \max\{T(r,c),T(r,a)\}+O(1). \end{aligned}$$

Similarly,

$$T(r,g)=\max\{T(r,c),T(r,b)\}+O(1),$$

and, hence, the theorem follows from the above estimates. □

The complex version of Theorem 7 was obtained by M. van Frankenhuysen [35]. In [36], Vaserstein considered the *abc*-problem for polynomials in characteristic $p > 0$. Boutabaa and Escassut [3] slightly improved the inequality in Theorem 7, generalized it in any characteristic, and shown application to polynomials. Vu and Doan [39] generalized the inequality in Theorem 7 to non-Archimedean holomorphic functions in several variables.

3. Diophantine approximation

Let κ be an algebraic number field. P. Vojta [37] translated Conjecture 2 into the following form:

Conjecture 5. *Let M be a nonsingular projective variety over κ. Let K be the canonical divisor of M, and let D be a normal crossings divisor on M. If $\varepsilon > 0$ and E is an almost ample divisor, then there exists a proper Zariski closed subset Z such that for all $x \in M(\kappa) - Z$,*

$$m(x, D) + h_K(x) \leq d(x) + \varepsilon h_E(x) + O(1),$$

where $M(\kappa)$ is the set of κ-valued points of M.

Influenced by Theorem 4, and considerations of Szpiro and Frey, Masser and Oesterlé formulated the *abc*-conjecture for integers as follows:

Conjecture 6. *Let a be a non-zero integer. Define the radical of a to be*

$$r(a) = \prod_{p|a} p$$

i.e. the product of the distinct primes dividing a. Given $\varepsilon > 0$, there exists a number $C(\varepsilon)$ having the following property. For any nonzero relatively prime integers a, b, c such that $a + b = c$, we have

$$\max\{|a|, |b|, |c|\} \leq C(\varepsilon) r(abc)^{1+\varepsilon}.$$

This conjecture also is a consequence of the Vojta's Conjecture (see Vojta [37]). Let a be a non-zero integer. Then

$$|a| = p_1^{i_1} \cdots p_n^{i_n}$$

holds for distinct primes $p_1, ..., p_n$. For a positive integer k, define

$$r_k(a) = \prod_{\nu=1}^{n} p_\nu^{\min\{i_\nu, k\}}.$$

Conjecture 7 ([13]). *Let $a_j (j = 0, \cdots, k)$ be nonzero integers such that the greatest common factor of $a_0, ..., a_k$ is 1,*

$$a_1 + \cdots + a_k = a_0, \tag{6}$$

and no proper subsum of (6) is equal to 0. Then for $\varepsilon > 0$, there exists a number $C(k, \varepsilon)$ such that

$$\max_{0 \le j \le k} \{|a_j|\} \le C(k, \varepsilon) \left(\prod_{i=0}^{k} r_{k-1}(a_i) \right)^{1+\varepsilon},$$

$$\max_{0 \le j \le k} \{|a_j|\} \le C(k, \varepsilon) r_{\frac{k(k-1)}{2}} (a_0 \cdots a_k)^{1+\varepsilon}.$$

If $k = 2$, this corresponds to the well known *abc*-conjecture. Conjecture 3 or Theorem 5 was translated the following problem by Hu-Yang:

Conjecture 8 ([13]). *Take $\varepsilon > 0$. Assume that for $\rho \in S$, the family*

$$\mathcal{A}_\rho = \{a_{\rho,0}, ..., a_{\rho,q}\} \subset \mathbb{P}(\amalg_m V^*) \ (q \ge n)$$

is admissible. Then there exists a finite set $b_1, ..., b_s$ of $\mathbb{P}(\amalg_m V^*)$ such that the inequality

$$\frac{1}{[\kappa : \mathbb{Q}]} \sum_{\rho \in S} \sum_{j=0}^{q} \log \frac{1}{\|x^{\amalg m}, a_{\rho,j}\|_\rho} < m(n + 1 + \varepsilon) h(x) + O(1)$$

holds for all $x \in \mathbb{P}(V) - \bigcup_i \ddot{E}[b_i]$.

Conjecture 8 follows from the following more stronger:

Conjecture 9 ([13]). *For $\rho \in S$, $i = 0, ..., n$, take $\alpha_{\rho,i} \in \amalg_m V^* - \{0\}$ such that the system*

$$\langle \xi^{\amalg m}, \alpha_{\rho,i} \rangle = 0, \quad i = 0, ..., n$$

has only the trivial solution $\xi = 0$ in V. Then for any $\varepsilon > 0$ there exists a finite set $b_1, ..., b_s$ of $\mathbb{P}(\amalg_m V^*)$ such that the inequality

$$\prod_{\rho \in S} \prod_{i=0}^{n} \frac{1}{\|\langle \xi^{\amalg m}, \alpha_{\rho,i} \rangle\|_\rho} \le \left\{ \max_{\rho \in S} \|\xi\|_\rho \right\}^\varepsilon$$

holds for all S-integral points $\xi \in \mathcal{O}_{V,S} - \bigcup_i E[b_i]$.

Conjecture 10 ([13]). *Take $\varepsilon > 0$. Assume that the family*

$$\mathcal{A} = \{a_0, ..., a_q\} \subset \mathbb{P}(\amalg_m V^*) \ (q \ge n)$$

is admissible. Then the set of points of $\mathbb{P}(V) - \bigcup_j \ddot{E}[a_j]$ *satisfying*

$$\sum_{j=0}^{q} m_{x^{\amalg m}}(a_j) \geq m(2n+\varepsilon)h(x) + O(1)$$

is finite.

If $m = 1$, Conjecture 9, 8 and 10 are well known (cf. [28]), which mainly correspond to Schmidt's subspace theorems.

4. Some progresses of abc-problem

Let κ be an algebraically closed field of characteristic zero, complete for a non-trivial non-Archimedean absolute value. The analogue of Conjecture 7 for entire functions defined over κ is the following (cf. [16]):

Theorem 8. *Let* $f_j (j = 0, \cdots, k)$ *be entire functions on* κ *such that* $f_0, \ldots, f_k$ *have no common zeros,* f_j $(j = 1, \cdots, k)$ *be linearly independent on* κ *and*

$$f_1 + \cdots + f_k = f_0. \tag{7}$$

Then the Nevanlinna functions $T(r, f_j)$ *and* $N_h(r, f_j)$ *satisfy*

$$\max_{0\leq j\leq k}\{T(r,f_j)\} \leq \sum_{i=0}^{k} N_{k-1}\left(r, \frac{1}{f_i}\right) - \frac{k(k-1)}{2}\log r + O(1)$$

$$\max_{0\leq j\leq k}\{T(r,f_j)\} \leq N_{\frac{k(k-1)}{2}}\left(r, \frac{1}{f_0\cdots f_k}\right) - \frac{k(k-1)}{2}\log r + O(1).$$

If f is a polynomial, it is easy to show

$$\deg(f) = \lim_{r\to\infty}\frac{T(r,f)}{\log r}, \quad r_k(f) := \lim_{r\to\infty} n_k\left(r, \frac{1}{f}\right) = \lim_{r\to\infty}\frac{N_k\left(r,\frac{1}{f}\right)}{\log r}.$$

As a direct consequence of Theorem 8, we obtain

Theorem 9. *Let* $f_j (j = 0, \cdots, k)$ *be polynomials on* κ *such that* $f_0, \ldots, f_k$ *have no common zeros,* f_j $(j = 1, \cdots, k)$ *are linearly independent on* κ *and*

$$f_1 + \cdots + f_k = f_0. \tag{8}$$

Then

$$\max_{0\leq j\leq k}\{\deg(f_j)\} \leq \sum_{i=0}^{k} r_{k-1}(f_i) - \frac{k(k-1)}{2},$$

$$\max_{0\leq j\leq k}\{\deg(f_j)\} \leq r_{\frac{k(k-1)}{2}}(f_0\cdots f_k) - \frac{k(k-1)}{2}.$$

When $k = 2$, Theorem 9 reduces to a Mason's theorem. If $k \geq 2$, the following example

$$f_0(z) = (z+1)^{k-1}, \quad f_{i+1}(z) = \binom{k-1}{i} z^i \; (i = 0, ..., k-1),$$

which obviously satisfy the conditions in Theorem 9, shows that the inequalities in the theorem, in fact, become equality for this example. For any positive integer k and any polynomial f on κ, note that

$$r_k(f) \leq k r_1(f).$$

Theorem 9 yields immediately the following:

Theorem 10. *Let $f_1, f_2, \cdots, f_k$ $(k \geq 2)$ be linearly independent polynomials in κ. Put $f_0 = f_1 + f_2 + \cdots + f_k$ and assume that $f_0, ..., f_k$ have no common zeros. Then the following inequalities*

$$\max_{0 \leq j \leq k} \{\deg(f_j)\} \leq (k-1) \sum_{i=0}^{k} r_1(f_i) - \frac{k(k-1)}{2}, \tag{9}$$

$$\max_{0 \leq j \leq k} \{\deg(f_j)\} \leq \frac{k(k-1)}{2} (r_1(f_0 \cdots f_k) - 1), \tag{10}$$

hold.

The inequality (10) was obtained independently by J. F. Voloch [38], W. D. Brownawell and D. Masser [5]. Earlier R. C. Mason [25] derived this estimate with $\frac{1}{2}k(k-1)$ replaced by 4^{k-1}. J. Browkin and J. Brzeziński [4] conjectured that the value $\frac{1}{2}k(k-1)$ in (10) would be replaced by $2k-3$. If the restriction on the linear independence of polynomials $f_1, ..., f_k$ is removed, we have

Theorem 11. *For fixed integer $k \geq 1$, let f_j $(j = 0, ..., k)$ be non-zero polynomials on κ such that $f_1 + \cdots + f_k = f_0$. Assume also that not all the f_j are constants, and the f_j are pairwise relatively prime. Then*

$$\max_{0 \leq j \leq k} \{\deg(f_j)\} \leq (d-1)(r_1(f_0 \cdots f_k) - 1), \tag{11}$$

where d is the dimension of the vector space spanned by the f_i over κ.

Proof of Theorem 8 is based on the non-Archimedean analogue of a result of Nevanlinna [26]:

Lemma 1 ([13], [17]). *Let $f_j (j = 1, \cdots, k)$ be linearly independent meromorphic functions on κ such that*

$$f_1 + \cdots + f_k = 1. \tag{12}$$

Then

$$\begin{aligned} T(r,f_j) \quad &< \sum_{i=1}^{k} N\left(r,\frac{1}{f_i}\right) - \sum_{i\neq j} N(r,f_i) + N(r,\mathbf{W}) \\ &-N\left(r,\frac{1}{\mathbf{W}}\right) - \frac{k(k-1)}{2}\log r + O(1), \quad 1 \le j \le k \quad (13) \end{aligned}$$

where $\mathbf{W}$ *is the Wronskian of* $f_1, ..., f_k$.

5. Generalized abc-problems for functions

Main results in this section are true for functions of several complex variables (cf. [18]). For simple, we only discuss the functions of one complex variable.

Theorem 12. *Let* $f_1, f_2, \cdots, f_k$ *be linearly independent entire functions in* $\mathbb{C}$. *Put* $f_0 = f_1 + f_2 + \cdots + f_k$ *and assume that* $f_0, ..., f_k$ *have no common zeros. Then for* $j = 1, 2, \cdots, k$, *the following inequalities*

$$T\left(r,\frac{f_j}{f_0}\right) < \sum_{i=0}^{k} N_{k-1}\left(r,\frac{1}{f_i}\right) - N_j(r) + l\log\left\{\frac{\rho T(R)}{r(\rho-r)}\right\} + O(1),$$

$$T\left(r,\frac{f_j}{f_0}\right) < N_l\left(r,\frac{1}{f_0\cdots f_k}\right) - N_j(r) + l\log\left\{\frac{\rho T(R)}{r(\rho-r)}\right\} + O(1),$$

hold for $r_0 < r < \rho < R$, *where* $l = \frac{k(k-1)}{2}$, N_j *is the valence function counting the common zero of* (f_0, f_j), *and*

$$T(r) = \max_{1\le i\le k}\left\{T\left(r,\frac{f_i}{f_0}\right)\right\}.$$

Theorem 13. *Let* $f_1, f_2, \cdots, f_k$ $(k \ge 2)$ *be linearly independent polynomials in* $\mathbb{C}$. *Put* $f_0 = f_1 + f_2 + \cdots + f_k$ *and assume that* $f_0, ..., f_k$ *have no common zeros. Then the following inequalities*

$$\max_{0\le j\le k}\{\deg(f_j)\} \le \sum_{i=0}^{k} r_{k-1}(f_i) - l, \tag{14}$$

$$\max_{0\le j\le k}\{\deg(f_j)\} \le r_l(f_0\cdots f_k) - l, \tag{15}$$

hold, where $l = \frac{k(k-1)}{2}$.

Theorem 9 and Theorem 13 imply directly

Theorem 14. *Given polynomials $f_1, f_2, \cdots, f_k$ $(k \geq 2)$ in an algebraically closed field K of characteristic 0 and positive integers $l_j (1 \leq j \leq k)$ such that*

(a) $f_1^{l_1}, f_2^{l_2}, \cdots, f_k^{l_k}$ *are linearly independent over K;*

(b) $f_0, f_1^{l_1}, f_2^{l_2}, \cdots, f_k^{l_k}$ *have not common zeros, where*

$$f_0 = \sum_{j=1}^{k} f_j^{l_j}.$$

Then the following inequality

$$\left\{1 - \sum_{j=1}^{k} \frac{k-1}{l_j}\right\} \max_{1 \leq j \leq k} \deg\left(f_j^{l_j}\right) \leq r_{k-1}(f_0) - \frac{k(k-1)}{2} \tag{16}$$

holds.

For the case

$$n = 2, \quad l_1 = 2, \quad l_2 = 3, \tag{17}$$

and

$$f_1 = f, \quad f_2 = -g, \tag{18}$$

the inequality (16) yields

$$\frac{1}{2} \deg(g) \leq \deg\left(f^2 - g^3\right) - 1, \tag{19}$$

which was proved by Davenport [12]. In fact, Davenport proved that (19) is true as long as $f^2 - g^3 \neq 0$ (also see [1], [34]).

The analogue of the inequality (19) in number theory is just Hall's conjecture (cf. [7]), namely, for any $\varepsilon > 0$ there exists a constant $C(\varepsilon)$ such that

$$|y|^{\frac{1}{2}} < C(\varepsilon)\left|x^2 + y^3\right|^{1+\varepsilon} \tag{20}$$

hold for integers x, y with $x^2 + y^3 \neq 0$. We conjecture that the analogue of Theorem 14 in number theory would be the following:

Conjecture 11. *Take positive integers $k, l_j (1 \leq j \leq k)$ such that*

$$k \geq 2, \quad \alpha = 1 - \sum_{j=1}^{k} \frac{k-1}{l_j} > 0.$$

Let $a_j (j = 1, \cdots, k)$ be nonzero integers such that $\gcd(a_1, ..., a_k) = 1$. *If there exist $\delta_j \in \{-1, 1\}$ $(j = 1, ..., k)$ such that*

$$\delta_1 a_1^{l_1} + \cdots + \delta_k a_k^{l_k} \neq 0, \tag{21}$$

and no proper subsum of (21) is equal to 0, then for $\varepsilon > 0$, there exists a number $C = C(k, \varepsilon)$ depending on k and ε such that

$$\max_{1 \le j \le k} \{|a_j|^{\alpha l_j}\} \le C r_{k-1} \left(\sum_{j=1}^{k} \delta_j a_j^{l_j} \right)^{1+\varepsilon}. \tag{22}$$

Conjecture 11 is a generalization of Hall's conjecture, and follows easily from Conjecture 7. In particular, for the case (17), Conjecture 11 claims that given $\varepsilon > 0$, there exists a constant $C(\varepsilon)$ such that

$$\max \left\{ |x|^{\frac{1}{3}}, |y|^{\frac{1}{2}} \right\} \le C(\varepsilon) r_1 \left(x^2 + y^3 \right)^{1+\varepsilon} \tag{23}$$

holds for nonzero integers x, y with $x^2 + y^3 \neq 0$, $\gcd(x, y) = 1$. Obviously, the inequality is stronger than (20) under above stronger conditions.

6. Borel theorem and a generalized Fermat conjecture

Take $a \in \mathbb{P}^1$. For a meromorphic function f on $\mathbb{C}^m$, let $\mu_f^a(z)$ be the a-valued multiplicity of f at a point $z \in \mathbb{C}^m$. For a positive integer k, define

$$\mu_{f)k}^a(z) = \begin{cases} \mu_f^a(z) & : \text{if } \mu_f^a(z) \le k, \\ 0 & : \text{otherwise.} \end{cases}$$

The following theorem is a variant of the classical Borel's theorem.

Theorem 15 ([19]). *Let $f_0, ..., f_n$ $(n \ge 2)$ be non-zero entire functions on $\mathbb{C}^m$ such that there are constants $a_i \in \mathbb{C}$ satisfying the following equation:*

$$\sum_{i=0}^{n} a_i f_i = 0.$$

Assume that there exist positive integers d_i $(i = 0, ..., n)$ such that

$$\mu_{f_i)d_i - 1}^0 = 0, \quad i = 0, ..., n.$$

If $1 - \sum_{j=0}^{n} \frac{n-1}{d_j} > 0$, then there exists a partition of indices

$$\{0, 1, ..., n\} = I_0 \cup I_1 \cup \cdots \cup I_k$$

such that $I_\alpha \neq \emptyset$ $(\alpha = 0, 1, ..., k)$, $I_\alpha \cap I_\beta = \emptyset$ $(\alpha \neq \beta)$,

$$\sum_{i \in I_\alpha} a_i f_i = 0, \quad \alpha = 0, 1, ..., k,$$

and f_i/f_j *is constant for any* $i, j \in I_\alpha$. *In particular, if* $a_i \neq 0$ *for* $i = 0, 1, ..., n$, *each* I_α *contains at least two indices.*

For a non-zero integer a and a prime p, we define an integer $v_p(a)$ as follows:

$$a = p^{v_p(a)} a', \quad p \nmid a'.$$

Based on Theorem 15, we suggested a problem in number theory as follows:

Conjecture 12 ([19]). *Let* $a_0, ..., a_n$ $(n \geq 2)$ *be non-zero integers satisfying the following equation:*

$$\sum_{i=0}^{n} a_i = 0.$$

Assume that there exist positive integers d_i $(i = 0, ..., n)$ *such that for each* $i \in \{0, 1, ..., n\}$, *there is not any prime* p *satisfying*

$$0 < v_p(a_i) < d_i.$$

If $1 - \sum_{j=0}^{n} \frac{n-1}{d_j} > 0$, *then there exists a partition of indices*

$$\{0, 1, ..., n\} = I_0 \cup I_1 \cup \cdots \cup I_k$$

such that $I_\alpha \neq \emptyset$ $(\alpha = 0, 1, ..., k)$, $I_\alpha \cap I_\beta = \emptyset$ $(\alpha \neq \beta)$,

$$\sum_{i \in I_\alpha} a_i = 0, \quad \alpha = 0, 1, ..., k,$$

and $a_i/a_j \in \{-1, 1\}$ *for any* $i, j \in I_\alpha$. *In particular, each* I_α *contains at least two indices.*

If $d_i = d$ for $i = 0, ..., n$, this conjecture becomes the following:

Conjecture 13 ([19]). *Let* $a_0, ..., a_n$ *be non-zero integers satisfying the following equation:*

$$\sum_{i=0}^{n} a_i = 0.$$

Assume that there exists a positive integer $d \geq n^2$ *such that for each* $i \in \{0, 1, ..., n\}$, *there is not any prime* p *satisfying*

$$0 < v_p(a_i) < d.$$

Then there exists a partition of indices

$$\{0,1,...,n\} = I_0 \cup I_1 \cup \cdots \cup I_k$$

such that $I_\alpha \neq \emptyset$ $(\alpha = 0,1,...,k)$, $I_\alpha \cap I_\beta = \emptyset$ $(\alpha \neq \beta)$,

$$\sum_{i \in I_\alpha} a_i = 0, \quad \alpha = 0,1,...,k,$$

and $a_i/a_j \in \{-1,1\}$ *for any* $i,j \in I_\alpha$. *In particular, each* I_α *contains at least two indices.*

Conjecture 14 ([19]). *Let* $a_0,...,a_n$ *be non-zero integers satisfying the following equation:*

$$\delta_0 a_0^d + \cdots + \delta_n a_n^d = 0,$$

where $\delta_j \in \{-1,1\}$ $(j = 0,...,n)$. *If* $d \geq n^2$, *there exists a partition of indices*

$$\{0,1,...,n\} = I_0 \cup I_1 \cup \cdots \cup I_k$$

such that $I_\alpha \neq \emptyset$ $(\alpha = 0,1,...,k)$, $I_\alpha \cap I_\beta = \emptyset$ $(\alpha \neq \beta)$,

$$\sum_{i \in I_\alpha} \delta_i a_i^d = 0, \quad \alpha = 0,1,...,k,$$

and $a_i/a_j \in \{-1,1\}$ *for any* $i,j \in I_\alpha$. *In particular, each* I_α *contains at least two indices.*

Obviously, Conjecture 14 implies the following generalized Fermat conjecture:

Conjecture 15 ([19]). *If* n *is even and if* $d \geq n^2$, *then for any given* $\delta_j \in \{-1,1\}$ $(j = 0,...,n)$, *the following equation*

$$\delta_0 a_0^d + \cdots + \delta_n a_n^d = 0,$$

has no non-zero integer solutions $a_0,...,a_n$.

References

[1] Birch, B. J., Chowla, S., M. Hall Jnr. and Schinzel, A., *On the difference* $x^3 - y^2$, Norske Vid. Selsk. Forh. (Trondheim) 38 (1965), 65-69.

[2] Bloch, A., *Sur les systèmes de fonctions uniformes satisfaisant à l'équation d'une variété algébrique dont l'irrégularité dépasse la dimension*, J. de Math. 5(1926), 19-66.

[3] Boutabaa, A. and Escassut, A., *Nevanlinna theory in characteristic p, and applications*, preprint.

[4] Browkin, J. and Brzezinski, J., *Some remarks on the abc-conjecture*, Mathematics of Computation 62 (1994), 931-939.

[5] Brownawell, W. D. and Masser, D., *Vanishing sums in function fields*, Math. Proc. Cambridge Philos. Soc. 100(1986), 427-434.

[6] Davenport, H., *On* $f^3(t) - g^2(t)$, Norske Vid. Selsk. Forh. (Trondheim) 38 (1965), 86-87.

[7] Green, M. & Griffiths, Ph., *Two applications of algebraic geometry to entire holomorphic mappings*, The Chern Symposium 1979, Proc. Internal. Sympos., Berkeley, CA., 1979, Springer-Verlag, New York, 1980, 41-74.

[8] Griffiths, Ph., *Holomorphic mappings: Survey of some results and discussion of open problems*, Bull. Amer. Math. Soc. **78** (1972),374-382.

[9] Griffiths, Ph. and King, J., *Nevanlinna theory and holomorphic mappings between algebraic varieties*, Acta Math. 130(1973),145-220.

[10] Hu, P. C., *Holomorphic mappings between spaces of different dimensions I*, Math. Z. **214**(1993), 567-577; II, Math. Z. **215**(1994), 187-193.

[11] Hu, P. C. and Yang, C. C., *Notes on second main theorems, Complex Variables* **37** (1-4) (1998), 251-277.

[12] Hu, P. C. & Yang, C. C., *A note on Shiffman's conjecture*, The report on 2-th ISAAC's Congress, Japan, 1999.

[13] Hu, P. C. & Yang, C. C., *Meromorphic Functions over non-Archimedean Fields*, Kluwer Academic Publishers, 2000.

[14] Hu, P. C. & Yang, C. C., *The "abc" conjecture over function fields*, Proc. Japan Acad. 76, Ser. A(2000), 118-120.

[15] Hu, P. C. & Yang, C. C., *Some progresses in non-Archimedean analysis*, Proc. of Joint AMS-HKMS Conference 2000, AMS Contemporary Math. Series, 2002.

[16] Hu, P. C. & Yang, C. C., *Notes on a generalized abc-conjecture over function fields*, Annales Mathematiques Blaise Pascal 8(1)(2001), 61-71.

[17] Hu, P. C. & Yang, C. C., *A generalized abc-conjecture over function fields*, to appear in J. Number Theory.

[18] Hu, P. C. & Yang, C. C., *A note on the abc-conjecture*, preprint.

[19] Hu, P. C. & Yang, C. C., *Borel theorem and a generalized Fermat conjecture*, preprint.

[20] Lang, S., *Introduction to complex hyperbolic spaces*, Springer-Verlag, 1987.

[21] Lang, S., *Number theory III*, Encyclop. Math. Sc. vol. **60** (1991), Springer-Verlag.

[22] Mason, R. C., *Equations over function fields*, Lecture Notes in Math. 1068 (1984), 149-157, Springer.

[23] Mason, R. C., *Diophantine equations over function fields*, London Math. Soc. Lecture Note Series, Vol. 96, Cambridge University Press, United Kingdom, 1984.

[24] Mason, R. C., *The hyperelliptic equation over function fields*, Math. Proc. Cambridge Philos. Soc. 93(1983), 219-230.

[25] Mason, R. C., *Norm form equations I*, J. Number Theory 22 (1986), 190-207.

[26] Nevanlinna, R., *Le théorème de Picard-Borel et la théorie des fonctions meromorphes*, Gauthier Villars, Paris, 1929; Reprint, Chelsea, New York, 1974.

[27] Ru, M., *A note on p-adic Nevanlinna theory*, Proc. Amer. Math. Soc. 129 (5) (2001), 1263-1269.

[28] Schmidt, W. M., *Diophantine approximation*, Lecture Notes in Math. 785(1980), Springer.

[29] Schmidt, W. M., *Diophantine approximations and Diophantine equations*, Lecture Notes in Math. **1467** (1991), Springer-Verlag.

[30] Shiffman, B., *Holomorphic curves in algebraic manifolds*, Bull. Amer. Math. Soc. **83** (1977), 553-568.

[31] Shiffman, B., *On holomorphic curves and meromorphic maps in projective space*, Indiana Math. J. **28** (1979), 627-641.

[32] Shiffman, B., *Uniqueness of entire and meromorphic functions sharing finite sets*, Complex Variables **43** (2001), 433-449.

[33] Siu, Y. T. and Yeung, S. K., *A generalized Bloch's theorem and the hyperbolicity of the complement of an ample divisor in an abelian variety*, Math. Ann. **306** (4) (1996), 743-758.

[34] Stothers, W. W., *Polynomial identities and Hauptmoduln*, Quart. J. Math. Oxford Ser. (2) 32 (1981), no. 127, 349-370.

[35] van Frankenhuysen, M., *Hyperbolic spaces and the abc conjecture*, Thesis, Katholieke Universiteit Nijmegen, 1995.

[36] Vaserstein, L. N., *Quantum (abc)-Theorems*, J. Number Theory 81 (2000), 351-358.

[37] Vojta, P., *Diophantine approximation and value distribution theory*, Lecture Notes in Math. 1239, Springer, 1987.

[38] Voloch, J. F., *Diagonal equations over function fields*, Bol. Soc. Brasil. Mat. 16 (1985), 29-39.

[39] Vu, H. A. & Doan, Q. M., *The "abc" conjecture for p-adic holomorphic functions of several variables*, preprint.

2

IN RAMANUJAN'S SHADOW

Irwin Kra*
State University of New York at Stony Brook
Stony Brook, NY 11794
E-mail: irwin@math.sumysb.edu

1. Introduction

One variable θ-functions and θ-constants are powerful tools in the study of function theory, combinatorics and number theory. This is the main theme of this article and our recently published book [2]. We emphasize a unified approach to a diverse set of problems that include uniformizations of Riemann surfaces represented by subgroups of the modular group, applications to combinatorial number theory and the derivation of beautiful identities. In this expository note we provide a sample of the methods and results of my joint work with H.M Farkas. We outline some proofs and supply complete references to results not appearing in [2].

2. One variable θ-functions and their basic properties

The main analytic tools in our study are the one variable theta functions $\theta\begin{bmatrix}\epsilon\\\epsilon'\end{bmatrix}(\zeta,\tau)$. These are functions of a *characteristic* $\begin{bmatrix}\epsilon\\\epsilon'\end{bmatrix}\in\mathbb{R}^2$, a *variable* $\zeta\in\mathbb{C}$ and a *parameter* $\tau\in\mathbb{H}^2=\{\tau\in\mathbb{C};\Im\tau>0\}$, defined

*This is an expanded version of an expository talk at the 9th International Conference on Finite or Infinite Dimensional Complex Analysis held in Hanoi, Vietnam in August of 2001. It is based mostly on joint work with H.M. Farkas of The Hebrew University of Jerusalem.

by absolutely and uniformly (on appropriate compact sets) convergent series as

$$\theta\begin{bmatrix} \epsilon \\ \epsilon' \end{bmatrix}(\zeta,\tau) = \sum_{n=-\infty}^{\infty} \exp 2\pi\imath\left(\frac{1}{2}\left(n+\frac{\epsilon}{2}\right)^2\tau + \left(n+\frac{\epsilon}{2}\right)\left(\zeta+\frac{\epsilon'}{2}\right)\right).$$

All the material of this manuscript except for the contents of §§6 and 7 comes from [2], where all the unreferenced claims made here are verified. We usually fix the characteristic and study the resulting holomorphic function $\theta\begin{bmatrix} \epsilon \\ \epsilon' \end{bmatrix}(\cdot,\cdot)$ on $\mathbb{C}\times\mathbb{H}^2$. *Theta constants*

$$\theta\begin{bmatrix} \epsilon \\ \epsilon' \end{bmatrix} = \theta\begin{bmatrix} \epsilon \\ \epsilon' \end{bmatrix}(0,\tau)$$

and *theta constant derivatives*

$$\theta'\begin{bmatrix} \epsilon \\ \epsilon' \end{bmatrix} = \frac{\partial}{\partial\zeta}\theta\begin{bmatrix} \epsilon \\ \epsilon' \end{bmatrix}(\zeta,\tau) \text{ evaluated at } \zeta=0$$

are holomorphic functions on $\mathbb{H}^2$ that play a central role in our investigations. There is "only" one θ-function, the one with characteristic $0\in\mathbb{Z}^2$,

$$\theta = \theta\begin{bmatrix} 0 \\ 0 \end{bmatrix}.$$

The identity (easily derived) connecting the general θ-function to the "universal" θ-function is

$$\theta\begin{bmatrix} \epsilon \\ \epsilon' \end{bmatrix}(\zeta,\tau) = e^{2\pi\imath\left\{\frac{1}{8}\epsilon^2\tau+\frac{1}{2}\epsilon\zeta+\frac{1}{4}\epsilon\epsilon'\right\}}\theta\left(\zeta+\frac{\epsilon'}{2}+\frac{\epsilon}{2}\tau,\tau\right).$$

The characteristic $\chi = \begin{bmatrix} \epsilon \\ \epsilon' \end{bmatrix}$, however, is a useful book-keeping device.

What are probably the two most interesting groups in mathematics are key players in our picture. The first of these is the free commutative rank two group $\mathbb{Z}\oplus\mathbb{Z}$ acting by translations on $\mathbb{C}$ (the "first" nontrivial abelian group); the second is $\mathrm{SL}(2,\mathbb{Z})$ and its analytic incarnation, the modular group, $\Gamma = \mathrm{PSL}(2,\mathbb{Z}) = \mathrm{SL}(2,\mathbb{Z})/\{\pm I\}$, acting by Möbius transformations on $\mathbb{H}^2$ (the "first" nontrivial nonabelian group). The second of these groups has many rich families of subgroups; three families are important in our approach. For $k\in\mathbb{Z}^+$, let

$$\Gamma_o(k) = \left\{\begin{bmatrix} a & b \\ c & d \end{bmatrix}\in\Gamma;\ k|c\right\},$$
$$\Gamma(k,k) = \{\gamma\in\Gamma_o(k);\ k|b\}$$

and
$\Gamma(k) = \{\gamma \in \Gamma(k,k);\ a \equiv 1 \mod k \equiv d\}$.
In most of our applications (and in this paper) k is prime. It is quite routine to establish, by elementary manipulations of convergent series, *the variable transformation formula*: for all $(n,m) \in \mathbb{Z} \oplus \mathbb{Z}$, all $\zeta \in \mathbb{C}$ and all $\tau \in \mathbb{H}^2$,

$$\theta\begin{bmatrix} \epsilon \\ \epsilon' \end{bmatrix}(\zeta + n + m\tau, \tau) = e^{2\pi\imath\left\{\frac{n\epsilon - m\epsilon'}{2} - m\zeta - \frac{m^2}{2}\tau\right\}}\theta\begin{bmatrix} \epsilon \\ \epsilon' \end{bmatrix}(\zeta, \tau).$$

Fix an $N \in \mathbb{Z}^+$. An entire function f is an *N-th order θ-function with characteristic* $\begin{bmatrix} \epsilon \\ \epsilon' \end{bmatrix}$ *(and period τ)* if for all $(n,m) \in \mathbb{Z} \oplus \mathbb{Z}$ and all $\zeta \in \mathbb{C}$,

$$f(\zeta + n + m\tau) = e^{2\pi\imath\left\{\frac{n\epsilon - m\epsilon'}{2} - N\left(m\zeta + \frac{m^2}{2}\tau\right)\right\}} f(\zeta).$$

These form a vector space $\mathcal{F}_N\begin{bmatrix} \epsilon \\ \epsilon' \end{bmatrix}$.

The function $\theta\begin{bmatrix} \epsilon \\ \epsilon' \end{bmatrix}(\cdot, \tau)$ has precisely one simple zero in a fundamental polygon for the torus with periods 1 and τ; located at the point

$$\zeta = \frac{1-\epsilon'}{2} + \frac{1-\epsilon}{2}\tau.$$

Therefore

$$0 \neq \theta'\begin{bmatrix} \epsilon \\ \epsilon' \end{bmatrix}\left(\frac{1-\epsilon'}{2} + \frac{1-\epsilon}{2}\tau, \tau\right) = e^{\frac{\pi\imath\epsilon}{4}(2+2\tau-\epsilon\tau)}\theta'\begin{bmatrix} 1 \\ 1 \end{bmatrix}(0, \tau).$$

It follows from this formula that N-th order theta functions with more or less prescribed zeros can be constructed from ordinary (that is, $N = 1$) theta functions and that

$$\dim \mathcal{F}_N\begin{bmatrix} \epsilon \\ \epsilon' \end{bmatrix} = N.$$

The *parameter transformation formula*

$$\frac{e^{\pi\imath\left\{\frac{-c\zeta^2}{c\tau+d}\right\}}\theta\begin{bmatrix} \epsilon \\ \epsilon' \end{bmatrix}\left(\frac{\zeta}{c\tau+d}, \frac{a\tau+b}{c\tau+d}\right)}{\theta\begin{bmatrix} a\epsilon + c\epsilon' - ac \\ b\epsilon + d\epsilon' + bd \end{bmatrix}(\zeta, \tau)} = \kappa\left(\begin{bmatrix} \epsilon \\ \epsilon' \end{bmatrix}, \gamma\right)(c\tau + d)^{\frac{1}{2}}$$

belongs to a deeper part of the theory. Here $\zeta \in \mathbb{C}$, $\tau \in \mathbb{H}^2$, $\chi = \begin{bmatrix} \epsilon \\ \epsilon' \end{bmatrix} \in \mathbb{R}^2$ and $\gamma = \begin{bmatrix} a & b \\ c & d \end{bmatrix} \in \mathrm{SL}(2,\mathbb{Z})$. The proof of the formula uses the heat equation (easily seen to be satisfied by θ-functions), the residue theorem and lots of calculations.

Let χ_1 and $\chi_2 \in \mathbb{R}^2$. We define an *equivalence relation* on characteristics by

$$\chi_1 \equiv \chi_2 \text{ iff } \chi_1 - \chi_2 \in 2\mathbb{Z}^2.$$

The equivalence relation makes

$$\chi \mapsto \chi\gamma = \begin{bmatrix} a\epsilon + c\epsilon' - ac \\ b\epsilon + d\epsilon' + bd \end{bmatrix}$$

a group action of the modular group on equivalence classes of characteristics. As a bonus, we obtain that θ-functions with equivalent characteristics differ by very little. The mysterious constant $\kappa = \kappa(\chi, \gamma)$ depends only on the characteristic $\chi \in \mathbb{R}^2$ and the motion $\gamma \in \mathrm{SL}(2, \mathbb{Z})$. It is of absolute value 1, a root of unity if $\chi \in \mathbb{Q}^2$ and an 8-th root of unity if $\chi \in \mathbb{Z}^2$.

The last formula simplifies considerable for θ-constants ($\zeta = 0$):

$$\frac{\theta[\chi](0, \gamma(\tau))}{\theta[\chi\gamma](0, \tau)} = \kappa(\chi, \gamma)(\gamma'(\tau))^{-\frac{1}{4}}.$$

The corresponding formula for θ-constant derivatives reads

$$\frac{\theta'[\chi](0, \gamma(\tau))}{\theta'[\chi\gamma](0, \tau)} = \kappa(\chi, \gamma)(\gamma'(\tau))^{-\frac{3}{4}}.$$

Thus if

$$\chi\gamma = \chi$$

for some $\chi \in \mathbb{Q}^2$ and all $\gamma \in G$, for some subgroup $G \subset \Gamma$ (this can be achieved "morally" for many groups G), then sufficiently high powers of $\theta[\chi]$ and $\theta'[\chi]$ are meromorphic automorphic forms for G; many times modular or even cusp forms for G.

Fix $k \in \mathbb{Z}^+$ (usually, but not necessarily, prime). A characteristic $\chi \in \mathbb{Q}^2$ is *good* if it is of the form $\begin{bmatrix} \frac{m}{k} \\ \frac{m'}{k} \end{bmatrix}$ with m and $m' \in \mathbb{Z}$ of the same parity as k. By passing to equivalent characteristics we may assume that for every good characteristic,

$$0 \le m \le k \text{ and } 0 \le m' \begin{cases} < 2k \text{ if } 0 \ne m \ne k \\ \le k \text{ if } m = 0 \text{ or } k \end{cases},$$

which we assume throughout. We denote the set of equivalence classes of good characteristics by $X(k)$. This is precisely the set of fixed points of the action of $\Gamma(k)$ on the set of equivalence classes of characteristics.

We also distinguish a set of equivalence classes of *better* characteristics that we denote by $X_o(k)$. These are the characteristics in $X(k)$ that do not belong to any $X(k')$ with k' a proper divisor (which includes 1 but excludes k, of course) of k.

Fix an $r \in \mathbb{Q}$. A holomorphic function f on $\mathbb{H}^2$ is an r-form for G provided

$$\frac{f(\gamma(\tau))\gamma'(\tau)^r}{f(\tau)} = c_\gamma$$

for all $\tau \in \mathbb{H}^2$ and all $\gamma \in G$, where c_γ is a nonzero constant independent of τ, and f extends meromorphically to all the cusps of G. The map

$$c : \gamma \mapsto c_\gamma$$

defines a homomorphism of G into $\mathbb{C}^* = \mathbb{C} - \{0\}$, whenever $r \in \mathbb{Z}$. We will call c, the *homomorphism induced by* f, even when $r \notin \mathbb{Z}$.

3. Evaluations of trigonometric sums

Let $k \in \mathbb{Z}^+$. For $\chi \in X(k)$, $\theta[\chi]$ is a $\frac{1}{4}$-form for $\Gamma(k)$ (it is identically zero if $\chi \in X(1)$) and $\theta'[\chi]$ is a $\frac{3}{4}$-form for $\Gamma(k)$ (it is identically zero if $\chi \in X_o(2)$). Hence for $\chi \in X_o(k)$, $k > 2$,

$$\left(\frac{\theta'[\chi]}{\theta[\chi]}\right)^2$$

is a 1-form for $\Gamma(k)$ with trivial homomorphism (it projects to an abelian differential of the third kind on $\mathbb{H}^2/\Gamma(k)$ with simple poles at the punctures and no other singularities).

The sum of the residues of an abelian differential on a compact Riemann surface is 0. This gives a proof of the four trigonometric identities appearing below.

There is an alternate approach that yields more. Let $G \subset H$ be subgroups of Γ with G of finite index in H. If $r \in \mathbb{Z}$ and φ is an r-form for G with trivial homomorphism, then the *relative Poincaré series*

$$(\Theta(\varphi))(\tau) = \frac{1}{[H:G]} \sum_{\gamma \in G \backslash H} \varphi(\gamma(\tau))\gamma'(\tau)^r, \tau \in \mathbb{H}^2,$$

is an r-form for H. The operator Θ preserves holomorphicity, modular forms and cusp forms. If we take $G = \Gamma(k)$, $H = \Gamma$ and $\varphi = \left(\frac{\theta'[\chi]}{\theta[\chi]}\right)^2$ with $\chi \in X_o(k)$, we have $r = 1$ and we hence obtain that $\Theta(\varphi)$ is a modular 1-form for Γ; hence identically zero. Further, we have the nice formula

$$\Theta(\varphi) = \frac{1}{n(k)} \sum_{\chi \in X_o(k)} \left(\frac{\theta'[\chi]}{\theta[\chi]} \right)^2, \tag{1}$$

where $n(k)$ is the number of punctures on $\mathbb{H}^2/\Gamma(k)$ (which is the same as the cardinality of $X_o(k)$). Evaluating the constant term of the Fourier series expansion of the last function leads to the following identities.

(a) For all integers $k \geq 2$,

$$\sum_{m=1}^{k-1} \tan^2 \frac{\pi m}{2k} = \frac{(k-1)(2k-1)}{3}.$$

(b) For each odd integer $k \geq 3$,

$$\sum_{l=0,1,\ldots,\frac{k-3}{2},\ (1+2l,k)=1} \tan^2 \frac{\pi(1+2l)}{2k} = \frac{1}{k} \sum_{l=0}^{\frac{k-3}{2}} \frac{(1+2l)^2}{(k,1+2l)} \varphi((k,1+2l))$$

and

$$\sum_{l=0}^{\frac{k-3}{2}} \tan^2 \frac{\pi(1+2l)}{2k} = \frac{(k-2)(k-1)}{6}.$$

(c) For even integers $k \geq 2$,

$$\sum_{l=0,1,\ldots,\frac{k-2}{2},\ (1+2l,k)=1} \tan^2 \frac{\pi(1+2l)}{2k} = \frac{2}{k} \sum_{l=1}^{k-1} \frac{l^2}{(k,l)} \varphi((k,l)).$$

(d) For odd integers $k \geq 3$,

$$\sum_{m=1,2,\ldots,\frac{k-1}{2},\ (m,k)=1} \tan^2 \frac{\pi m}{k} = \frac{1}{k} \sum_{l=1}^{k-1} (2,l) \frac{l^2}{(k,l)} \varphi((k,l)).$$

In the above, we use standard notation. Thus (a,b) represent the gcd of the integers a and b, and $\varphi(n)$ is the number of positive integers $\leq n$ that are relatively prime to $n \in \mathbb{Z}^+$. The first of the above identities states, in particular, that for all even integers $k \geq 4$,

$$\sum_{m=1}^{\frac{k}{2}-1} \tan^2 \frac{\pi m}{k} = \frac{(k-1)(k-2)}{6},$$

which is a positive integer provided $k \not\equiv 0 \mod 3$.

The above identities **(a)**, **(b)**, **(c)** and **(d)** are consequences of the fact that the 0-th Fourier coefficient of the function defined by (1) is zero. Since the function is identically zero, all Fourier coefficients vanish. The resulting identities are too complicated for simple interpretations. See [1] for more details.

4. Interesting θ-constant identities

We can use four main tools for obtaining identities. They can be briefly described as follows.

(a) Using theta constants and/or theta constant derivatives, construct on surfaces represented by subgroups of the modular group, especially $\Gamma_o(k)$, the same meromorphic function or meromorphic differential in more than one way.

(b) Using N-th order theta functions, construct an elliptic function. The sum of the residues of an elliptic function vanishes giving many identities.

(c) Construct $N+1$ N-th order theta functions with the same characteristic. Since the space of such functions has dimension N, we must have at least one linear dependence.

(d) In analogy to the last method, we can exploit the finite dimensionality of spaces of modular and/or cusp forms for subgroups of Γ.

It is convenient to introduce into the picture the classical η-function defined by

$$\eta(\tau) = x^{\frac{1}{24}} \prod_{n=1}^{\infty} (1 - x^n) = x^{\frac{1}{24}} \sum_{n=-\infty}^{\infty} (-1)^n x^{\frac{n(3n+1)}{2}},$$

where $x = e^{2\pi\imath\tau}$ (this local coordinate is used throughout the paper). It is known and not hard to show that η is a $\frac{1}{4}$-form for Γ.

The connection between η and θ-constants is provided by the derivative formula

$$\theta' \begin{bmatrix} 1 \\ 1 \end{bmatrix} (0, \tau) = 2\pi\imath\theta^3 \begin{bmatrix} \frac{1}{3} \\ 1 \end{bmatrix} (0, 3\tau)$$

and the identity

$$\eta(\tau) = e^{-\frac{\pi\imath}{6}} \theta \begin{bmatrix} \frac{1}{3} \\ 1 \end{bmatrix} (0, 3\tau).$$

The last two formulae are easy consequences of method **(d)** described above after observing that $\eta^{24}(\tau)$, $\theta'^8 \begin{bmatrix} 1 \\ 1 \end{bmatrix} (0, \tau)$ and $\theta^{24} \begin{bmatrix} \frac{1}{3} \\ 1 \end{bmatrix} (0, 3\tau)$

are all cusp 6-forms for the full modular Γ and that by the Riemann-Roch theorem, this space is one dimensional.

5. Partition Identities. Formuale of Ramanujan

For the classical partition function $P : \mathbb{Z}^+ \to \mathbb{Z}^+$ defined by

$$\prod_{n=1}^{\infty} \frac{1}{1-x^n} = \sum_{n=0}^{\infty} P(n)x^n$$

Ramanujan obtained two beautiful identities:
(a) Identity for the prime 5.

$$\sum_{m=1}^{\infty} P(5m-1)x^m = 5\; x\, \frac{\prod_{n=1}^{\infty}(1-x^{5n})^5}{\prod_{n=1}^{\infty}(1-x^n)^6}.$$

(b) Identity for the prime 7.

$$\sum_{m=1}^{\infty} P(7m-2)x^m = 7\; x\, \frac{\prod_{n=1}^{\infty}(1-x^{7n})^3}{\prod_{n=1}^{\infty}(1-x^n)^4} \left[7\; x\, \frac{\prod_{n=1}^{\infty}(1-x^{7n})^4}{\prod_{n=1}^{\infty}(1-x^n)^4} + 1\right].$$

Since for $n \in \mathbb{Z}^+$, $P(n)$ represents the total number of partitions of the positive integer n, the combinatorial content of the last two identities is that for all $n \in \mathbb{Z}^+$,

$$P(5n-1) \equiv 0 \mod 5 \text{ and } P(7n-2) \equiv 0 \mod 7.$$

Different constructions of the same automorphic meromorphic function for $\Gamma_o(k)$ yield proofs of the last two identities. We start with the fact that

$$\tau \mapsto \frac{\eta(k\tau)}{\eta\left(\frac{\tau}{k}\right)}, \tau \in \mathbb{H}^2,$$

projects as a holomorphic function to $\mathbb{H}^2/\Gamma(k,k)$ that extend meromorphically to its compactification $\overline{\mathbb{H}^2/\Gamma(k,k)}$. Averaging this function over $\Gamma(k,k)\backslash\Gamma_o(k)$ produces a $\Gamma_o(k)$ invariant function whose divisor can readily be computed. Alternate descriptions of these functions lead us to the following two relations. For $k=5$, where $\mathbb{H}^2/\Gamma_o(5)$ is a sphere, we get

$$\frac{1}{5}\sum_{l=0}^{4} \frac{\eta(5(\tau+l))}{\eta(\frac{\tau+l}{5})} = 5\left(\frac{\eta(5\tau)}{\eta(\tau)}\right)^6.$$

For $k=7$, where $\mathbb{H}^2/\Gamma_o(7)$ is a torus, we are lead to the more complicated formula

$$\frac{1}{7}\sum_{l=0}^{6} \frac{\eta(7(\tau+l))}{\eta(\frac{\tau+l}{7})} = 7\left(\frac{\eta(7\tau)}{\eta(\tau)}\right)^4 \left(7\left(\frac{\eta(7\tau)}{\eta(\tau)}\right)^4 + 1\right).$$

Translation of the last two equations to local coordinate $x = e^{2\pi\imath\tau}$ yield the two Ramanujan identities.

In the same spirit as the two formulae of Ramanujan is the following relation between products and sums:

$$\sum_{n=-\infty}^{\infty} (6n+1)x^{\frac{n(3n+1)}{2}} = \prod_{n=1}^{\infty}(1-x^n)^3(1-x^{2n-1})^2.$$

To obtain its combinatorial interpretation, we consider partitions of positive integers using 3 sets of even positive integers and 5 sets of odd positive integers and do not allow repetitions. As usual, in this case, the parity of a partition is the parity of the number of its parts. For these partitions the difference between the number of even and odd partitions of $m \in \mathbb{Z}^+$ is 0 if m is not a (generalized) pentagonal number (that is, not of the form $\frac{n(3n+1)}{2}$) and equals $6n+1$ for $m = \frac{n(3n+1)}{2}$. The proof of the last identity is a study of the third order even theta function for the characteristic $\begin{bmatrix} 1 \\ 1 \end{bmatrix}$,

$$\theta\begin{bmatrix} \frac{1}{3} \\ 1 \end{bmatrix}(3\zeta, 3\tau) + \theta\begin{bmatrix} \frac{1}{3} \\ 1 \end{bmatrix}(-3\zeta, 3\tau)$$

$$= c(\tau)\theta\begin{bmatrix} 0 \\ 0 \end{bmatrix}(\zeta,\tau)\theta\begin{bmatrix} 0 \\ 1 \end{bmatrix}(\zeta,\tau)\theta\begin{bmatrix} 1 \\ 0 \end{bmatrix}(\zeta,\tau),$$

where

$$c(\tau) = \frac{2\theta\begin{bmatrix} \frac{1}{3} \\ 1 \end{bmatrix}(0,3\tau)}{\theta\begin{bmatrix} 0 \\ 0 \end{bmatrix}(0,\tau)\theta\begin{bmatrix} 0 \\ 1 \end{bmatrix}(0,\tau)\theta\begin{bmatrix} 1 \\ 0 \end{bmatrix}(0,\tau)}.$$

More complicated relations between sums are given by

$$\sum_{n=-\infty}^{\infty} (-1)^n x^{n(5n+1)} \left(\sum_{n=-\infty}^{\infty} (-1)^n x^{(n+1)(5n+3)} + \sum_{n=-\infty}^{\infty} (-1)^n x^{n(5n+2)} \right) =$$

$$\sum_{n=-\infty}^{\infty} (-1)^n x^{n(5n+3)} \left(\sum_{n=-\infty}^{\infty} (-1)^n x^{5n^2+6n+2} + \sum_{n=-\infty}^{\infty} (-1)^n x^{5n^2+4n+1} \right)$$

and

$$\sum_{n=-\infty}^{\infty} (-1)^n x^{n(5n+1)} \left(\sum_{n=-\infty}^{\infty} (-1)^n x^{(5n+3)(n-1)} + \sum_{n=-\infty}^{\infty} (-1)^n x^{n(5n-8)} \right) =$$

$$\sum_{n=-\infty}^{\infty}(-1)^n x^{n(5n+3)}\left(\sum_{n=-\infty}^{\infty}(-1)^n x^{5n^2-4n-2}+\sum_{n=-\infty}^{\infty}(-1)^n x^{5n^2-6n-1}\right).$$

The last two identities obviously have rather complicated combinatorial interpretations. They are the special cases $\zeta = 1$ and -1 of the Septuple Product Identity discussed below.

None of our formula are quite as elegant as the identities

$$\prod_{n=1}^{\infty}(1-x^n)=\sum_{n=-\infty}^{\infty}(-1)^n x^{\frac{n(3n+1)}{2}}$$

and

$$\prod_{n=1}^{\infty}(1-x^n)^3=\sum_{n=0}^{\infty}(-1)^n(2n+1)x^{\frac{n(n+1)}{2}}$$

of Euler and Jacobi that suggest an investigation of the partition functions $P_N : \mathbb{Z}^+ \to \mathbb{Z}$, $N \in \mathbb{Z}$, defined by

$$\prod_{n=1}^{\infty}(1-x^n)^{-N}=\sum_{n=0}^{\infty}P_N(n)x^n.$$

The only known closed forms for P_N are the cases $N = -1$ and -3 described above.

6. The Jacobi Quartic

This classical formula reads

$$\theta^4\begin{bmatrix}0\\0\end{bmatrix}=\theta^4\begin{bmatrix}1\\0\end{bmatrix}+\theta^4\begin{bmatrix}0\\1\end{bmatrix}.$$

An equivalent form is thus

$$\theta^4\begin{bmatrix}0\\0\end{bmatrix}-\theta^4\begin{bmatrix}1\\0\end{bmatrix}-\left(\theta^4\begin{bmatrix}0\\1\end{bmatrix}-\theta^4\begin{bmatrix}1\\1\end{bmatrix}\right)=0.$$

A generalization of the last identity places it in a family of equations. For all a and $b \in \mathbb{R}$,

$$\theta^4\begin{bmatrix}a\\b\end{bmatrix}-\theta^4\begin{bmatrix}a+1\\b\end{bmatrix}-e^{-2\pi i a}\left(\theta^4\begin{bmatrix}a\\b+1\end{bmatrix}-\theta^4\begin{bmatrix}a+1\\b+1\end{bmatrix}\right)=0.$$

The last formula is a special case of a more general theorem established in [4].

To prove the last identity, we fix τ and define

$$f(\zeta) = \frac{\theta\begin{bmatrix} a \\ b \end{bmatrix}(\zeta,\tau)\,\theta\begin{bmatrix} -a \\ -b \end{bmatrix}(\zeta,\tau)\,\theta\begin{bmatrix} c \\ d \end{bmatrix}(\zeta,\tau)\,\theta\begin{bmatrix} -c \\ -d \end{bmatrix}(\zeta,\tau)}{\theta\begin{bmatrix} 0 \\ 0 \end{bmatrix}(\zeta,\tau)\ \theta\begin{bmatrix} 0 \\ 1 \end{bmatrix}(\zeta,\tau)\ \theta\begin{bmatrix} 1 \\ 0 \end{bmatrix}(\zeta,\tau)\ \theta\begin{bmatrix} 1 \\ 1 \end{bmatrix}(\zeta,\tau)}.$$

The function f is elliptic with periods 1 and τ. Its sum of residues is zero. This gives us a θ-constant identity. The desired formula follows by setting $c = a$ and $d = b$ (after some algebraic simplification using properties of theta functions).

7. The Jacobi derivative formula

This famous formula

$$\theta'\begin{bmatrix} 1 \\ 1 \end{bmatrix} = -\pi\theta\begin{bmatrix} 0 \\ 0 \end{bmatrix}\theta\begin{bmatrix} 1 \\ 0 \end{bmatrix}\theta\begin{bmatrix} 0 \\ 1 \end{bmatrix}.$$

has a far reaching generalization.

The following conditions are equivalent:

(a) $\theta'[\chi]$ is a finite nontrivial product of powers of theta constants.

(b) $\theta'[\chi]$ has no zeros on $\mathbb{H}^2$.

(c) $\chi \in X_o(k)$ for $k = 1$, 4 or 6.

(d) $\theta'[\chi]$ has an infinite product expansion (in terms of x valid on the disc $|x| < 1$).

We refer to [3] for a proof. A sample of the consequences are the identities

$$\theta'\begin{bmatrix} 1 \\ \frac{1}{2} \end{bmatrix} = \frac{\pi}{2}\frac{\theta^2\begin{bmatrix} 1 \\ 0 \end{bmatrix}\theta^2\begin{bmatrix} 0 \\ \frac{1}{2} \end{bmatrix}}{\theta\begin{bmatrix} 1 \\ \frac{1}{2} \end{bmatrix}}$$

and

$$\theta'\begin{bmatrix} 1 \\ \frac{2}{3} \end{bmatrix} = c\frac{\theta^3\begin{bmatrix} 1 \\ \frac{1}{3} \end{bmatrix}\theta'\begin{bmatrix} 1 \\ 1 \end{bmatrix}}{\theta\begin{bmatrix} 1 \\ 0 \end{bmatrix}\theta^2\begin{bmatrix} 1 \\ \frac{2}{3} \end{bmatrix}}$$

(for some nonzero constant c that can easily be determined). The power series versions, after some algebraic manipulations, of these formulae

$$\sum_{n=-\infty}^{\infty}(4n+1)(-1)^n x^{n(2n+1)} = \prod_{n=1}^{\infty}(1-x^{2n})^3(1+x^{2n-1})^3$$

and

$$\sum_{n=-\infty}^{\infty}\left((6n+1)x^{\frac{3n(3n+1)}{2}}-(6n+3)x^{\frac{(3n+1)(3n+2)}{2}}\right)$$

$$=\prod_{n=1}^{\infty}\frac{(1-x^n)^3(1-x^{2n-1})^2(1+x^n+x^{2n})^3}{(1-x^n+x^{2n})^2}$$

are two more identities that obviously have combinatorial content for partitions.

8. Products and sums

We end with three identities relating products and sums which leave a big unanswered question.

(a) The first is the Jacobi Triple Product

$$\frac{1+z}{z}\prod_{n=1}^{\infty}(1-x^n)(1+x^nz)\left(1+\frac{x^n}{z}\right)=\sum_{n=-\infty}^{\infty}x^{\frac{n(n+1)}{2}}z^n$$

that involves a study of zeros of θ.

(b) The second is the Quintuple Product Identity

$$(1+z)\prod_{n=1}^{\infty}(1-x^n)(1+x^nz)\left(1+\frac{x^n}{z}\right)(1-x^{2n-1}z^2)\left(1-\frac{x^{2n-1}}{z^2}\right)$$

$$=\sum_{n=-\infty}^{\infty}(-1)^nx^{\frac{n(3n-1)}{2}}z^{3n}+z\sum_{n=-\infty}^{\infty}(-1)^nx^{\frac{n(3n+1)}{2}}z^{3n}$$

whose proof can be based on a study of the even third order θ-function

$$\theta\begin{bmatrix}\frac{1}{3}\\1\end{bmatrix}(3\zeta,3\tau)+\theta\begin{bmatrix}\frac{1}{3}\\1\end{bmatrix}(-3\zeta,3\tau).$$

(c) The last is the Septuple Product Identity

$$(1+z)(1-z)^2\prod_{n=1}^{\infty}(1-x^n)^2(1-x^nz)\left(1-\frac{x^n}{z}\right)(1-x^nz^2)\left(1-\frac{x^n}{z^2}\right)$$

$$=\sum_{n=-\infty}^{\infty}(-1)^nx^{\frac{n(5n+1)}{2}}(I)-\sum_{n=-\infty}^{\infty}(-1)^nx^{\frac{n(5n+3)}{2}}(II)$$

whose proof is a study of the even fifth order θ-function, in $\mathcal{F}_5\begin{bmatrix}1\\1\end{bmatrix}$,

$$\theta\begin{bmatrix}0\\0\end{bmatrix}(\zeta,\tau)\theta\begin{bmatrix}0\\1\end{bmatrix}(\zeta,\tau)\theta\begin{bmatrix}1\\0\end{bmatrix}(\zeta,\tau)\theta^2\begin{bmatrix}1\\1\end{bmatrix}(\zeta,\tau)$$

where

$$I=\sum_{n=-\infty}^{\infty}(-1)^n x^{\frac{n(5n+3)}{2}}z^{5n+3}+\sum_{n=-\infty}^{\infty}(-1)^n x^{\frac{n(5n-3)}{2}}z^{5n},$$

$$II=\sum_{n=-\infty}^{\infty}(-1)^n x^{\frac{n(5n+1)}{2}}z^{5n+2}+\sum_{n=-\infty}^{\infty}(-1)^n x^{\frac{n(5n-1)}{2}}z^{5n+1}.$$

An open problem is the identification of unifying good generalizations of the last three identities, if such exist.

References

[1] H.M. Farkas and I. Kra, *Identities in the theory of theta constants*, to appear in Contemporary Math.

[2] ———, *Theta Constants, Riemann Surfaces and the Modular Group*, Graduate Studies in Mathematics, vol. 37, American Mathematical Society, 2001.

[3] I. Kra, *On theta constant identities*, in preparation.

[4] ———, *Theta constant identities involving arbitrary characteristics*, in preparation.

3

HOLOMORPHIC FUNCTIONS ON THE LIE BALL AND RELATED TOPICS

Mitsuo Morimoto
Department of Mathematics
International Christian University
Tokyo 181-8585, Japan
E-mail: mitsuo@icu.ac.jp

Keiko Fujita
Faculty of Culture and Education
Saga University
Saga 840-8502, Japan
E-mail: keiko@cc.saga-u.ac.jp

Introduction

In this note, we wish to report our recent results related with the Lie ball (See [7], [8], and [3]).

The Lie ball appeared to us in our research [6] of hyperfunctions on the (real) sphere.

Let $r > 0$. We call

$$B(r) = \{x \in \mathbb{R}^{n+1}; \|x\| < r\}$$

the (real) ball of radius r and

$$\tilde{B}(r) = \{z \in \mathbb{C}^{n+1}; L(z) < r\}$$

the Lie ball of radius r, where $\|x\| = \sqrt{x^2}$ is the Euclidean norm,

$$L(z) = \sqrt{\|z\|^2 + \sqrt{\|z\|^4 - |z^2|^2}} \tag{1}$$

is the Lie norm, $z^2 = z_1^2 + z_2^2 + \cdots + z_{n+1}^2$ and $\|z\|^2 = |z_1|^2 + |z_2|^2 + \cdots + |z_{n+1}|^2$ for $z = (z_1, z_2, \cdots, z_{n+1}) \in \mathbb{C}^{n+1}$. It is convenient to introduce the canonical bilinear form on $\mathbb{C}^{n+1}$

$$z \cdot w = z_1 w_1 + z_2 w_2 + \cdots + z_{n+1} w_{n+1}$$

for $z = (z_1, z_2, \cdots, z_{n+1})$ and $w = (w_1, w_2, \cdots, w_{n+1}) \in \mathbb{C}^{n+1}$. Then we have $z^2 = z \cdot z$ and $\|z\|^2 = z \cdot \bar{z}$.

We denote by $B[r]$ the closure of $B(r)$ and will be called the closed real ball. The Lie ball $\tilde{B}(r)$ is one of the classical bounded domain of E. Cartan and was studied by Hua Loo Keng. The closure of $\tilde{B}(r)$ is denoted by $\tilde{B}[r]$ and will be called the closed Lie ball.

Via analytic continuation every harmonic function on the ball $B(r)$ can be extended uniquely to a complex harmonic function on the Lie ball $\tilde{B}(r)$.

Next, via boundary values every harmonic function on the ball $B(r)$ can be associated with a hyperfunction on the sphere $\mathbb{S}_r = \partial B(r)$.

Therefore, complex harmonic functions on the Lie ball and hyperfunctions on the sphere can be corresponded naturally in a one-to-one manner.

1. Definitions

To be precise, we introduce notation. We denote by $\mathcal{A}(B(r))$ the space of real analytic functions on $B(r)$ and by $\mathcal{A}(B[r])$ the space of germs of real analytic functions on $B[r]$. We call

$$\mathcal{A}_\Delta(B(r)) = \{f \in \mathcal{A}(B(r)); \Delta_x f = 0\}$$

the space of harmonic functions on $B(r)$, where Δ_x denotes the Laplacian operator. $\mathcal{A}_\Delta(B[r])$ has the similar meaning.

We denote by $\mathcal{O}(\tilde{B}(r))$ the space of holomorphic functions on $\tilde{B}(r)$ and by $\mathcal{O}(\tilde{B}[r])$ the space of germs of holomorphic functions on $\tilde{B}[r]$. We call

$$\mathcal{O}_\Delta(\tilde{B}(r)) = \{f \in \mathcal{O}(\tilde{B}(r)); \Delta_z f = 0\}$$

the space of complex harmonic functions on $\tilde{B}(r)$, where

$$\Delta_z = \frac{\partial^2}{\partial z_1^2} + \frac{\partial^2}{\partial z_2^2} + \cdots + \frac{\partial^2}{\partial z_{n+1}^2}$$

is the complex Laplacian. $\mathcal{O}_\Delta(\tilde{B}[r])$ has the similar meaning.

Theorem 1 (Siciak). *The restriction mapping gives a topological linear isomorphism:*

$$\mathcal{O}_\Delta(\tilde{B}(r)) \xrightarrow{\sim} \mathcal{A}_\Delta(B(r)).$$

The Lie ball $\tilde{B}(r)$ is the harmonic hull of the real ball $B(r)$.

Let

$$\mathbb{S}_r = \{x \in \mathbb{R}^{n+1}; \|x\| = r\}$$

be the (real) sphere of radius $r > 0$. We denote by $\mathcal{A}(\mathbb{S}_r)$ the space of real analytic functions on $\mathbb{S}_r$ and by $\mathcal{B}(\mathbb{S}_r)$ the dual space of $\mathcal{A}(\mathbb{S}_r)$, which we call the space of hyperfunctions on $\mathbb{S}_r$.

Theorem 2 (M. Sato). *The boundary value operator maps $\mathcal{A}_\Delta(B(r))$ onto $\mathcal{B}(\mathbb{S}_r)$ linearly and topologically:*

$$\mathcal{A}_\Delta(B(r)) \xrightarrow{\sim} \mathcal{B}(\mathbb{S}_r).$$

Combining Theorems 1 and 2 we can conclude

$$\mathcal{O}_\Delta(\tilde{B}(r)) \xrightarrow{\sim} \mathcal{B}(\mathbb{S}_r).$$

The Poisson integral formula is closely connected with the above theorem.

Theorem 3 (Poisson formula). (i) *For $f \in \mathcal{O}_\Delta(\tilde{B}[r])$ we have*

$$f(z) = \int_{\mathbb{S}_r} K_r(z, x) f(x) \dot{dx}, \quad z \in \tilde{B}(r),$$

where $K_r(z, w) = K_1(z/r, w/r)$,

$$K_1(z, w) = \frac{1 - z^2\bar{w}^2}{(1 - 2z \cdot \bar{w} + z^2\bar{w}^2)^{(n+1)/2}}$$

is the Poisson kernel, and $\dot{dx}$ denotes the normalized invariant mesure on $\mathbb{S}_r$.

(ii) *The restriction mapping maps $\mathcal{O}_\Delta(\tilde{B}[r])$ onto $\mathcal{A}(\mathbb{S}_r)$ and the Poisson integral gives the inverse mapping:*

$$\mathcal{O}_\Delta(\tilde{B}[r]) \xrightarrow{\sim} \mathcal{A}(\mathbb{S}_r).$$

2. Cauchy-Hua integral formula

Let $\lambda \in \mathbb{C}$. The complex variety

$$\tilde{\mathbb{S}}_\lambda = \{z \in \mathbb{C}^{n+1}; z^2 = \lambda^2\}$$

is called the complex sphere of radius λ. We also call it the complex light cone if $\lambda = 0$. We call

$$\mathbb{S}_\lambda = \{\lambda x; x \in \mathbb{R}^{n+1}, x^2 = 1\}$$

the real sphere of complex radius λ and

$$\Sigma_r = \{e^{i\theta}x; \theta \in \mathbb{R}, x \in \mathbb{S}_r\}.$$

the Lie sphere of radius r. Note that $\dim_{\mathbb{R}} \tilde{\mathbb{S}}_\lambda = 2n$, $\dim_{\mathbb{R}} \mathbb{S}_\lambda = n$ and $\dim_{\mathbb{R}} \Sigma_r = n+1$. (We studied the Lie sphere first in [5].)

The Lie sphere Σ_r is the Shilov boundary of the Lie ball $\tilde{B}[r]$. A continuous function on $\tilde{B}[r]$ which is holomorphic in $\tilde{B}(r)$ takes the maximum value on Σ_r.

Theorem 4 (Cauchy-Hua formula). *For $f \in \mathcal{O}(\tilde{B}[r])$ we have*

$$f(z) = \int_{\Sigma_r} H_r(z,w) f(w) \dot{dw}, \quad z \in \tilde{B}(r),$$

where $H_r(z,w) = H_1(z/r, w/r)$,

$$H_1(z,w) = \frac{1}{(1 - 2z \cdot \bar{w} + z^2 \bar{w}^2)^{(n+1)/2}}$$

is the Cauchy-Hua kernel, and $\dot{dw}$ denotes the normalized invariant measure on Σ_r.

Note the relation

$$H_r(z,w) = \frac{K_r(z,w)}{1 - (z/r)^2 (\bar{w}/r)^2}. \tag{2}$$

3. Double series expansion

Let $f \in \mathcal{O}(\tilde{B}[r])$. Define the k-homogeneous polynomial f_k by

$$f_k(z) = \frac{1}{2\pi} \int_0^{2\pi} f(e^{i\theta} z) e^{-ik\theta} d\theta. \tag{3}$$

Then we have the homogeneous polynomial expansion of f:

$$f(z) = \sum_{k=0}^{\infty} f_k(z), \quad z \in \tilde{B}[r],$$

the convergence being uniform in a neighborhood of $\tilde{B}[r]$.

We denote by $\mathcal{P}^k_\Delta(\mathbb{C}^{n+1})$ the space of homogeneous harmonic polynomials of degree k on $\mathbb{C}^{n+1}$ and by $N(k,n)$ the dimension of $\mathcal{P}^k_\Delta(\mathbb{C}^{n+1})$:

$$N(k,n) = \frac{(2k+n-1)(k+n-2)!}{k!(n-1)!}.$$

We denote by $P_{k,n}(t)$ the Legendre (Gegenbauer) polynomial and define the harmonic extended Legendre polynomial $\tilde{P}_{k,n}(z,w)$ by

$$\tilde{P}_{k,n}(z,w) = (\sqrt{z^2})^k(\sqrt{\bar{w}^2})^k P_{k,n}(\frac{z}{\sqrt{z^2}} \cdot \frac{\bar{w}}{\sqrt{\bar{w}^2}}).$$

Then $\Delta_z \tilde{P}_{k,n}(z,w) = 0$,

$$|\tilde{P}_{k,n}(z,w)| \le L(z)^k L(w)^k,$$

and $\tilde{P}_{k,n}(z,w) = \overline{\tilde{P}_{k,n}(w,z)}$ (see, for example, [6]).

$\tilde{P}_{k,n}(z,w)$ is the reproducing kernel for $\mathcal{P}^k_\Delta(\mathbb{C}^{n+1})$. For $f_k \in \mathcal{P}^k_\Delta(\mathbb{C}^{n+1})$ we have

$$f_k(z) = \frac{N(k,n)}{r^{2k}} \int_{\mathbb{S}_r} \tilde{P}_{k,n}(z,x) f_k(x) \dot{dx}, \tag{4}$$

where $\dot{dx}$ is the normalized invariant measure on the real sphere $\mathbb{S}_r$. Note also that $\mathcal{P}^k_\Delta(\mathbb{C}^{n+1})$'s are mutually orthogonal:

$$\mathcal{P}^k_\Delta(\mathbb{C}^{n+1}) \perp \mathcal{P}^l_\Delta(\mathbb{C}^{n+1}), \quad k \neq l.$$

Because the k-homogeneous component f_k of $f \in \mathcal{O}_\Delta(\tilde{B}[r])$ is a harmonic polynomial, putting

$$K_r(z,w) = \sum_{k=0}^{\infty} \frac{N(k,n)}{r^{2k}} \tilde{P}_{k,n}(z,w),$$

we have the Poisson integral formula (Theorem 3).

For a k-homogeneous polynomial f_k, define $f_{k,k-2l} \in \mathcal{P}^{k-2l}_\Delta(\mathbb{C}^{n+1})$ by

$$f_{k,k-2l}(z) = \frac{N(k-2l,n)}{r^{2k-2l}} \int_{\mathbb{S}_r} \tilde{P}_{k-2l,n}(z,x) f_k(x) \dot{dx}. \tag{5}$$

Then we have

$$f_k(z) = \sum_{l=0}^{[k/2]} (z^2)^l f_{k,k-2l}(z). \tag{6}$$

Thus $f \in \mathcal{O}(\tilde{B}[r])$ can be expanded by means of homogeneous harmonic polynomials as follows:

$$f(z) = \sum_{k=0}^{\infty} \sum_{l=0}^{[k/2]} (z^2)^l f_{k,k-2l}(z). \tag{7}$$

We call the right-hand side in (7) the double series expansion of f and $f_{k,k-2l}$ the $(k, k-2l)$-harmonic component of f.

By (3) and (5) we have

$$\begin{aligned} f_{k,k-2l}(z) &= \\ &= \frac{N(k-2l,n)}{r^{2k-2l}} \int_{\mathbb{S}_r} \tilde{P}_{k-2l,n}(z,x) f_k(x) \dot{dx} \\ &= \frac{N(k-2l,n)}{r^{2k-2l}} \int_{\mathbb{S}_r} e^{i(k-2l)\theta} \tilde{P}_{k-2l,n}(z, e^{i\theta}x) \frac{1}{2\pi} \int_0^{2\pi} f(e^{i\theta}x) e^{-ik\theta} d\theta \dot{dx} \\ &= \frac{N(k-2l,n)}{r^{2k-2l}} \int_{\Sigma_r} e^{-i2l\theta} \tilde{P}_{k-2l,n}(z,w) f(w) \dot{dw} \\ &= \frac{N(k-2l,n)}{r^{2k}} \int_{\Sigma_r} (\bar{w}^2)^l \tilde{P}_{k-2l,n}(z,w) f(w) \dot{dw}, \end{aligned}$$

where $w = e^{i\theta}x$, and $\dot{dw} = \frac{1}{2\pi} d\theta \dot{dx}$ is the normalized invariant measure on the Lie sphere Σ_r. Putting

$$H_r(z,w) = \sum_{k=0}^{\infty} \sum_{l=0}^{[k/2]} \frac{N(k-2l,n)}{r^{2k}} (z^2)^l (\bar{w}^2)^l \tilde{P}_{k-2l,n}(z,w)$$

we have the Cauchy-Hua formula (Theorem 4).

The holomorphic function on the Lie ball can be characterized by the Cauchy-Hadamard growth condition of its $(k, k-2l)$-harmonic coefficients:

Theorem 5 ([5]). *Let (7) be the double series expansion of $f(z)$. Then we have*

$$f \in \mathcal{O}(\tilde{B}(r)) \text{ iff } \limsup_{k\to\infty} \left(\|f_{k,k-2l}\|_{C(\mathbb{S})} \right)^{\frac{1}{k}} \le \frac{1}{r},$$

where $\| \cdot \|_{C(\mathbb{S})}$ denotes the supremum norm of f on the unit sphere $\mathbb{S}$.

4. Truncated Lie ball

Suppose $0 < |\lambda| < r$ and consider the truncated Lie balls:

$$\tilde{B}(\lambda, r) = \{z \in \tilde{B}(r); |z^2| < |\lambda|^2\},$$

$$\tilde{B}[\lambda, r] = \{z \in \tilde{B}[r]; |z^2| \le |\lambda|^2\}.$$

Put

$$\tilde{\mathbb{S}}_{\lambda,r} = \partial(\tilde{\mathbb{S}}_\lambda \cap \tilde{B}[r]) = \{z \in \tilde{\mathbb{S}}_\lambda; L(z) = r\},$$

$$\Sigma_{\lambda,r} = \{e^{i\theta} z; \theta \in \mathbb{R}, z \in \tilde{\mathbb{S}}_{\lambda,r}\}.$$

$\tilde{\mathbb{S}}_{\lambda,r}$ and $\Sigma_{\lambda,r}$ are real analytic manifolds,

$$\dim_{\mathbb{R}} \tilde{\mathbb{S}}_{\lambda,r} = 2n - 1 \text{ and } \dim_{\mathbb{R}} \Sigma_{\lambda,r} = 2n.$$

If $|\lambda| \to r$, then $\tilde{\mathbb{S}}_{\lambda,r}$ reduces to $\mathbb{S}_\lambda$ and $\Sigma_{\lambda,r}$ reduces to Σ_r.

Let $f_k \in \mathcal{P}^k_\Delta(\mathbb{C}^{n+1})$. We can shift the integral region in (4) into the complex neighborhood as follows:

$$f_k(z) = \frac{N(k,n)}{L_{k,\lambda,r}} \int_{\tilde{\mathbb{S}}_{\lambda,r}} \tilde{P}_{k,n}(z,w) f_k(w) d\dot{w},$$

where

$$L_{k,\lambda,r} = \lambda^{2k} P_{k,n}\left(\frac{1}{2}\left(\frac{\lambda^2}{r^2} + \frac{r^2}{\lambda^2}\right)\right).$$

and $d\dot{w}$ is the normalized invariant measure of $\tilde{\mathbb{S}}_{\lambda,r}$. We define the generalized Poisson kernel $K_{\lambda,r}(z,w)$ by

$$K_{\lambda,r}(z,w) = \sum_{k=0}^{\infty} \frac{N(k,n)}{L_{k,\lambda,r}} \tilde{P}_{k,n}(z,w).$$

Then we can generalize the Poisson integral (Theorem 3):

Theorem 6 ([7]). *For $f \in \mathcal{O}_\Delta(\tilde{B}[r])$ we have*

$$f(z) = \int_{\tilde{\mathbb{S}}_{\lambda,r}} K_{\lambda,r}(z,w) f(w) d\dot{w}, \quad z \in \tilde{B}(r).$$

If

$$f(z) = \sum_{k=0}^{\infty} \sum_{l=0}^{[k/2]} (z^2)^l f_{k,k-2l}(z) \in \mathcal{O}(\tilde{B}[\lambda, r]),$$

then $f_{k,k-2l}$ is represented by an integral over $\Sigma_{\lambda,r}$

$$f_{k,k-2l}(z) = \frac{N(k-2l,n)}{\lambda^{4l} L_{k-2l,\lambda,r}} \int_{\Sigma_{\lambda,r}} (\bar{w}^2)^l \tilde{P}_{k-2l,n}(z,w) f(w) d\dot{w},$$

where $\dot{dw}$ is the normalized invariant measure on $\Sigma_{\lambda,r}$. Putting

$$H_{\lambda,r}(z,w) = \sum_{k=0}^{\infty}\sum_{l=0}^{[k/2]} \frac{N(k-2l,n)}{\lambda^{4l}L_{k-2l,\lambda,r}}(z^2)^l(\bar{w}^2)^l\tilde{P}_{k-2l,n}(z,w),$$

we can generalize the Cauchy-Hua formula (Theorem 4):

Theorem 7 ([7]). *For $f \in \mathcal{O}(\tilde{B}[\lambda,r])$, we have*

$$f(z) = \int_{\Sigma_{\lambda,r}} H_{\lambda,r}(z,w)f(w)\dot{dw}, \quad z \in \tilde{B}(\lambda,r).$$

Note that the relation (2) between the Poisson kernel and the Cauchy-Hua kernel can be extended as follows:

$$H_{\lambda,r}(z,w) = \frac{K_{\lambda,r}(z,w)}{1-(z/\lambda)^2(\bar{w}/\lambda)^2}.$$

5. N_p norms

The dual Lie norm

$$L^*(z) = \sup\{|z\cdot\zeta|;\ L(\zeta)\le 1\}$$

has the following expression:

$$L^*(z) = \sqrt{(\|z\|^2+|z^2|)/2} = \frac{1}{2}\left(L(z)+M(z)\right), \tag{8}$$

where

$$M(z) = \frac{|z^2|}{L(z)} = \sqrt{\|z\|^2 - \sqrt{\|z\|^4-|z^2|^2}}$$

(see [6]).

For $p \ge 1$, we define the function $N_p(z)$ on $\mathbb{C}^{n+1}$ as follows:

$$N_p(z) = \left\{\frac{1}{2}\left(L(z)^p + M(z)^p\right)\right\}^{1/p}.$$

Note that $N_2(z) = \|z\|$, $N_1(z) = L^*(z)$ by (8), $N_p(z)$ is monotone increasing in p, and $\lim_{p\to\infty} N_p(z) = L(z)$ by (1). We interpret $N_\infty(z) = L(z)$.

If $n=1$, then $N_p(z)$ is equivalent to the Lebesgue L^p norm. In fact, for $z=(z_1,z_2)\in\mathbb{C}^2$ we have $L(z) = \max\{|z_1 \pm iz_2|\}$, $M(z) = \min\{|z_1\pm iz_2|\}$, and

$$N_p(z) = \left\{\frac{1}{2}\left(|z_1+iz_2|^p + |z_1-iz_2|^p\right)\right\}^{1/p}.$$

More generally, we have the following

Theorem 8 ([4],[8]). (i) *$N_p(z)$ are norms on $\mathbb{C}^{n+1}$ for $p \geq 1$.*

(ii) *$N_q(z)$ is the dual norm of $N_p(z)$, where $1/p + 1/q = 1$.*

For $r > 0$ we define the N_p-ball as follows:

$$\tilde{B}_p(r) = \{z \in \mathbb{C}^{n+1}; N_p(z) < r\}$$

$\tilde{B}_1(r)$ is the dual Lie ball, $\tilde{B}_2(r)$ is the Euclidean ball, and $\tilde{B}_\infty(r)$ is interpreted as the Lie ball.

Because $N_p(x) = \|x\|$ for $x \in \mathbb{R}^{n+1}$, we have $\tilde{B}_p(r) \cap \mathbb{R}^{n+1} = B(r)$.

The N_p-ball is a union of truncated Lie balls:

$$\tilde{B}_p(r) = \bigcup \{\tilde{B}[\lambda, R]; 0 < \lambda < R, R^p + \frac{\lambda^{2p}}{R^p} < 2r^p\}.$$

Using this fact we can characterize holomorphic functions f in $\mathcal{O}(\tilde{B}_p(r))$ by the Cauchy-Hadamard growth behavior of their harmonic components $f_{k,k-2l}$. The following theorem generalizes Theorem 5 and theorems for the dual Lie ball or the complex Euclidean ball in [1] or [2]:

Theorem 9 ([3]). *Let (7) be the double series expansion of f and $1 \leq p \leq \infty$.*

$$f \in \mathcal{O}(\tilde{B}_p(r)) \text{ iff } \limsup_{k \to \infty} \left(\left(\frac{2^k l!(k-l)!}{k!} \right)^{\frac{1}{p}} \|f_{k,k-2l}\|_{C(S)} \right)^{\frac{1}{k}} \leq \frac{1}{r}.$$

References

[1] K.Fujita, *On the double series expansion with harmonic components*, Finite or Infinite Dimensional Complex Analysis: Lecture Notes in Pure and Applied Mathematics, Marcel Dekker, New York, 2000, 77–86.

[2] K.Fujita and M.Morimoto, *Holomorphic functions on the dual Lie ball*, Proceedings of the Second ISAAC Congress, Vol 1, Kluwer Academic Publishers, 2000, pp 771–780.

[3] K.Fujita and M.Morimoto, *On the double series expansion of holomorphic functions*, submitted.

[4] T.Kimura, *Norm related to the Lie norm*, to appear in Tokyo J. Math.

[5] M.Morimoto, *Analytic functionals on the Lie sphere*, Tokyo J. Math., 3(1980), 1–35

[6] M.Morimoto, *Analytic Funtionals on the Sphere*, Translation of Mathematical Monographs, AMS, 1998.

[7] M.Morimoto, *A generalization of the Cauchy-Hua integral formula on the Lie ball*, Tokyo J. Math., 22(1999), 177-192.

[8] M.Morimoto and K.Fujita, *Between Lie norm and dual Lie norm*, to appear in Tokyo J. Math., 24(2001)

4

INVARIANCE OF PLURIGENERA AND TORSION-FREENESS OF DIRECT IMAGE SHEAVES OF PLURICANONIAL BUNDLES

Yum-Tong Siu*
Department of Mathematics
Harvard University
Cambridge, MA 02138, U.S.A
E-mail: siu@math.harvard.edu

Abstract: The deformational invariance of plurigenera proved recently by the author is a special case of the torsion-free property of the first direct image sheaf of the pluricanonical line bundle when the target space of the proper surjective holomorphic map is the open unit 1-disk with nonsingular fibers. This article discusses the adaptation of the techniques used in the proof of the deformational invariance of plurigenera to the general problem of proving the torsion-free property of the first direct image sheaf of the pluricanonical line bundle. The discussion covers also the more general case of the torsion-free property of the first direct image of the pluricanonical line bundle after twisting by a Hermitian holomorphic line bundle with semipositive curvature current and by its multiplier ideal sheaf. A number of results are obtained for the torsion-free problems by the adaptation of the techniques of the proof of the deformational invariance of plurigenera.

AMS-classification: 32J25, 32G05, 32C35, 32L10, 32L20, 32U05, 32U25

Table of Contents.

§0. Introduction.
§1. Multiplier ideal sheaves and their global generation.

---

*Partially supported by a grant from the National Science Foundation.

§2. Normal ideal sheaf of hypersurface.
§3. Extension theorem of Ohsawa-Takegoshi type.
§4. Induction argument and extension of twisted sections.
§5. Extension from singular fiber without twisting.
§6. Induction technique on multiplicity of multiple point.

0. Introduction

The deformational invariance of plurigenera proved in [8, 9] is a special case of the torsion-free property of the first direct image sheaf of the *pluricanonical* line bundle. The special case is when the target space of the proper surjective holomorphic map is the open unit 1-disk with nonsingular fibers. Kollar in [3] proved, among a number of other related results, the following torsion-free property of the direct image sheaves of the *canonical* line bundle. Let X and Y be projective varieties and assume that X is smooth. Let $f : X \to Y$ be a surjective map. Then $R^i f_* (\mathcal{O}_X (K_X))$ is torsion-free for $i \geq 0$, where K_X is the canonical line bundle of X. ([3], Th.2.1 (i)).

In this article we discuss the problem of the torsion-free property of the first direct image of the pluricanonical line bundle. Since the verification of the nonexistence of torsion involves only the use of a single holomorphic function, without loss of generality we can confine ourselves to the case when the target manifold is the open unit 1-disk. Let Δ be the open unit 1-disk $\{t \in \mathbf{C} \mid |t| < 1\}$. Unless explicitly mentioned to the contrary, t will be used as the coordinate for $\mathbf{C}$ and, in particular, for Δ. The canonical line bundle of a complex manifold Y is denoted by K_Y. The structure sheaf of a complex space Z is denoted by $\mathcal{O}_Z$. We use the terminology of a ringed space to highlight the structure sheaf of a possibly unreduced complex space, *i.e.*, a complex space whose structure sheaf may contain nonzero nilpotent elements. A ringed space $(Z, \mathcal{O}_Z)$, used to denote a complex space, is a pair whose first component is the underlying space Z and whose second component is the structure sheaf $\mathcal{O}_Z$ which may admit nonzero nilpotent elements.

Problem (0.1). Let X be a complex manifold and $\pi : X \to \Delta$ be a proper surjective holomorphic map (whose fibers may be singular and may have unreduced complex structure). Suppose that there exists some positive line bundle over X (*i.e.*, $\pi : X \to \Delta$ is a family of projective algebraic complex spaces). Let m be an integer ≥ 2. Is the first direct image sheaf $R^1\pi_* (\mathcal{O}_X (mK_X))$ of the m-th power of the canonical line bundle of X torsion-free on Δ ? An equivalent formulation is whether the restriction map from $\Gamma (X, mK_X)$ to $\Gamma \left(X, \left(\mathcal{O}_X/\mathcal{O}_X w\right) (mK_X)\right)$ is surjective for any holomorphic function w on Δ.

The special case of Problem (0.1) where the fibers of $\pi : X \to \Delta$ are all nonsingular (with multiplicity 1) is equivalent to the following result on the deformational invariance of the plurigenera proved in [9].

Theorem (0.2). Let $\pi : X \to \Delta$ be a holomorphic family of compact complex projective algebraic manifolds over Δ with fiber X_t $(t \in \Delta)$. Let m be any positive integer. Then the restriction map

$$\Gamma(X, mK_X) \to \Gamma(X_t, mK_{X_t})$$

is surjective for $t \in \Delta$ and, in particular, the complex dimension of $\Gamma(X_t, m K_{X_t})$ is independent of t for $t \in \Delta$.

Theorem (0.2) is a special case of the following theorem proved in [9] which allows twisting in the deformational invariance of the plurigenera.

Theorem (0.3). Let $\pi : X \to \Delta$ be a holomorphic family of compact complex projective algebraic manifolds over Δ with fiber X_t $(t \in \Delta)$. Let L be a holomorphic line bundle over X with a (possibly singular) metric $e^{-\varphi}$ whose curvature current $\frac{\sqrt{-1}}{2\pi}\partial\bar{\partial}\varphi$ is semi-positive on X such that $e^{-\varphi}|_{X_0}$ is locally integrable on X_0. Let m be any positive integer. Then any element $s \in \Gamma(X_0, m K_{X_0} + L)$ with $|s|^2 e^{-\varphi}$ locally bounded on X_0 can be extended to an element $\tilde{s} \in \Gamma(X, m K_X + L)$ in the sense that $\tilde{s}|_{X_0} = s \wedge \pi^*(dt)$.

Theorem (0.3) motivates the following more general version of Problem (0.1).

Problem (0.4). Let X be a complex manifold and $\pi : X \to \Delta$ be a proper surjective holomorphic map. Suppose that there exists some positive line bundle over X. Let L be a holomorphic line bundle over X with a (possibly singular) metric $e^{-\varphi}$ whose curvature current $\frac{\sqrt{-1}}{2\pi}\partial\bar{\partial}\varphi$ is semi-positive on X. Let $\mathcal{I}_\varphi$ be the multiplier ideal sheaf for the metric $e^{-\varphi}$ in the sense that $\mathcal{I}_\varphi$ consists of all holomorphic function germs f on X with $|f|^2 e^{-\varphi}$ locally integrable on X. Let m be an integer ≥ 2. Is the first direct image sheaf $R^1\pi_*(\mathcal{I}_\varphi(mK_X + L))$ torsion-free on Δ ?

Theorem (0.3) is not yet strong enough to imply even the special case of Problem (0.4) where the fibers of $\pi : X \to \Delta$ are all nonsingular (with multiplicity 1). That special case under the additional assumption of $\mathcal{I}_\varphi \cap (t\,\mathcal{O}_X) = t\,\mathcal{I}_\varphi$ would be a consequence of the following more general conjecture which is still open.

Conjecture (0.5). Let $\pi : X \to \Delta$ be a holomorphic family of compact complex projective algebraic manifolds over Δ with fiber X_t $(t \in \Delta)$. Let L be a holomorphic line bundle over X with a (possibly singular) metric

$e^{-\varphi}$ whose curvature current $\frac{\sqrt{-1}}{2\pi}\partial\bar{\partial}\varphi$ is semi-positive on X. Let m be any positive integer. Let $s \in \Gamma(X_t, mK_{X_t} + L)$ with $|s|^2 e^{-\varphi}$ locally integrable on X_t. Then s can be extended to $\tilde{s} \in \Gamma(X, mK_X + L)$ with $|\tilde{s}|^2 e^{-\varphi}$ locally integrable on X.

In Theorems (0.1), (0.2), (0.3), Problems (0.1), (0.4), and Conjecture (0.5), we assume the existence of a positive line bundle on X. When we replace that assumption by the assumption that X is Kähler, the situation becomes much more difficult and the analogs of Theorems (0.1), (0.2), (0.3) for the Kähler setting still remain open. We will not discuss the Kähler case here and also we will not discuss the higher direct image sheaves $R^q\pi_*(\mathcal{O}_X(mK_X))$ and $R^q\pi_*(\mathcal{I}_\varphi(mK_X + L))$ for $q \geq 2$.

In this article we will present partial results toward the solutions of Problems (0.1) and (0.4) which we can obtain by refining and adapting the methods of [8, 9]. The methods of [8, 9], though useful and powerful for the problem of the deformatonal invariance of plurigenera, have their limitations when applied to Problems (0.1) and (0.4). Theorem (0.6) and Theorem (0.7) below respectively give our partial results toward the solutions of Problem (0.1) and Problem (0.4). Some crucial techniques for the complete solutions of Problems (0.1) and (0.4) are still lacking.

Theorem (0.6). Let $\pi : X \to \Delta$ be a holomorphic family of compact complex projective algebraic manifolds whose fibers are of complex dimension n. Let A be a holomorphic line bundle on X such that for every point P of X there are a finite number of elements of $\Gamma(X, E)$ which all vanish to order at least $n+1$ at P and which do not simultaneously vanish outside P. Let a be a positive integer such that $aK_X - A$ admits a non identically zero global holomorphic section s_D on X which is non identically zero on X_0 and whose divisor in X is D. Let L be a holomorphic line bundle over X with a (possibly singular) metric $e^{-\kappa}$ with semi-positive curvature current on X. Let $\ell \geq 2$ be an integer such that $e^{-\frac{\ell}{\ell-1}\frac{a}{2\ell+a}\kappa}|s_D|^{\frac{-2}{\ell-1}}$ is locally integrable on X. Let m_0 be any positive integer. Then any $s \in \Gamma(X_0, m_0K_{X_0} + L)$ with $\left(|s|^2 e^{-\kappa}\right)^{\frac{p}{m_0}} e^{-\ell\kappa}$ locally integrable on X_0 for $1 \leq p \leq \ell m_0$ can be extended to an element of $\Gamma(X, m_0K_X + L)$.

Theorem (0.7). Let X be a complex manifold of complex dimension $n+1$ and $\pi : X \to \Delta$ be a proper surjective holomorphic map. Let t be the coordinate of Δ. Assume that the set of points of $X_0 := X \cap \{t = 0\}$ where dt vanishes is of complex codimension at least 1 in X_0. Let $\mathcal{J}$ be the ideal sheaf generated by all germs of holomorphic $n+1$ forms on X of the form $\omega \wedge dt$ such that ω is a local holomorphic n-form defined on some open subset U of X whose pullback to the intersection of U and the regular part of X_0 is square integrable. Then every element of $\Gamma\left(X_0, \left(\mathcal{J}^m / \left(\mathcal{J}^m \cap \mathcal{O}_X t\right)\right)(mK_X)\right)$ can be extended to an element of $\Gamma(X, mK_X)$.

Remark (0.8). Theorem (0.6) is an improvement over Theorem (0.3) only in the following two points. The improvement is obtained at the expense of the additional assumption of general type. The first point is the removal of the assumption of local integrability of $e^{-\varphi}$ from Theorem (0.3). The second point is the replacement of the assumption of the local supremum bound of $|s|^2 e^{-\varphi}$ by the local integrability of $|s|^{2k} e^{-\ell\varphi}$ for some finite positive numbers k and ℓ. The significance of the difference between the the supremum bound and the local integrability is that the latter condition defines a coherent ideal sheaf for the holomorphic function germs s while the sheaf defined by the former is in general not coherent (see §1 below).

Remark (0.9). Twisting by a holomorphic line bundle L on X with metric $e^{-\varphi}$ and with φ locally plurisubharmonic can be added to Theorem (0.7) in the same way that twisting is added to Theorem (0.2) to yield Theorem (0.3). We will not give a proof of the statement with the twisting for Theorem (0.67), because the modifications needed for it are quite straightforward and could be readily seen from [9] and the arguments presented in this article.

Remark (0.10). The ideal sheaf $\mathcal{J}$ introduced in Theorem (0.7) is coherent. A proof of its coherence as well as the definitions for it from other perspectives will be given in §2.

For the invariance of plurigenera the main argument is in [8]. The argument by induction on m extends m-canonical sections from a fiber to the family and it relies on the two following ingredients: (i) the global generation of multiplier ideal sheaves on a compact complex manifold after twisting by a sufficiently ample line bundle A which is independent of the multiplier ideal sheaves; (ii) an extension theorem of L^2 holomorphic top-degree forms with respect to a weight $e^{-\varphi}$, with φ plurisubharmonic, from the fiber to the family when $e^{-\varphi}$ is defined on the family.

The induction argument on m works because an m-canonical section can be regarded as a top-degree form with coefficient in the $(m-1)$-canonical bundle. It is based on the following naive motivation. If one could write an element $s^{(m)}$ of $\Gamma(X_0, m K_{X_0})$ as a sum of terms, each of which is the product of an element $s^{(1)}$ of $\Gamma(X_0, K_{X_0})$ and an element $s^{(m-1)}$ of $\Gamma(X_0, (m-1) K_{X_0})$, then one can extend $s^{(m)}$ to an element of $\Gamma(X, m K_X)$ by induction on m. While in general it is clearly impossible to so express $s^{(m)}$ as a sum of such products, one could successfully implement a modified form of the naive motivation, in which $s^{(1)}$ is only a local holomorphic section and $s^{(m-1)}$ is an element of $\Gamma\left(X_0, (m-1)K_{X_0} + A\right)$ instead of $\Gamma\left(X_0, (m-1)K_{X_0}\right)$, where A is an ample line bundle on X independent of m and is sufficiently ample for the global generation of multiplier ideal sheaves.

The argument by induction first enables one to extend a holomorphic m-canonical section from a fiber to the family after multiplying it by a holomorphic section of A. To get rid of the contribution from A, one uses the technique of first raising the section s to a power ℓ for some large ℓ before extension and then uses the ℓ-th root of the absolute value of s to get the correct weight $e^{-\varphi}$ for the application of the extension theorem. To avoid the limiting process of $\ell \to \infty$, the technical assumption of general type is used in [8]. The limiting process of $\ell \to \infty$ poses a convergence problem for the construction of the weight $e^{-\varphi}$ which is needed for the application of the extension theorem.

In [8] the convergence problem of the limiting process of $\ell \to \infty$ is overcome by (i) introducing a Hermitian metric h as singular possible for the canonical line bundle of the fiber with respect to which the m-canonical section on the fiber to be extended is L^2 and (ii) using the concavity of the logarithm in the estimate. The metric h as singular as possible is used to guarantee the uniform bound of the dimension of the space of all L^2 holomorphic ℓm-canonical sections with respect to h on the fiber as $\ell \to \infty$. Such a uniform bound is essential for the estimate by the concavity of the logarithm.

It is the need to use the supremum bound in the handling of the convergence problem that makes impossible a direct adaptation of the method for the proof of Theorem (0.3) to the proof of Conjecture (0.5). The additional assumption of general type in Theorem (0.6) enables us to avoid the limiting process so that we can replace the use of the supremum bound by an L^p bound for some finite p. However, the improvement is still not enough to enable us to stay with the L^2 norm to get Conjecture (0.6).

In his recent preprint [11] posted on the web Tsuji introduced a new generalized Bergman kernel for the pluricanonical line bundle to handle the problem of convergence in the deformational invariance of plurigenera in the argument of [8] when the assumption of general type is removed. Tsuji's generalized Bergman metric is an elegant and natural approach to the convergence problem. Unfortunately one crucial estimate is still lacking and seems unlikely to be establishable, as explained in [9].

Since the arguments in this article are adaptations of those in [8, 9], the emphasis is on the presentation of the adaptations and reformulations. Many details which are in [8, 9] are therefore not provided here in their entirety.

Prior to [8, 9] which uses induction on m to prove the surjectivity of $\Gamma(X, mK_X) \to \Gamma(X_0, mK_{X_0})$, Levine [4] used the degeneracy of the Hodge to DeRham spectral sequences to extend, in the Kähler setting, elements of $\Gamma(X_0, mK_{X_0})$ to over a double point at the origin of Δ. By induction on the multiplicity of the multiple point at the origin of Δ over which elements of $\Gamma(X_0, mK_{X_0})$ can be extended, he also verified the deformational invariance of the plurigenera for the Kähler setting, provided that the general section of the m-canonical bundle defines a nonsingular divisor or a divisor with only "mild" singularities.

In applying the methods of [8, 9] to the invariance of twisted plurigenera and to Problem (0.4) the main difficulty is the sufficiently ample line bundle A which is required for the global generation of multiplier ideal sheaves. It is possible to combine the methods of multiplier ideal sheaves and the process, introduced by Levine [4], of induction on the multiplicity of multiple points in the base over which extension of the section is possible. Such a combination has the advantage that no sufficiently ample line bundle A is needed. However, there are still obstacles beyond the stage of the double points in this combination of techniques. The reason is that the fibers over the multiple points of the base are unreduced complex spaces and there are still no vanishing theorems of Kawamata-Viehweg-Nadel [2, 12, 5] available for such unreduced complex spaces.

At the end of this article (§7) we will discuss the combination of the methods of multiplier ideal sheaves and the process of induction on the multiplicity of the multiple point. We will illustrate this combination of techniques in the proof of a result on the extension of twisted pluricanonical sections on a reduced fiber to the unreduced fiber over the double point of the base.

1. Multiplier Ideal Sheaves and Their Global Generation

For a plurisubharmonic function φ on an open subset Ω of $\mathbf{C}^n$, the multiplier ideal sheaf $\mathcal{I}_\varphi$ is defined as the sheaf of all holomorphic function germs f on Ω such that $|f|^2 e^{-\varphi}$ is locally integrable on Ω. The multiplier ideal sheaf $\mathcal{I}_\varphi$ is coherent and was first introduced in [5].

Multiplier ideal sheaves are defined by L^2 conditions with respect to $e^{-\varphi}$. One can generalize the notion to L^p multiplier ideal sheaves for other values of $p \geq 1$. The L^p multiplier ideal sheaf $\mathcal{I}_{\varphi,L^p}$ is defined as the sheaf of all holomorphic function germs f such that $|f|^p e^{-\varphi}$ is locally integrable on Ω. The multiplier ideal sheaves $\mathcal{I}_\varphi$ introduced by Nadel are the L^2 multiplier ideal sheaves $\mathcal{I}_{\varphi,L^2}$.

When $1 \leq p < \infty$, the L^p multiplier ideal sheaf $\mathcal{I}_{\varphi,L^p}$ is also coherent. Since we will not use the coherence of $\mathcal{I}_{\varphi,L^p}$ for $1 \leq p < \infty$ in this article, we will only very briefly sketch here the arguments for the coherence without providing all the details. We first consider the special case when $\varphi = \alpha \log |g|^2$ for some $\alpha > 0$ and some holomorphic function g on Ω. For that special case, we take a resolution $\pi : Y \to \Omega$ of the embedded singularity of the zero-set of g and a collection of nonsingular hypersurfaces $\{E_j\}_j$ in Y in normal crossing so that the divisor of $\pi^* g$ is of the form $\sum_j a_j E_j$ and $K_Y - \pi^* K_\Omega = \sum_j b_j E_j$. Then $\mathcal{I}_{\varphi,L^p}$ is the zeroth direct image, under π, of the ideal sheaf on Y for the divisor $\sum_j \left\lceil \frac{1}{p} \max\left(0, \lfloor \alpha\, a_j - b_j \rfloor\right)\right\rceil$, where $\lfloor u \rfloor$ (respectively $\lceil u \rceil$) means the largest integer not exceeding u (respectively the smallest integer no less than u). For the general case, one uses the coherence criterion of generation by global sections. Take a local section f. We argue that the local integrability $|f|^p e^{-\varphi}$ is an open condition in the sense that $|f|^p e^{-(1+\varepsilon)\varphi}$ is still locally integrable for some sufficiently small $\varepsilon > 0$. Locally one finds a sequence of real numbers α_ν and a sequence of holomorphic functions g_ν such that the difference of the Lelong number of $\sqrt{-1}\partial\bar\partial \left(\alpha_\nu \log |g_\nu|^2\right)$ and the Lelong number of $\sqrt{-1}\partial\bar\partial\left((1+\varepsilon)\varphi\right)$ goes to zero uniformly on compact subsets as $\nu \to \infty$ (see *e.g.*, [1], p.346, Proposition 9.1). Take $0 < \varepsilon' < \varepsilon$. Then locally for ν sufficiently large,

$$\mathcal{I}_{(1+\varepsilon)\varphi,L^p} \subset \mathcal{I}_{\alpha_\nu (1+\varepsilon') \log|g_\nu|^2, L^p} \subset \mathcal{I}_{\varphi,L^p}.$$

When $p = \infty$, we can also define $\mathcal{I}_{\varphi,L^\infty}$ as the sheaf of all holomorphic function germs f on Ω such that $|f|^2 e^{-\varphi}$ is locally bounded on Ω. However, the ideal sheaf $\mathcal{I}_{\varphi,L^\infty}$ in general is not coherent. An easy example is the plurisubharmonic function $\varphi = \sum_{j=1}^\infty \varepsilon_j \log|z - c_j|$ on $\Omega = \mathbf{C}$ with $c_j \in \mathbf{C}$ and $\varepsilon_j > 0$ such that $c_j \to 0$ and $\varepsilon_j \to 0$ sufficiently rapidly. Then the stalk of $\mathcal{I}_{\varphi,L^\infty}$ at 0 is zero, but its stalk and at any $c \neq 0$ near 0 is nonzero. The difference between the case of $1 \leq p < \infty$ and the case of $p = \infty$ is that the condition for $|f|^2 e^{-\varphi}$ to be locally bounded is not an open condition and we cannot conclude the local boundedness of $|f|^2 e^{-(1+\varepsilon)\varphi}$ for any sufficiently small $\varepsilon > 0$ from the local boundedness of $|f|^2 e^{-\varphi}$.

For the proof of Theorem (0.6) we need only the following original statement on the global generation of multiplier ideal sheaves in [8].

Proposition (1.1)(Global Generation of Multiplier Ideal Sheaves). Let L be a holomorphic line bundle over an n-dimensional compact complex manifold Y with a Hermitian metric which is locally of the form $e^{-\xi}$ with ξ plurisubharmonic. Let $\mathcal{I}_\xi$ be the multiplier ideal sheaf of the Hermitian metric $e^{-\xi}$. Let A be an ample holomorphic line bundle over Y such that for every point P of Y there are a finite number of elements of $\Gamma(Y, A)$ which all vanish to order at least $n+1$ at P and which do not simultaneously vanish outside P. Then $\Gamma(Y, \mathcal{I}_\xi \otimes (L + A + K_Y))$ generates $\mathcal{I}_\xi \otimes (L + A + K_Y)$ at every point of Y.

For the proof of Theorem (0.7) we need the effective version of the global generation of multiplier ideal sheaves. To keep the notations as simple as possible, we restate the effective version in [[9], p.234, Theorem 2.1] in a simpler form with less precise constants of estimation but still sufficient for our purpose. We state it here for a general line bundle L rather just for the line bundle $(m-1) K_X$ in [[9], p.234, Theorem 2.1]. The arguments for a general line bundle is completely analogous to the special case $(m-1) K_X$. Actually for the proof of Theorem (0.7) only the special case $L = (m-1) K_X$ is needed. We still state our effective version here for the general line bundle L, because the notations are easy to follow for the general case.

Proposition (1.2)(Effective Global Generation of Multiplier Ideal Sheaves) Let L be a holomorphic line bundle over an n-dimensional compact complex manifold Y with a Hermitian metric which is locally of the form $e^{-\xi}$ with ξ plurisubharmonic. Let $\mathcal{I}_\xi$ be the multiplier ideal sheaf of the Hermitian metric $e^{-\xi}$. Let A be an ample holomorphic line bundle over Y such that for every point P of Y there are a finite number of elements of $\Gamma(Y, A)$ which all vanish to order at least $n+1$ at P and which do not simultaneously vanish outside P. Let $\mathfrak{C}$ be a finite open cover of Y. Let N be the complex dimension of the subspace of all elements

$$s \in \Gamma(Y, K_Y + L + A)$$

such that

$$\int_Y |s|^2 \, e^{-\xi} \, h_A < \infty \, .$$

Then there exists a constant $C^\flat$ and there exist open subsets U_λ, U_λ^* of Y $(1 \le \lambda \le \Lambda)$ such that U_λ is relatively compact in U_λ^* and $\bigcup_{\lambda=1}^{\Lambda} U_\lambda = Y$ and each U_λ^* is contained in some member of $\mathfrak{C}$ and the following property is satisfied. There exist

$$\sigma_1, \cdots, \sigma_N \in \Gamma\left(Y, K_Y + L + A\right)$$

with

$$\int_Y \left|\sigma_k\right|^2 e^{-\xi} h_A \leq 1$$

$(1 \leq k \leq N)$ such that, if $1 \leq \lambda \leq J$ and if $s \in \Gamma\left(U_\lambda^*, m K_Y + A\right)$ with

$$\int_{U_\lambda^*} |s|^2 e^{-\xi} h_A = C_s < \infty ,$$

then one can find holomorphic functions b_k on U_λ such that

$$s = \sum_{k=1}^{N} b_k\, \sigma_k^{(m)}$$

on U_λ and

$$\sup_{U_\lambda} \sum_{k=1}^{N} \left|b_k\right|^2 \leq C^b\, C_s \, .$$

Definition (1.3). To facilitate the description of the global generation of multiplier ideal sheaves, we introduce the following definition to describe the condition on the ample line bundle A in Propositions (1.1) and (1.2). Let Y be a complex manifold of complex dimension n. A positive holomorphic line bundle A on Y is said to be *sufficiently ample for the global generation of multiplier ideal sheaves on* Y if for every point P of Y there are a finite number of elements of $\Gamma(Y, A)$ which all vanish to order at least $n+1$ at P and which do not simultaneously vanish outside P.

2. Normal Ideal Sheaf of Hypersurface

We now describe the ideal sheaf $\mathcal{J}$ of Theorem (0.7) from other perspectives and prove its coherence.

Let Z be a complex submanifold of complex codimension 1 in a complex manifold Y of complex dimension n defined by $w = 0$ for some holomorphic function w on Y so that dw is nonzero at every point of Z. Let T_Z (respectively T_Y) be the tangent bundle of Z (respectively Y) and $N_{Z|Y}$ be the normal bundle of Z in Y. The sequence

$$(2.0.1) \qquad 0 \rightarrow T_Z \rightarrow T_Y|_Z \rightarrow N_{Z|Y} \rightarrow 0$$

is exact over Z. The bundle-homomorphism $T_Y|Z \rightarrow Z \times \mathbf{C}$ defined by evaluation by dw induces a bundle-isomorphism $\Psi : N_{Z|Y} \rightarrow Z \times \mathbf{C}$

between $N_{Z|Y}$ and the trivial line bundle $Z \times \mathbf{C}$ over Z. From the exact sequence (2.0.1) and the isomorphism Ψ we have an isomorphism $\Phi : K_Z \to K_Y|_Z$ defined as follows. For a local $(n-1)$-form ω on Z we extend it to a local $(n-1)$-form $\tilde{\omega}$ on Y. Then $\Phi(\omega) = \tilde{\omega} \wedge dw$. We denote $\Phi^{-1}(s)$ by $s \sqcap dw$ for any local section s of K_Y over Z.

Another description of $s \sqcap dw$ is as follows. Let

$$s \in \Gamma\left(Z, \left(\mathcal{O}_Y / \mathcal{O}_Y w\right)(K_Y)\right).$$

Then there exists a unique holomorphic $(n-1)$-form on ω on Z such that $\omega \wedge dw = s$ on Z in the sense that every point Q of Z admits an open neighborhood U in Z and $\tilde{\omega} \in \Gamma\left(U, \bigwedge^{n-1} T_Y^*\right)$ which is pulled back to ω on U and which satisfies $\tilde{\omega} \wedge dw = s$ at every point of $Z \cap U$. (Here $T{*}_Y$ is the dual bundle of T_Y and $\bigwedge^{n-1} T_Y^*$ is the exterior product of $(n-1)$ copies of T_Y^*.) For, if $\tilde{\omega}_1, \tilde{\omega}_2 \in \Gamma\left(U, \bigwedge^{n-1} T_Y^*\right)$ with $\tilde{\omega}_j \wedge dw = s$ at every point of $Z \cap U$ for $j = 1, 2$, then

$$\tilde{\omega}_1 - \tilde{\omega}_2 = \tau \wedge dw$$

at every point of U' for some open neighborhood U' of Q in U and for some $\tau \in \Gamma\left(U', \bigwedge^{n-2} T_Y^*\right)$, as one can readily see by expressing $\tilde{\omega}_j$ in terms of a local holomorphic basis of T_Y^* with dw as a member. We can now define $s \sqcap dw$ as ω.

Definition (2.1). Let Ω be an open subset of $\mathbf{C}^n$ and w be a holomorphic function on Ω. Let E be the set of points of Ω where $w = 0$ and $dw \neq 0$. Assume that E is of complex codimension ≥ 2 in Ω. Let $\mathcal{I}_{\sqcap dw}$ be the ideal sheaf such that its stalk at $Q \in \Omega$ consists of all of holomorphic functions f on Ω such that

$$\int_{\{w=0\} \cap (U-E)} \left| f\,(dz_1 \wedge \cdots \wedge dz_n) \sqcap dw \right|^2 < \infty$$

for some open neighborhood U of Q in Ω. We call the ideal sheaf $\mathcal{I}_{\sqcap dw}$ the *normal ideal sheaf* of the hypersurface defined by w.

Remark (2.2). It is by a considerable abuse of language and for lack of a better name that the ideal sheaf $\mathcal{I}_{\sqcap dw}$ is called the normal ideal sheaf of the hypersurface defined by w. When $Y = \Omega - E$ and $Z = Y \cap \{w = 0\}$ with the continuation of the above notations, the normal bundle $N_{Z|Y}$ is naturally isomorphic to $\mathrm{Hom}\,(K_Y, K_Z)$ through the pairing map

$$(\omega, g) \mapsto \tilde{\omega} \wedge g\, dw,$$

where ω is a local $(n-1)$-form on Z and g is a local holomorphic function on Z so that $g\,dw$ defines a local section of the dual $N^*_{Z|Y}$ of $N_{Z|Y}$ and $\tilde{\omega}\wedge g\,dw$ is a local n-form on Y. Since K_Y has a global basis $dz_1\wedge\cdots\wedge dz_n$, the natural isomorphism between $N_{Z|Y}$ and $\text{Hom}\,(K_Y, K_Z)$ identifies a local section of $N_{Z|Y}$ as a local section of K_Z. The ideal sheaf $\mathcal{I}_{\sqcap dw}$ on Ω is the subsheaf of all holomorphic function germs g defined on U such that ω is a square integrable $(n-1)$-forms on $Z\cap U$ when $\tilde{\omega}\wedge g\,dw = dz_1\wedge\cdots\wedge dz_n$. The designation of $\mathcal{I}_{\sqcap dw}$ as the normal ideal sheaf of $w=0$ is motivated by the fact that the equation $\tilde{\omega}\wedge g\,dw = dz_1\wedge\cdots\wedge dz_n$ makes the sheaf of holomorphic function germs g naturally isomorphic with $\mathcal{O}_\Omega\left(N_{Z|Y}\right)$ on $\Omega - E$ when the transition functions of the two underlying bundles on $\Omega - E$ are compared.

Lemma (2.3). Let $w = z_1\cdots z_k\, z_{k+1}^{\nu_{k+1}}\cdots z_n^{\nu_n}$ in $\mathbf{C}^n$, where $\nu_j \geq 2$ for $1\leq j\leq n$. Then normal ideal sheaf $\mathcal{I}_{\sqcap dw}$ on $\mathbf{C}^n$ of the hypersurface defined by w is equal to the ideal sheaf on $\mathbf{C}^n$ generated by

$$z_1\cdots z_{\ell-1}z_{\ell+1}\cdots z_k\, z_{k+1}^{\nu_{k+1}}\cdots z_n^{\nu_n}\quad (1\leq \ell\leq k).$$

Proof. For $1\leq\ell\leq k$ let $E_\ell = \{z_\ell = 0\}\cap\left(\bigcup_{j\neq\ell}\{z_j=0\}\right)$. Consider a holomorphic n-form

$$\omega := f\,(z_1,\cdots,z_n)\,dz_1\wedge\cdots\wedge dz_n,$$

where $f\,(z_1,\cdots,z_n)$ is a local holomorphic function germ on $\mathbf{C}^n$ at 0. Let

$$\omega\sqcap dw = \sum_{j=1}^n g_j\,(z_1,\cdots,z_n)\,dz_1\wedge\cdots\wedge dz_{j-1}\wedge dz_{j+1}\wedge\cdots\wedge dz_n.$$

Then on E_ℓ we have

$$\begin{aligned} &f\,(z_1,\cdots,z_n)\,dz_1\wedge\cdots\wedge dz_n\\ &= \sum_{j=1}^n g_j dz_1\wedge\cdots\wedge dz_{j-1}\wedge dz_{j+1}\wedge\cdots\wedge dz_n\wedge dw\\ &= g_\ell dz_1\wedge\cdots\wedge dz_{\ell-1}\wedge dz_{\ell+1}\wedge\cdots\wedge dz_n\wedge\left(z_1\cdots z_{\ell-1}z_{\ell+1}\cdots z_k z_{k+1}^{\nu_{k+1}}\cdots z_n^{\nu_n}dz_\ell\right)\end{aligned}$$

and

$$g_\ell = \frac{(-1)^{\ell-n}f}{z_1\cdots z_{\ell-1}z_{\ell+1}\cdots z_k z_{k+1}^{\nu_{k+1}}\cdots z_n^{\nu_n}}.$$

For $\omega\sqcap dw$ to be locally square integrable on E_ℓ at the origin, one must have the divisibility of $f|_{E_\ell}$ by

$$z_1\cdots z_{\ell-1}z_{\ell+1}\cdots z_k z_{k+1}^{\nu_{k+1}}\cdots z_n^{\nu_n}$$

on E_ℓ at the origin as holomorphic function germs. For this to hold for every $1 \leq \ell \leq k$, it is necessary and sufficient f belongs to the ideal generated by

$$z_1 \cdots z_{\ell-1} z_{\ell+1} \cdots z_k z_{k+1}^{\nu_{k+1}} \cdots z_n^{\nu_n} \quad (1 \leq \ell \leq k).$$

Q.E.D.

Remark (2.4). The normal ideal sheaf $\mathcal{I}_{\sqcap dw}$ can also be described by the finiteness of the limit of integrals of tubular neighborhoods. Let Ω be an open subset of $\mathbf{C}^n$ and w be a holomorphic function on Ω. Let E be the set of points of Ω where $w = 0$ and $dw \neq 0$. Assume that E is of complex codimension ≥ 2 in Ω. If $f \in (\mathcal{O}_\Omega)_Q$ is a holomorphic function germ on Ω at $Q \in \Omega$ which is defined on some open neighborhood U of Q in Ω, then f belongs to the stalk of $\mathcal{I}_{\sqcap dw}$ at Q if and only if

$$\lim_{\varepsilon \to 0+} \frac{1}{\varepsilon^2} \int_{U' \cap \{|w| < \varepsilon\}} |f \, dz_1 \wedge \cdots \wedge dz_n|^2 < \infty$$

for some open neighborhood U' of Q in U.

According to Lemma (2.3), even in the case $n = 2$ and $w = z_1 z_2$ where the singularity of the zero-set of w in $\mathbf{C}^2$ is a singularity of normal crossing at the origin, the ideal sheaf $\mathcal{I}_{\sqcap dw}$ at the origin is different from the unit ideal and is equal to the maximum ideal of $\mathbf{C}^2$. When one uses the description of $\mathcal{I}_{\sqcap dw}$ by the finiteness of the limit of integrals of tubular neighborhoods, one can understand why $\mathcal{I}_{\sqcap dw}$ is different from the unit ideal at the origin by computing the limiting behavior of the volume V_ε of the regular part of

$$\{|z_1 z_2| < \varepsilon, |z_1| < 1, |z_2| < 1\}$$

as a function of $\varepsilon > 0$ when $\varepsilon \to 0$. Let $\eta_j = |z_j|^2$ for $j = 1, 2$. Then

$$\begin{aligned} V_\varepsilon &= \pi^2 \int_{\eta_1=0}^{1} \int_{\eta_2=0}^{\max\left(\frac{\varepsilon^2}{\eta_1}, 1\right)} d\eta_1 d\eta_2 = \pi^2 \int_{\eta_1=0}^{1} \max\left(\frac{\varepsilon^2}{\eta_1}, 1\right) d\eta_1 \\ &= \pi\varepsilon^2 + \pi \int_{\eta_1=\varepsilon^2}^{1} \frac{\varepsilon^2}{\eta_1} d\eta_1 = \pi\varepsilon^2 \left(1 + \log \frac{1}{\varepsilon^2}\right). \end{aligned}$$

Hence

$$\lim_{\varepsilon \to 0+} \frac{1}{\varepsilon^2} V_\varepsilon = \lim_{\varepsilon \to 0+} \frac{1}{\varepsilon^2} \left(\pi\varepsilon^2 \left(1 + \log \frac{1}{\varepsilon^2}\right)\right) = \infty.$$

Lemma (2.5). Let Ω be an open subset of $\mathbf{C}^n$ and w be a holomorphic function on Ω. Let E be the set of points of Ω where $w = 0$ and $dw \neq 0$. Assume that E is of complex codimension ≥ 2 in Ω. Then the normal ideal sheaf $\mathcal{I}_{\sqcap dw}$ of the hypersurface defined by w is coherent on Ω.

Proof. Let $\pi : Y \to \Omega$ be a resolution of the embedded singularities of $\{w = 0\}$ such that

(i) $\pi|_{Y-\pi^{-1}(E)} : Y - \pi^{-1}(E) \to \Omega - E$ is a biholomorphism,

(ii) the pullback of the ideal $\mathcal{O}_\Omega w$ to Y is the ideal of a divisor $\sum_{j\in J} a_j V_j$, where a_j is a positive integer and $\{V_j\}_{j\in J}$ is a collection of nonsingular hypersurfaces in Y in normal crossing, and

(iii) $K_Y - \pi^* K_\Omega = \sum_{j\in J} b_j V_j$, where b_j is a nonnegative integer.

Let J_0 be the subset of J such that $j \in J_0$ if and only if π maps V_j biholomorphically onto a branch of the zero-set of w. Then $a_j = 1$ and $b_j = 0$ for $j \in J_0$. Let $c_j = \max(0, a_j - b_j)$ for $j \in J$. Then $c_j = 1$ for $j \in J_0$. Let ζ_j be the local defining function of V_j for $j \in J$. Let $\mathcal{J}$ be the ideal sheaf on Y generated by the local functions

$$\prod_{j\in J, j\neq \ell} (\zeta_j)^{c_j} \quad \text{for } \ell \in J_0.$$

By the argument in the proof of Lemma (2.3) the ideal sheaf $\mathcal{I}_{\sqcap dw}$ is equal to zeroth direct image of $\mathcal{J}$ under π and hence is coherent. Q.E.D.

Remark (2.6). The normal ideal sheaf $\mathcal{I}_{\sqcap dw}$ in some cases is the unit ideal at points where dw vanishes. The Fermat hypersurface cone $w = z_1^k + \cdots + z_n^k$ for different values of k and n is a good example for illustration. We blow up the origin of $\mathbf{C}^n$ to get $\pi : Y \to \mathbf{C}^n$ with exceptional fiber $E = \pi^{-1}(0)$. Then E is biholomorphic to $\mathbf{P}_{n-1}$ and Y is biholomorphic to the total space of the line bundle $\mathcal{O}_{\mathbf{P}_{n-1}}(-1)$ over $E = \mathbf{P}_{n-1}$. Let $\theta : Y \to E$ be the bundle map $\mathcal{O}_{\mathbf{P}_{n-1}}(-1) \to \mathbf{P}_{n-1}$. Let V be the Fermat hypersurface in $\mathbf{P}_{n-1}$ consisting of all points whose homogeneous coordinates $[z_1, \cdots, z_n]$ satisfy $z_1^k + \cdots + z_n^k = 0$. Let $\tilde{V} = \theta^{-1}(V)$. Then the divisor of $\pi^* w$ is equal to $k\,E + \tilde{V}$. The two hypersurfaces E and $\tilde{V}$ of Y intersect normally and $K_Y = (n-1)\,E$. Let $\mathcal{J}$ be the ideal sheaf on Y for the divisor $\max(0, k - (n-1))\,E$. From the argument in the proof of Lemma (2.3), we conclude that the ideal sheaf $\mathcal{I}_{\sqcap dw}$ is equal to zeroth direct image of $\mathcal{J}$ under π. So the ideal sheaf $\mathcal{I}_{\sqcap dw}$ is the unit ideal if and only if $k \leq n-1$. This example also shows that, when $k \geq n$, the holomorphic $(n-1)$-form $(dz_1 \wedge \cdots \wedge dz_n) \sqcap dw$ on the regular part of the affine hypersurface $z_1^k + \cdots + z_n^k = 0$ is not square integrable on the regular part of $\{z_1^k + \cdots + z_n^k = 0\} \cap U$ for any open neighborhood U of the origin in $\mathbf{C}^n$, even though the affine hypersurface $z_1^k + \cdots + z_n^k = 0$ is normal when $n \geq 3$.

3. Extension Theorem of Ohsawa-Takegoshi Type.

Ohsawa-Takegoshi [7] and Ohsawa [6] developed the theory of extension of holomorphic L^2 sections from a submanifold to an ambient manifold with respect to a weight $e^{-\varphi}$ with φ plurisubharmonic on the ambient manifold. The form of such an extension theorem which we need is the following.

Theorem (3.1) (Extension Theorem of Ohsawa-Takegoshi Type). Let Y be a complex manifold of complex dimension n. Let w be a bounded holomorphic function on Y with nonsingular zero-set Z so that dw is nonzero at any point of Z. Let L be a holomorphic line bundle over Y with a (possibly singular) metric $e^{-\kappa}$ whose curvature current is semi-positive. Assume that there exists a hypersurface V in Y such that $V \cap Z$ is a subvariety of codimension at least 1 in Z and $Y - V$ is the union of a sequence of Stein subdomains Ω_ν of smooth boundary and Ω_ν is relatively compact in $\Omega_{\nu+1}$. If f is an L-valued holomorphic $(n-1)$-form on Z with

$$\int_Z |f|^2 \, e^{-\kappa} < \infty \, ,$$

then $f dw$ can be extended to an L-valued holomorphic n-form F on Y such that

$$\int_Y |F|^2 \, e^{-\kappa} \leq 8\pi \, e \sqrt{2 + \frac{1}{e}} \left(\sup_Y |w|^2 \right) \int_Z |f|^2 \, e^{-\kappa} \, .$$

Remark (3.2). Theorem (3.1) is given and proved in [[9], p.241, Theorem 3.1]. In its proof there are some misprints on page 246, lines 7 and 8 from the bottom, in [9] where the formula should correctly read

$$\begin{aligned} |(u, g_\varepsilon)_{\Omega,\psi}|^2 &\leq \left(\int_\Omega |\langle u, g_\varepsilon \rangle| \, e^{-\psi} \right)^2 \\ &= \left(\int_\Omega \left| \left\langle u, \frac{\tilde{f} \wedge dw}{\varepsilon^2} \varrho' \left(\frac{|w|^2}{\varepsilon^2} \right) d\bar{w} \right\rangle \right| e^{-\psi} \right)^2 . \end{aligned}$$

4. Induction Argument and Extension of Twisted Sections

We give now the induction argument for the extension of twisted pluricanonical sections from a fiber to the family which is used in the proof of Theorem (0.6).

Proposition (4.1). Let $\pi : X \to \Delta$ be a holomorphic family of compact complex projective algebraic manifolds. Let $m \geq 2$ be an integer. Let E be a holomorphic line bundle over X with a (possibly singular) metric $e^{-\varphi_E}$ with semi-positive curvature current on X. Let $e^{-\varphi_p}$ be a metric for $p K_{X_0} + E$ over X_0 with semi-positive curvature current on X_0 for $0 \leq p < m$. Let A be a holomorphic line bundle on X which is sufficiently ample for the global generation of multiplier ideal sheaves on X. Assume that

(i) $\mathcal{I}_{\varphi_p} \subset \mathcal{I}_{\varphi_{p-1}}$ on X_0 for $0 < p < m$, and

(ii) $\mathcal{I}_{\varphi_E|X_0}$ agrees with $\mathcal{I}_{\varphi_0}$ on X_0 (where $\mathcal{I}_{\varphi_E|X_0}$ means the multiplier ideal sheaf on X_0 for the metric $e^{-\varphi|X_0}$ of $E|X_0$ on X_0).

Let f be an element of $\Gamma\left(X_0, m K_{X_0} + E + A\right)$ which locally on X_0 belongs to the multiplier ideal sheaf $\mathcal{I}_{\varphi_{m-1}}$. Then f can be extended to an element $\tilde{f}$ of $\Gamma\left(X, m K_X + E + A\right)$.

Proof. By the assumption on A, we can apply Proposition (1.1) to obtain the following open cover for X_0. We cover X_0 by a finite number of open subsets U_λ $(1 \leq \lambda \leq \Lambda)$ so that, for some nowhere zero element $\xi_\lambda \in \Gamma\left(U_\lambda, -K_{X_0}\right)$, we can write

$$\xi_\lambda\, f|U_\lambda = \sum_{k=1}^{N_{m-1}} b_k^{(m-1,\lambda)} s_k^{(m-1)},$$

where $b_k^{(m-1,\lambda)}$ is a holomorphic function on U_λ and

$$s_k^{(m-1)} \in \Gamma\left(X_0, \mathcal{I}_{\varphi_{m-2}}\left((m-1)\, K_{X_0} + E + A\right)\right)$$

for $1 \leq k \leq N_{m-1}$. We can further assume that the open subsets U_λ are chosen small enough so that, for $1 \leq p \leq m-2$, inductively we can write

$$\xi_\lambda\, s_j^{(p+1)}|U_\lambda = \sum_{k=1}^{N_p} b_{j,k}^{(p,\lambda)} s_k^{(p)},$$

where $b_{j,k}^{(p,\lambda)}$ is a holomorphic function on U_λ and

$$s_k^{(p)} \in \Gamma\left(X_0, \mathcal{I}_{\varphi_{p-1}}\left(p\, K_{X_0} + E + A\right)\right)$$

for $1 \leq k \leq N_p$. We are going to verify the following claim by induction on $1 \leq p < m$.

Claim (4.1.1).

$$s_j^{(p)} \in \Gamma\left(X_0, p K_{X_0} + E + A\right)$$

can be extended to

$$\tilde{s}_j^{(p)} \in \Gamma\left(X, p K_X + E + A\right)$$

for $1 \leq j \leq N_p$ and $1 \leq p < m$.

The case $p = 1$ follows from assumption on the agreement between $\mathcal{I}_{\varphi_E | X_0}$ and $\mathcal{I}_{\varphi_0}$ on X_0 and Theorem (3.1). Now we go from Step p to Step $p+1$. Let

$$\chi_p = \log \sum_{k=1}^{N_p} \left| \tilde{s}_k^{(p)} \right|^2$$

so that $e^{-\chi_p}$ is a metric of $p K_X + E + A$. Since $\tilde{s}_j^{(p)} | X_0 = s_j^{(p)}$ for $1 \leq p \leq N_p$ by Step p, the germ of $s_j^{(p+1)}$ at any point of X_0 belongs to $\mathcal{I}_{\chi_p | X_0}$ for $1 \leq j \leq N_{p+1}$. By Theorem (3.1),

$$s_j^{(p+1)} \in \Gamma\left(X_0, (p+1) K_{X_0} + E + A\right)$$

can be extended to

$$\tilde{s}_j^{(p+1)} \in \Gamma\left(X, (p+1) K_X + E + A\right),$$

finishing the verification of Step $p+1$ of Claim (4.1.1).

Since $\tilde{s}_j^{(m-1)} | X_0 = s_j^{(m-1)}$ for $1 \leq p \leq N_{m-1}$ from Claim (4.1.1), the germ of f at any point of X_0 belongs to $\mathcal{I}_{\chi_{m-1} | X_0}$. By Theorem (3.1),

$$f \in \Gamma\left(X_0, m K_{X_0} + E + A\right)$$

can be extended to

$$\tilde{f} \in \Gamma\left(X, m K_X + E + A\right).$$

Q.E.D.

(4.2) *Proof of Theorem (0.6).* Let $m = \ell m_0$. Let $E = \ell L + D$ and $\varphi_E = \ell \kappa + \log |s_D|^2$. Let

$$\varphi_p = \log |s|^{\frac{2p}{m_0}} + \frac{m-p}{m_0} \kappa + \log |s_D|^2$$

for $1 \leq p \leq m$ and $\varphi_0 = \varphi_E$. Then $e^{-\varphi_p}$ is a metric for $pK_{X_0} + \ell L + D = pK_{X_0} + E$. We are going to verify the following claim.

Claim (4.2.1). $\mathcal{I}_{\varphi_p} \subset \mathcal{I}_{\varphi_{p-1}}$ for $1 \leq p \leq m$.

To verify Claim (4.2.1), we take a local holomorphic function g and assume that g belongs locally to $\mathcal{I}_{\varphi_p}$ for some $1 \leq p \leq m$ so that

$$\int \frac{|g|^2}{|s_D|^2 |s|^{\frac{2p}{m_0}}} e^{-\frac{m-p}{m_0}\kappa} < \infty. \tag{4.2.2}$$

We have to verify that g belongs locally to $\mathcal{I}_{\varphi_{p-1}}$ so that

$$\int \frac{|g|^2}{|s_D|^2 |s|^{\frac{2(p-1)}{m_0}}} e^{-\frac{m-p+1}{m_0}\kappa} < \infty.$$

Since

$$\frac{e^{-\frac{m-p}{m_0}\kappa}}{|s|^{\frac{2p}{m_0}}}$$

is bounded from below by a positive number, it follows from (4.2.2) that the local meromorphic function $\frac{g}{s_D}$ is L^2 and hence is locally equal to a holomorphic function G. So we need only show that

$$\int \frac{|G|^2}{|s|^{\frac{2p}{m_0}}} e^{-\frac{m-p}{m_0}\kappa} < \infty$$

implies

$$\int \frac{|G|^2}{|s|^{\frac{2(p-1)}{m_0}}} e^{-\frac{m-p+1}{m_0}\kappa} < \infty. \tag{4.2.3}$$

We now take any $0 < \eta < 1$ and rewrite the left-hand side of (4.2.3) as

$$\begin{aligned}
&\int \frac{|G|^2}{|s|^{\frac{2(p-1)}{m_0}}} e^{-\frac{m-p+1}{m_0}\kappa} \\
&= \int |G|^2 \left(\frac{e^{-\frac{p-\eta}{p}\frac{m-p}{m_0}\kappa}}{|s|^{\frac{2(p-\eta)}{m_0}}} \right) \left(|s|^{\frac{2(1-\eta)}{m_0}} e^{-\left(\frac{m-p+1}{m_0} - \frac{p-\eta}{p}\frac{m-p}{m_0}\right)\kappa} \right) \\
&= \int |G|^2 \left(\frac{e^{-\frac{p-\eta}{p}\frac{m-p}{m_0}\kappa}}{|s|^{\frac{2(p-\eta)}{m_0}}} \right) \left(|s|^{\frac{2(1-\eta)}{m_0}} e^{-\frac{\eta m+(1-\eta)p}{pm_0}\kappa} \right).
\end{aligned}$$

We now apply Hölder's inequality with $q = \frac{p}{p-\eta}$ and $q' = \frac{q}{q-1} = \frac{p}{\eta}$ so that $\frac{1}{q} + \frac{1}{q'} = 1$. We get

$$\int |G|^2 \left(\frac{e^{-\frac{p-\eta}{p}\frac{m-p}{m_0}\kappa}}{|s|^{\frac{2(p-\eta)}{m_0}}} \right) \left(|s|^{\frac{2(1-\eta)}{m_0}} e^{-\frac{\eta m+(1-\eta)p}{p m_0}\kappa} \right)$$

$$\leq \left(\int |G|^2 \left(\frac{e^{-\frac{p-\eta}{p}\frac{m-p}{m_0}\kappa}}{|s|^{\frac{2(p-\eta)}{m_0}}} \right)^{\frac{p}{p-\eta}} \right)^{\frac{p-\eta}{p}} \left(\int |G|^2 \left(|s|^{\frac{2(1-\eta)}{m_0}} e^{-\frac{\eta m+(1-\eta)p}{p m_0}\kappa} \right)^{\frac{p}{\eta}} \right)^{\frac{\eta}{p}}$$

$$\leq \left(\int |G|^2 \frac{e^{-\frac{m-p}{m_0}\kappa}}{|s|^{\frac{2p}{m_0}}} \right)^{\frac{p-\eta}{p}} \left(\sup |G|^2 \right)^{\frac{\eta}{p}} \left(\int \left(|s|^2 e^{-\kappa} \right)^{\frac{p}{\eta}\frac{1-\eta}{m_0}} e^{-\ell\kappa} \right)^{\frac{\eta}{p}}.$$

We set $\eta = \frac{1}{2}$ and get

$$\int \left(|s|^2 e^{-\kappa} \right)^{\frac{p}{\eta}\frac{1-\eta}{m_0}} e^{-\ell\kappa} = \int \left(|s|^2 e^{-\kappa} \right)^{\frac{p}{m_0}} e^{-\ell\kappa}.$$

We now use our assumption that

$$\left(|s|^2 e^{-\kappa} \right)^{\frac{p}{m_0}} e^{-\ell\kappa}$$

is integrable for $1 \leq p \leq \ell m_0$. This finishes the verification of Claim (4.2.1).

We now apply Proposition (4.1). Since clearly $s^\ell s_D$ locally belongs to $\mathcal{I}_{\varphi_m}$, it follows from $\mathcal{I}_{\varphi_m} \subset \mathcal{I}_{\varphi_{m-1}}$ that locally $s^\ell s_D$ locally belongs to $\mathcal{I}_{\varphi_{m-1}}$. Let

$$s_{A,1}, \cdots, s_{A,q_A} \in \Gamma(X, A)$$

so that

$$s_{A,1}|X_0, \cdots, s_{A,q_A}|X_0 \in \Gamma(X_0, A)$$

form a basis of $\Gamma(X_0, A)$ over **C**. By Proposition (4.1) we can extend each

$$s^\ell s_D s_{A,j} \in \Gamma(X_0, mK_{X_0} + \ell L + D + A)$$

to an element

$$\tilde{s}_j \in \Gamma(X, mK_X + \ell L + D + A)$$

for $1 \leq j \leq q_A$. Let

$$\chi_m = \log \sum_{j=1}^{q_A} |\tilde{s}_j|^2$$

so that $e^{-\chi_m}$ is a metric for $mK_X + \ell L + D + A = (m+a) K_X + \ell L$.

We now use Hölder's inequality. Let h_D be a smooth metric for D without any curvature condition. Let h_L be a smooth metric for L without any curvature condition. Let $h_{K_{X_0}}$ be a smooth metric for K_{X_0} without any curvature condition. Then $dV_{X_0} := \left(h_{K_{X_0}}\right)^{-1}$ defines a smooth volume form of X_0. Let $\ell' = \frac{\ell}{\ell-1}$ so that $\frac{1}{\ell}+\frac{1}{\ell'}=1$. By Hölder's inequality

$$\int_{X_0} |s|^2 e^{-\left(\frac{m_0}{m+a}\right)\chi_m - \frac{a}{m+a}\kappa}\, dV_{X_0} =$$

$$= \int_{X_0} \left(\left| s\,(s_D)^{\frac{1}{\ell}} \right|^2 h_D^{\frac{1}{\ell}}\, e^{-\left(\frac{m_0}{m+a}\right)\chi_m} h_L^{\frac{a}{m+a}} \right) \left(\frac{e^{-\frac{a}{m+a}\kappa}}{\left|(s_D)^{\frac{1}{\ell}}\right|^2 h_D^{\frac{1}{\ell}} h_L^{\frac{a}{m+a}}} \right) dV_{X_0}$$

$$\leq \left(\int_{X_0} \left| s^\ell s_D \right|^2 h_D\, e^{-\ell\left(\frac{m_0}{m+a}\right)\chi_m} h_L^{\frac{\ell a}{m+a}}\, dV_{X_0} \right)^{\frac{1}{\ell}} \times$$

$$\times \left(\int_{X_0} \frac{e^{-\frac{\ell}{\ell-1}\frac{a}{m+a}\kappa} dV_{X_0}}{\left(h_D \left| s_D \right|^2 h_L^{\frac{\ell a}{m+a}} \right)^{\frac{1}{\ell-1}}} \right)^{1-\frac{1}{\ell}}$$

$$\leq \left(\int_{X_0} \frac{\sum_j \left| s^\ell s_D s_{A,j} \right|^2 h_D\, e^{-\ell\left(\frac{m_0}{m+a}\right)\chi_m} h_L^{\frac{\ell a}{m+a}}\, dV_{X_0}}{\sum_j \left| s_{A,j} \right|^2} \right)^{\frac{1}{\ell}} \times$$

$$\times \left(\int_{X_0} \frac{e^{-\frac{\ell}{\ell-1}\frac{a}{m+a}\kappa} dV_{X_0}}{\left(h_D \left| s_D \right|^2 h_L^{\frac{\ell a}{m+a}} \right)^{\frac{1}{\ell-1}}} \right)^{1-\frac{1}{\ell}}$$

$$\leq \left(\sup_{X_0} \frac{e^{\left(\frac{a}{m+a}\right)\chi_m} h_D h_L^{\frac{\ell a}{m+a}}}{\sum_j \left| s_{A,j} \right|^2} \right)^{\frac{1}{\ell}} \left(\int_{X_0} \sum_j \left| s^\ell s_D s_{A,j} \right|^2 e^{-\chi_m} dV_{X_0} \right)^{\frac{1}{\ell}} \times$$

$$\times \left(\int_{X_0} \frac{e^{-\frac{\ell}{\ell-1}\frac{a}{m+a}\kappa} dV_{X_0}}{\left(h_D \left| s_D \right|^2 h_L^{\frac{\ell a}{m+a}} \right)^{\frac{1}{\ell-1}}} \right)^{1-\frac{1}{\ell}}$$

which is finite. Hence

$$\int_{X_0} |s|^2 e^{-\left(\frac{m_0-1}{m+a}\right)\chi_m - \frac{m_0-1}{m_0}\frac{a}{m+a}\kappa}\, dV_{X_0}$$

$$\leq \left(\sup_{X_0} \frac{e^{\left(\frac{1}{m+a}\right)\chi_m + \frac{1}{m_0}\frac{a}{m+a}\kappa}}{h_{K_{X_0}}}\right) \int_{X_0} |s|^2 e^{-\left(\frac{m_0}{m+a}\right)\chi_m - \frac{a}{m+a}\kappa} h_{K_{X_0}}\, dV_{X_0} < \infty.$$

Thus s can be extended to an element $\tilde{s} \in \Gamma(X, m_0 K_X + L)$. This concludes the proof of Theorem (0.6).

5. Extension from Singular Fiber Without Twisting

In this §5 we give the proof of Theorem (0.7). We use all the notations in Theorem (0.7).

(5.1) Choose a subvariety Z of complex codimension 1 in X such that $X - Z$ is Stein and Z does not contain any branch of any fiber of $\pi : X \to \Delta$ and $Z \cap X_0$ contains the singular set of X_0. We can assume without loss of generality that Z is contained in the zero-set of some holomorphic section s_A of some ample line bundle A over X. Let $Z_0 = Z \cap X_0$. Let $\tilde{\pi}_0 : \tilde{X}_0 \to X_0$ be a desingularization of X_0 so that $\tilde{X}_0$ is a compact complex manifold of complex dimension n and $\tilde{\pi}_0|_{\tilde{X}_0 - \tilde{\pi}_0^{-1}(Z_0)}$ maps $\tilde{X}_0 - \tilde{\pi}_0^{-1}(Z_0)$ biholomorphically onto $X_0 - Z_0$. We identify $X_0 - Z_0$ with $\tilde{X}_0 - \tilde{\pi}_0^{-1}(Z_0)$ through this biholomorphic map $\tilde{\pi}_0|_{\tilde{X}_0 - \tilde{\pi}_0^{-1}(Z_0)}$ so that $X_0 - Z_0$ is also regarded as an open subset of $\tilde{X}_0$. Let $\tilde{Z}_0 = \tilde{\pi}_0^{-1}(Z_0)$.

(5.2) In general the pullback $\tilde{A}_0$ of $A|X_0$ to $\tilde{X}_0$ under $\tilde{\pi}_0$ is not ample any more on $\tilde{X}_0$. We have to look at $\tilde{\pi}_0 : \tilde{X}_0 \to X_0$ so that we have a collection of nonsingular hypersurfaces H_j ($j \in J$) in normal crossing in $\tilde{X}_0$ so that $\tilde{A}_0 - \sum_{j \in J} \varepsilon_j H_j$ is ample on $\tilde{X}_0$ for a certain collection of sufficiently small positive rational numbers ε_j. Moreover, $\tilde{\pi}_0(H_j) \subset Z_0$. We can assume that $\tilde{\pi}_0^*(\operatorname{div} s_A) - \sum_{j \in J} \varepsilon_j H_j$ is effective by choosing ε_j sufficiently small. We choose a sufficiently large positive integer q such that the line bundle $B = q\left(\tilde{A}_0 - \sum_{j \in J} \varepsilon_j H_j\right)$ is sufficiently ample for the global generation of multiplier ideal sheaves on $\tilde{X}_0$. Let h_A (respectively h_B) be a smooth metric of A on X (respectively B on $\tilde{X}_0$) with positive curvature.

(5.3) Let m_0 be a positive integer. Take

$$f \in \Gamma\left(X_0, \left(\mathcal{J}^{m_0} / \left(\mathcal{J}^{m_0} \cap \mathcal{O}_X t\right)\right)(m_0 K_X)\right).$$

For the proof of Theorem (0.7) we have to extend f to an element of $\Gamma(X, m_0 K_X)$. The fibers of $\pi|_{X-Z} : X - Z \to \Delta$ are all manifolds. Hence $K_{X_0 - Z_0}$ can be naturally identified with $K_X|_{X_0-Z_0}$ by the map $\omega \to \omega \wedge dt$. This identification extends to a natural identification between $pK_{X_0-Z_0}$ and $pK_X|_{X_0-Z_0}$ for any positive integer p. Except when there is a specific need, we will make this natural identification without any explicit mention. By this natural identification we regard $f|_{X_0-Z_0}$ as an element of $\Gamma(X_0 - Z_0, m_0 K_{X_0-Z_0})$. Let

$$\tilde{f} \in \Gamma\left(\tilde{X}_0 - \tilde{Z}_0, m_0 K_{\tilde{X}_0}\right)$$

be the pullback of

$$f|_{X_0-Z_0} \in \Gamma(X_0 - Z_0, m_0 K_{X_0-Z_0})$$

under $\tilde{\pi}_0|_{\tilde{X}_0-\tilde{Z}_0}$.

First we verify that $\tilde{f}$ can be extended to an element of $\Gamma\left(\tilde{X}_0, m_0 K_{\tilde{X}_0}\right)$. For every point Q of Z_0 there exists some open neighborhood U of Q in X such that $f|_{X_0 \cap U}$ is induced by $f_1 \cdots f_{m_0}$ with $f_j \in \Gamma(U, \mathcal{J}(K_X))$ for $1 \leq j \leq m_0$. Let $g_j = f_j \sqcap dt$ for $1 \leq j \leq m_0$. Then

$$g_j|_{(X_0-Z_0)\cap U} \in \Gamma((X_0 - Z_0) \cap U, K_{X_0-Z_0})$$

is square integrable on $(X_0 - Z_0) \cap U$ as an $(n-1)$-form.

Because of the square integrability of $g_j|_{(X_0-Z_0)\cap U}$ on $(X_0 - Z_0) \cap U$, the pullback of $g_j|_{(X_0-Z_0)\cap U}$ to $\left(\tilde{X}_0 - \tilde{Z}_0\right) \cap \tilde{\pi}_0^{-1}(U)$ under $\tilde{\pi}_0|_{\tilde{X}_0-\tilde{Z}_0}$ is square integrable and therefore can be extended to an element of $\Gamma\left(\tilde{X}_0 \cap \tilde{\pi}_0^{-1}(U), K_{\tilde{X}_0}\right)$. So the pullback $\tilde{g}$ of $(g_1 \cdots g_{m_0})|_{(X_0-Z_0)\cap U}$ to $\left(\tilde{X}_0 - \tilde{Z}_0\right) \cap \tilde{\pi}_0^{-1}(U)$ under $\tilde{\pi}_0|_{\tilde{X}_0-\tilde{Z}_0}$ can be extended to an element of $\Gamma\left(\tilde{X}_0 \cap \tilde{\pi}_0^{-1}(U), m_0 K_{\tilde{X}_0}\right)$.

Through the natural identification between $m_0 K_{X_0-Z_0}$ and $m_0 K_X|_{X_0-Z_0}$, the section

$$\tilde{f} \in \Gamma\left(\tilde{X}_0 - \tilde{Z}_0, m_0 K_{\tilde{X}_0}\right)$$

agrees with

$$\tilde{g} \in \Gamma\left(\tilde{X}_0 - \tilde{Z}_0, m_0 K_{\tilde{X}_0}\right)$$

and hence can be extended to an element

$$s^{(m_0)} \in \Gamma\left(\tilde{X}_0, m_0 K_{\tilde{X}_0}\right).$$

(5.4) We consider

$$s^{(m_0)}\tilde{\pi}_0^*\left(s_A^q\right)\left(\prod_{j\in J} s_{H_j}^{-q\varepsilon_j}\right)\in\Gamma\left(\tilde{X}_0, m_0 K_{\tilde{X}_0}+B\right).$$

Since $\prod_{j\in J} s_{H_j}^{q\varepsilon_j}$ is a holomorphic section of the line bundle $q\tilde{A}_0 - B$, it follows that

$$\left|\prod_{j\in J} s_{H_j}^{q\varepsilon_j}\right|^2 \left(\tilde{\pi}_0^* h_A\right)^q h_B^{-1}$$

is a well-defined bounded nonnegative smooth function on $\tilde{X}_0$ and we denote this function by γ. Let C_γ be the supremum of γ on $\tilde{X}_0$.

Let $\psi = \frac{1}{m_0}\log\left|s^{(m_0)}\right|^2$ and $m_1 = \ell m_0$. For $1 \le p < m_1$ let N_p be the complex dimension of the subspace of all elements $s \in \Gamma\left(\tilde{X}_0, pK_{\tilde{X}_0}+B\right)$ such that

$$\int_{\tilde{X}_0} |s|^2\, e^{-(p-1)\psi}\, h_B < \infty.$$

Let $\tilde{N} = \sup_{1\le p<m_1} N_p$. Then $\tilde{N}$ is bounded independently of ℓ. This supremum bound being independent of ℓ is a key point in this proof.

(5.5) Let $\{U_\lambda\}_{1\le\lambda\le\Lambda}$ be a finite covering of $\tilde{X}_0$ by Stein open subsets such that

(i) for some U_λ^* which contains U_λ as a relatively compact subset, the assumptions and conclusions of Theorem (1.2) are satisfied when L with metric $e^{-\xi}$ is replaced by $pK_{\tilde{X}_0}$ with metric $e^{-p\psi}$ for $0 \le p \le m_1 - 1$ and when A is replaced by B, and

(ii) for $1 \le \lambda \le \Lambda$ there exist nowhere zero

$$\tau_{\lambda,A} \in \Gamma\left(U_\lambda^{**}, A\right),$$
$$\xi_\lambda \in \Gamma\left(U_\lambda^{**}, -K_{\tilde{X}_0}\right)$$

for some open subset U_λ^{**} of $\tilde{X}_0$ which contains U_λ^* as a relatively compact subset.

Let $\{\rho_\lambda\}_{1\le\lambda\le\Lambda}$ be a smooth partition of unity subordinate to the covering $\{U_\lambda\}_{1\le\lambda\le\Lambda}$ of $\tilde{X}_0$. Let

$$C^{(\%)} = \int_{\tilde{X}_0}\left|s^{(m_0)}\right|^{\frac{2}{m_0}}.$$

Let $C^{\natural}$ be the maximum of

$$\begin{cases} C^{\flat} C^{(\%)} \max_{1 \leq \lambda \leq \Lambda} \sup_{U_\lambda^*} \left(|\xi_\lambda|^2 e^{\psi} \right), \\ C^{\flat} C^{(\%)} \max_{1 \leq \lambda \leq \Lambda} \sup_{U_\lambda^*} \left(\left| \tilde{\pi}_0^* \left(s_A^q \right) \left(\prod_{j \in J} s_{H_j}^{-q\varepsilon_j} \right) h_B \right|^2 \right), \end{cases}$$

where $C^{\flat}$ is from Theorem (1.2).

Claim (5.6). There exist

(i)

$$s_1^{(p)}, \cdots, s_{N_p}^{(p)} \in \Gamma \left(\tilde{X}_0, p K_{\tilde{X}_0} + B \right)$$

for $1 \leq p \leq m_1 - 1$,

(ii) holomorphic functions $b_{j,k}^{(p,\lambda)}$ on U_λ, for $1 \leq p \leq m_1 - 2$, $1 \leq j \leq N_{p+1}$, $1 \leq k \leq N_p$, and $1 \leq \lambda \leq \Lambda$, and

(iii) holomorphic functions $b_k^{(m_1-1,\lambda)}$ on U_λ for $1 \leq k \leq N_{m_1-1}$ and $1 \leq \lambda \leq \Lambda$ such that, for $1 \leq \lambda \leq \Lambda$,

(a)

$$\xi_\lambda \left(s^{(m_0)} \right)^{\ell} \tilde{\pi}_0^* \left(s_A^q \right) \left(\prod_{j \in J} s_{H_j}^{-q\varepsilon_j} \right) = \sum_{k=1}^{N_{m_1-1}} b_k^{(m_1-1,\lambda)} s_k^{(m_1-1)}$$

on U_λ,

(b)

$$\xi_\lambda s_j^{(p+1)} = \sum_{k=1}^{N_p} b_{j,k}^{(p,\lambda)} s_k^{(p)}$$

on U_λ for $1 \leq p \leq m_1 - 2$ and $1 \leq j \leq N_{p+1}$,

(c)

$$\sup_{U_\lambda} \sum_{1 \leq k \leq N_{m_1-1}} \left| b_k^{(m_1-1,\lambda)} \right|^2 \leq C^{\natural},$$

(d)

$$\sup_{U_\lambda} \sum_{k=1}^{N_p} \left| b_{j,k}^{(p,\lambda)} \right|^2 \leq C^{\natural}$$

for $1 \leq j \leq N_{p+1}$ and $1 \leq k \leq N_p$ and $1 \leq p \leq m_1 - 2$, and

(e)

$$\int_{X_0} \left| s_k^{(p)} \right|^2 e^{-(p-1)\psi} h_B \leq 1$$

for $1 \leq k \leq N_p$ and $1 \leq p \leq m_1 - 1$.

The proof of Claim (5.6) follows immediately from Theorem (1.2) and descending induction on p. We have

$$\int_{\tilde{X}_0} \left| s_j^{(p)} \left(\prod_{j \in J} s_{H_j}^{q\varepsilon_j} \right) \right|^2 e^{-(p-1)\psi} \left(\tilde{\pi}_0^* h_A \right)^q$$
$$= \int_{\tilde{X}_0} \left| s_j^{(p)} \right|^2 \gamma\, e^{-(p-1)\psi} h_B \leq \sup_{\tilde{X}_0} \gamma = C_\gamma .$$

(5.7) By Theorem (3.1) there exists a positive constant $C^\heartsuit$ such that, for any holomorphic line bundle E over X with a (possibly singular) metric h_E of semi-positive curvature current and for any element

$$s \in \Gamma \left(X_0 - Z_0, E + K_{X_0 - Z_0} \right)$$

with

$$\int_{X_0 - Z_0} |s|^2 h_E < \infty ,$$

there exists

$$\tilde{s} \in \Gamma \left(X - Z, E + K_X \right)$$

such that $\tilde{s} = s \wedge dt$ on $X_0 - Z_0$ and

$$\int_{X - Z} |\tilde{s}|^2 h_E \leq C^\heartsuit \int_{X_0 - Z_0} |s|^2 h_E .$$

We now verify the following claim.

Claim (5.8). The element

$$s_j^{(p)} \left(\prod_{j \in J} s_{H_j}^{q\varepsilon_j} \right) \Bigg|_{X_0 - Z} \in \Gamma \left(X_0 - Z_0, p\, K_{X_0 - Z_0} + q\, A \right)$$

can be extended to

$$\tilde{s}_j^{(p)} \in \Gamma \left(X - Z, p\, K_X + qA \right)$$

for $1 \leq p < m_1$ and $1 \leq j \leq N_p$ such that

(i)

$$\int_{X-Z} \frac{\left|\tilde{s}_j^{(p+1)}\right|^2}{\max_{1\le k\le N_p}\left|\tilde{s}_k^{(p)}\right|^2} \le C^{\heartsuit} C^{\natural} C_\gamma \int_{X_0-Z} \sum_{\lambda=1}^{\Lambda} \frac{\rho_\lambda}{|\xi_\lambda|^2}$$

for $1 \le p \le m_1 - 2$ and $1 \le j \le N_{p+1}$,

(ii)

$$\int_{X-Z} \left|\tilde{s}_j^{(1)}\right|^2 h_A^q \le C^{\heartsuit} C_\gamma$$

for $1 \le j \le N_1$, and

(iii)

$$\int_{X_0-Z} \frac{\left|\left(s^{(m_0)}\right)^\ell \left(\prod_{j\in J} s_{H_j}^{q\varepsilon_j}\right) s_A^q\right|^2}{\max_{1\le k\le N_{m_1-1}}\left|\tilde{s}_k^{(m_1-1)}\right|^2} \le C^{\natural} C_\gamma \int_{X_0-Z} \sum_{\lambda=1}^{\Lambda} \frac{\rho_\lambda}{|\xi_\lambda|^2}.$$

The proof of Claim (5.8) is by induction on $1 \le p < m_1$ and by applying (5.7). With $\tilde{s}_k^{(p)}$ from the preceding induction step (and using the convention $\max_{1\le k\le N_0}\left|\tilde{s}_k^{(0)}\right|^2 = 1$), from (5.6)(b) we conclude that

$$\text{(5.8.1)} \qquad \int_{X_0-Z_0} \frac{\left|s_j^{(p+1)}\left(\prod_{j\in J} s_{H_j}^{q\varepsilon_j}\right)\right|^2}{\max_{1\le k\le N_p}\left|\tilde{s}_k^{(p)}\right|^2} \le C^{\natural} C_\gamma \int_{X_0-Z} \sum_{\lambda=1}^{\Lambda} \frac{\rho_\lambda}{|\xi_\lambda|^2}$$

for $0 \le p \le m_1 - 2$ and $1 \le j \le N_{p+1}$. We apply (5.7) to the line bundle $E = p\,K_X + q\,A$ with the metric

$$\frac{1}{\max_{1\le k\le N_p}\left|\tilde{s}_k^{(p)}\right|^2}.$$

The interpretation of the integrand on the left-hand side of (5.8.1) is as follows. The section $s_j^{(p+1)}\left(\prod_{j\in J} s_{H_j}^{q\varepsilon_j}\right)$ of $(p+1)\,K_{X_0-Z_0} + q\,A$ over $X_0 - Z_0$ is naturally identified with a $(p\,K_X + q\,A)$-valued $(n-1)$-form on $X_0 - Z_0$ so that

$$\frac{\left|s_j^{(p+1)}\left(\prod_{j\in J} s_{H_j}^{q\varepsilon_j}\right)\right|^2}{\max_{1\le k\le N_p}\left|\tilde{s}_k^{(p)}\right|^2}$$

is a (possibly singular) volume form of $X_0 - Z_0$. By (5.7) we obtain $\tilde{s}_k^{(p+1)}$. The statement (5.8)(iii) follows from (5.6)(a). The interpretation of the integrand on the left-hand side of (5.8)(iii) is analogous to that of (5.8.1) and is as follows. The expression

$$\frac{1}{\max_{1\leq k\leq N_{m_1-1}}\left|\tilde{s}_k^{(m_1-1)}\right|^2}$$

defines a metric for $(m_1-1)\,K_X+q\,A$. The section

$$\left(s^{(m_0)}\right)^\ell\left(\prod_{j\in J}s_{H_j}^{q\varepsilon_j}\right)\,s_A^q\in\Gamma\left(X_0-Z_0,m_1\,K_{X_0-Z_0}+q\,A\right)$$

is naturally identified with an $((m_1-1)\,K_X+q\,A)$-valued $(n-1)$-form on X_0-Z_0 so that

$$\frac{\left|\left(s^{(m_0)}\right)^\ell\left(\prod_{j\in J}s_{H_j}^{q\varepsilon_j}\right)\,s_A^q\right|^2}{\max_{1\leq k\leq N_{m_1-1}}\left|\tilde{s}_k^{(m_1-1)}\right|^2}$$

is a (possibly singular) volume form of X_0-Z_0. This finishes the verification of Claim (5.8)

By applying (5.7) and using (5.8)(iii), we conclude that the element

$$\left.\left(s^{(m_0)}\right)^\ell\left(\prod_{j\in J}s_{H_j}^{q\varepsilon_j}\right)\,s_A^q\right|_{X_0-Z}\in\Gamma\left(X_0-Z_0,m_1\,K_{X_0-Z_0}+q\,A\right)$$

can be extended to

$$\hat{s}^{(m_1)}\in\Gamma\left(X-Z,m_1K_X+qA\right)$$

such that

$$\int_{X-Z}\frac{\left|\hat{s}^{(m_1)}\right|^2}{\max_{1\leq k\leq N_{m_0-1}}\left|\tilde{s}_k^{(m_1-1)}\right|^2}\leq C^{\heartsuit}\,C^{\natural}\,C_\gamma\int_{X_0-Z}\sum_{\lambda=1}^{\Lambda}\frac{\rho_\lambda}{|\xi_\lambda|^2}.$$

(5.9) We now prepare to pass to limit as $\ell\to\infty$ and will use the concavity of the logarithmic function for the estimate of the limiting process together with the fact from (5.4) that $\tilde{N}$ is bounded independently of ℓ. Let $C^{\diamondsuit}$ be the maximum of $C^{\heartsuit}\,C_\gamma$ and

$$C^{\heartsuit}\,C^{\natural}\,C_\gamma\int_{X_0-Z}\sum_{\lambda=1}^{\Lambda}\frac{\rho_\lambda\left|s_A^q\right|^2}{\left|\xi_\lambda\,\tau_{\lambda,A}^q\right|^2}.$$

Let $0 < r_0 < r_1 < r_2 < 1$. Choose a finite number of coordinate charts $\tilde{W}_\lambda$ with coordinates

$$\left(z^{(\lambda)}, t\right) = \left(z_1^{(\lambda)}, \cdots, z_n^{(\lambda)}, t\right)$$

for $1 \leq \lambda \leq \hat{\Lambda}$ such that

(ii) each

$$W_\lambda'' := \left\{\left|z_1^{(\lambda)}\right| < r_2, \cdots, \left|z_n^{(\lambda)}\right| < r_2, |t| < r_2\right\}$$

is relatively compact in $\tilde{W}_\lambda$ for $1 \leq \lambda \leq \hat{\Lambda}$,

(ii) $X \cap \{|t| < r_2\} = \bigcup_{\lambda=1}^{\hat{\Lambda}} W_\lambda''$,

(iii) $X \cap \{|t| < r_0\} = \bigcup_{\lambda=1}^{\hat{\Lambda}} W_\lambda$, where

$$W_\lambda = \left\{\left|z_1^{(\lambda)}\right| < r_0, \cdots, \left|z_n^{(\lambda)}\right| < r_0, |t| < r_0\right\}$$

for $1 \leq \lambda \leq \hat{\Lambda}$, and

(iv) there exist nowhere zero

$$\hat{\tau}_{\lambda,A} \in \Gamma\left(W_\lambda'', A\right),$$
$$\hat{\xi}_\lambda \in \Gamma\left(W_\lambda'', -K_X\right)$$

for $1 \leq \lambda \leq \hat{\Lambda}$.

Let $dV_{z^{(\lambda)},t}$ be the Euclidean volume form in the coordinates system $\left(z^{(\lambda)}, t\right)$. Let

$$W_\lambda' = \left\{\left|z_1^{(\lambda)}\right| < r_1, \cdots, \left|z_n^{(\lambda)}\right| < r_1, |t| < r_1\right\}.$$

From the concavity of the logarithmic function we conclude that

$$\begin{aligned}
&\frac{1}{\left(\pi r_1^2\right)^{n+1}}\int_{W'_\lambda}\left(\log\max_{1\le j\le N_{p+1}}\left|\hat{\tau}_{\lambda,A}^{-q}\,\hat{\xi}_\lambda^{p+1}\,\tilde{s}_j^{(p+1)}\right|^2\right)dV_{z^{(\lambda)},t}\\
&\quad-\frac{1}{\left(\pi r_1^2\right)^{n+1}}\int_{W'_\lambda}\left(\log\max_{1\le k\le N_p}\left|\hat{\tau}_{\lambda,A}^{-q}\,\hat{\xi}_\lambda^{p}\,\tilde{s}_k^{(p)}\right|^2\right)dV_{z^{(\lambda)},t}\\
&=\frac{1}{\left(\pi r_1^2\right)^{n+1}}\int_{W'_\lambda}\left(\log\frac{\max_{1\le j\le N_{p+1}}\left|\hat{\tau}_{\lambda,A}^{-q}\,\hat{\xi}_\lambda^{p+1}\,\tilde{s}_j^{(p+1)}\right|^2}{\max_{1\le k\le N_p}\left|\hat{\tau}_{\lambda,A}^{-q}\,\hat{\xi}_\lambda^{p}\,\tilde{s}_k^{(p)}\right|^2}\right)dV_{z^{(\lambda)},t}\\
&\le\log\left(\frac{1}{\left(\pi r_1^2\right)^{n+1}}\int_{W'_\lambda}\frac{\max_{1\le j\le N_{p+1}}\left|\hat{\tau}_{\lambda,A}^{-q}\,\hat{\xi}_\lambda^{p+1}\,\tilde{s}_j^{(p+1)}\right|^2}{\max_{1\le k\le N_p}\left|\hat{\tau}_{\lambda,A}^{-q}\,\hat{\xi}_\lambda^{p}\,\tilde{s}_k^{(p)}\right|^2}\,dV_{z^{(\lambda)},t}\right)\\
&\le\log\left(\sup_{W'_\lambda}\left(\frac{1}{\left(\pi r_1^2\right)^{n+1}}\left|\hat{\xi}_\lambda\right|^2dV_{z^{(\lambda)},t}\right)\int_{W'_\lambda}\frac{\max_{1\le j\le N_{p+1}}\left|\tilde{s}_j^{(p+1)}\right|^2}{\max_{1\le k\le N_p}\left|\tilde{s}_k^{(p)}\right|^2}\right)\\
&\qquad\le\log\left(\tilde{N}\,C^{\diamondsuit}\sup_{W'_\lambda}\left(\frac{1}{\left(\pi r_1^2\right)^{n+1}}\left|\hat{\xi}_\lambda\right|^2dV_{z^{(\lambda)},t}\right)\right)
\end{aligned}$$

for $1\le p\le m_1-2$. Likewise

$$\begin{aligned}
&\frac{1}{\left(\pi r_1^2\right)^{n+1}}\int_{W'_\lambda}\left(\log\max_{1\le j\le N_{m_1}}\left|\hat{\tau}_{\lambda,A}^{-q}\,\hat{\xi}_\lambda^{m_1}\,\hat{s}_j^{(m_1)}\right|^2\right)dV_{z^{(\lambda)},t}\\
&-\frac{1}{\left(\pi r_1^2\right)^{n+1}}\int_{W'_\lambda}\left(\log\max_{1\le k\le N_{m_1-1}}\left|\hat{\tau}_{\lambda,A}^{-q}\,\hat{\xi}_\lambda^{m_1-1}\,\tilde{s}_k^{(m_1-1)}\right|^2\right)dV_{z^{(\lambda)},t}\\
&\qquad\le\log\left(\tilde{N}\,C^{\diamondsuit}\sup_{W'_\lambda}\left(\frac{1}{\left(\pi r_1^2\right)^{n+1}}\left|\hat{\xi}_\lambda\right|^2dV_{z^{(\lambda)},t}\right)\right)
\end{aligned}$$

Adding up, we get

$$\begin{aligned}
&\frac{1}{\left(\pi r_1^2\right)^{n+1}}\int_{W'_\lambda}\left(\log\max_{1\le j\le N_{m_1}}\left|\hat{\tau}_{\lambda,A}^{-q}\,\hat{\xi}_\lambda^{m_1}\,\hat{s}_j^{(m_1)}\right|^2\right)dV_{z^{(\lambda)},t}\\
&\quad\le\frac{1}{\left(\pi r_1^2\right)^{n+1}}\int_{W'_\lambda}\left(\log\max_{1\le k\le N_1}\left|\hat{\tau}_{\lambda,A}^{-q}\,\hat{\xi}_\lambda\,\tilde{s}_k^{(1)}\right|^2\right)dV_{z^{(\lambda)},t}\\
&\quad+(m_1-1)\log\left(\tilde{N}\,C^{\diamondsuit}\sup_{W'_\lambda}\left(\frac{1}{\left(\pi r_1^2\right)^{n+1}}\left|\hat{\xi}_\lambda\right|^2dV_{z^{(\lambda)},t}\right)\right).
\end{aligned}$$

(5.10) By the sub-mean-value property of plurisubharmonic functions

$$\sup_{W_\lambda} \log \left| \hat{\tau}_{\lambda,A}^{-q} \, \hat{\xi}_\lambda^{m_1} \, \hat{s}^{(m_1)} \right|^2$$

$$\leq \frac{1}{\left(\pi \left(r_1 - r_0\right)^2\right)^{n+1}} \int_{W'_\lambda} \left(\log \left| \hat{\tau}_{\lambda,A}^{-q} \, \hat{\xi}_\lambda^{m_1} \, \hat{s}^{(m_1)} \right|^2 \right) dV_{z^{(\lambda)},t}.$$

Choose a positive number $C^\spadesuit$ such that $\frac{1}{m_0} \log C^\spadesuit$ is no less than

$$\left(\frac{r_1}{r_1 - r_0} \right)^{2(n+1)} \log \left(\tilde{N} \, C^\diamondsuit \sup_{W'_\lambda} \left(\frac{1}{\left(\pi r_1^2\right)^{n+1}} \left| \hat{\xi}_\lambda \right|^2 dV_{z^{(\lambda)},t} \right) \right)$$

for every $1 \leq \lambda \leq \Lambda$ and $1 \leq p \leq m_1 - 1$. Since $\tilde{N}$ is bounded independently of ℓ, we can choose $C^\spadesuit$ to be independent of ℓ. Let $\hat{C}$ be defined by

$$\log \hat{C} = \frac{1}{\left(\pi r_1\right)^{n+1}} \int_{W'_\lambda} \left(\log \max_{1 \leq k \leq N_1} \left| \hat{\tau}_{\lambda,A}^{-q} \, \hat{\xi}_\lambda \, \tilde{s}_k^{(1)} \right|^2 \right) dV_{z^{(\lambda)},t}.$$

Then

$$\sup_{1 \leq \lambda \leq \hat{\Lambda}} \sup_{W_\lambda} \left| \hat{\tau}_{\lambda,A}^{-q} \, \hat{\xi}_\lambda^{\ell m_0} \, \hat{s}^{(\ell m_0)} \right|^2 \leq \hat{C} \left(C^\spadesuit \right)^{\ell m_0 - 1}.$$

(5.11) For $1 \leq \lambda \leq \hat{\Lambda}$ let χ_λ be the function on W_λ which is the upper semi-continuous envelope of

$$\limsup_{\ell \to \infty} \log \left| \hat{\xi}_\lambda^{\ell m_0} \, \hat{s}^{(\ell m_0)} \right|^{\frac{2}{\ell}}$$

which is $\leq m_0 \log C^\spadesuit$. From the definition of χ_λ, we have

$$\sup_{(X_0 - Z) \cap W_\lambda} \left(\left| \hat{\xi}_\lambda^{m_0} \, s^{(m_0)} \right|^2 e^{-\chi_\lambda} \right) \leq 1$$

for $1 \leq \lambda \leq \hat{\Lambda}$. Let $e^{-\chi} = \left| \hat{\xi}_\lambda \right|^2 e^{-\chi_\lambda}$ be the metric of K_X on $X \cap \{|t| < r\}$ so that the square of the pointwise norm of a local section σ of K_X on W_λ is $\left| \sigma \hat{\xi}_\lambda \right|^2 e^{-\chi_\lambda}$. Let $\{\hat{\rho}_\lambda\}_{1 \leq \lambda \leq \hat{\Lambda}}$ be a smooth partition of unity on X_0 subordinate to the covering $\{W_\lambda \cap X_0\}_{1 \leq \lambda \leq \hat{\Lambda}}$ of X_0. Then

$$\int_{X_0 - Z_0} \left| s^{(m_0)} \right|^2 e^{-\left(\frac{m_0 - 1}{m_0}\right)\chi} \leq C^\spadesuit \sum_{1 \leq \lambda \leq \Lambda} \int_{X_0 \cap W_\lambda} \frac{\hat{\rho}_\lambda}{\left| \hat{\xi}_\lambda \right|^2} < \infty.$$

Here the interpretation of the integrand on the left-hand side is as follows. The section $s^{(m_0)}$ of $m_0 K_{X_0 - Z_0}$ over $X_0 - Z_0$ is naturally identified with an $(m_0 - 1) K_X$-valued $(n-1)$-form on $X_0 - Z_0$ so that $\left|s^{(m_0)}\right|^2 e^{-\left(\frac{m_0 - 1}{m_0}\right)\chi}$ is a (possibly singular) volume form of $X_0 - Z_0$.

Finally by applying (5.7), we can extend $s^{(m_0)}|_{X_0 - Z}$ to an element

$$s^* \in \Gamma\big(X \cap \{|t| < r\} - Z, m_0 K_X\big)$$

with

$$\int_{X \cap \{|t|<r\} - Z} |s^*|^2 \, e^{-\left(\frac{m_0 - 1}{m_0}\right)\chi} < \infty.$$

By the removability of codimension-one singularities of L^2 holomorphic functions, we can extend s^* to an element

$$s^{**} \in \Gamma\big(X \cap \{|t| < r\}, m_0 K_X\big).$$

According to the general Stein theory for coherent sheaves on Δ, there exists an element $s^{\#}$ of $\Gamma(X, m_0 K_X)$ which induces the same element in $\Gamma\left(X, \left(\mathcal{O}_X / \mathcal{O}_X t\right)(m_0 K_X)\right)$ as s^{**}. Thus $s^{\#} \in \Gamma(m_0 K_X)$ extends

$$f \in \Gamma\left(X_0, \left(\mathcal{J}^{m_0} / \left(\mathcal{J}^{m_0} \cap \mathcal{O}_X t\right)\right)(m_0 K_X)\right).$$

This concludes the proof of Theorem (0.7).

6. Induction Technique on Multiplicity of Multiple Point

In this §6 we combine the method of multiplier ideal sheaves with the induction process of Levine [4] on the multiplicity of the multiple point of the base over which the sections can be extended. First we formulate the induction process of Levine [4] in the following lemma.

Lemma (6.1) Let $\pi : X \to \Delta$ be a holomorphic family of compact complex projective algebraic manifolds over Δ with fiber X_t ($t \in \Delta$). Let $\{U_i\}_i$ be an open cover of X such that each U_i is a coordinate chart with coordinates $\left(z^{(i)}, t\right) = \left(z_1^{(i)}, \cdots, z_n^{(i)}, t\right)$. Let $z^{(i)} = g^{(ij)}\left(z^{(j)}, t\right)$ be the coordinate transformation from the coordinate chart U_j to the coordinate chart U_i. Let E be a holomorphic line bundle over X with transition functions $\Xi_{ij}\left(z^{(j)}, t\right)$ on $U_i \cap U_j$. Let ℓ be a nonnegative integer. Let s be a global holomorphic section of E over the ringed space $\left(X_0, \mathcal{O}_X / \mathcal{O}_X t^{\ell+1}\right)$ such that s is given by a holomorphic function $s_i\left(z^{(i)}, t\right)$ on U_i. Let

$$\Theta_{ji} = \sum_{\nu=1}^{n} \frac{\partial s_i}{\partial z_\nu^{(i)}} \frac{\partial g_\nu^{(ij)}}{\partial t} - \frac{\partial \Xi_{ij}}{\partial t} \Xi_{ij}^{-1} s_i$$

on $U_i \cap U_j$. Then $\{\Theta_{ji}\}_{i,j}$ defines a cocycle for the covering $\{U_i\}$ with coefficients in E whose vanishing implies that there exists some global holomorphic section $\tilde{s}$ of E over the ringed space $\left(X_0, \mathcal{O}_X/\mathcal{O}_X t^{\ell+2}\right)$ whose restriction to $\left(X_0, \mathcal{O}_X/\mathcal{O}_X t\right)$ agrees with the restriction of s to $\left(X_0, \mathcal{O}_X/\mathcal{O}_X t\right)$.

Proof. By assumption on s and s_i, we have

$$s_i\left(g^{(ij)}(z^{(j)},t),t\right) \equiv \Xi_{ij}(z^{(j)},t)s_j(z^{(j)},t) \qquad \mathrm{mod}\ t^{\ell+1}$$

and

$$t^{\ell+1}\zeta_{ji} \equiv s_i\left(g^{(ij)}(z^{(j)},t),t\right) - \Xi_{ij}(z^{(j)},t)s_j(z^{(j)},t) \ \ \mathrm{mod}\ t^{\ell+2} \tag{6.1.1}$$

for some $\zeta_{ji} \in \Gamma\left(U_i \cap U_j, E\right)$. If we are able to write $t^\ell \zeta_{ji} = \sigma_i - \sigma_j$ modulo $t^{\ell+1}$ on $U_i \cap U_j$ for some holomorphic function σ_i on U_i, then the collection $\{s_j - t\sigma_j\}_j$ would define an element of $\left(X_0, \mathcal{O}_X/\mathcal{O}_X t^{\ell+2}\right)$ which extends s.

For fixed z_j we differentiate (6.1.1) with respect to t and we get

$$(\ell+1)t^\ell\zeta_{ji} \equiv \sum_{\nu=1}^{n} \frac{\partial s_i}{\partial z_\nu^{(i)}}\frac{\partial g_\nu^{(ij)}}{\partial t} + \frac{\partial s_i}{\partial t} - \frac{\partial \Xi_{ij}}{\partial t}s_j - \Xi_{ij}\frac{\partial s_j}{\partial t} \qquad \mathrm{mod}\ t^{\ell+1}$$

$$\equiv \left(\sum_{\nu=1}^{n} \frac{\partial s_i}{\partial z_\nu^{(i)}}\frac{\partial g_\nu^{(ij)}}{\partial t} - \frac{\partial \Xi_{ij}}{\partial t}\Xi_{ij}^{-1}s_i\right) + \left(\frac{\partial s_i}{\partial t} - \Xi_{ij}\frac{\partial s_j}{\partial t}\right) \qquad \mathrm{mod}\ t^{\ell+1}.$$

Since

$$\left\{\frac{\partial s_i}{\partial t} - \Xi_{ij}\frac{\partial s_j}{\partial t}\right\}_{i,j}$$

is a coboundary, it follows that

$$\left\{\sum_{\nu=1}^{n} \frac{\partial s_i}{\partial z_\nu^{(i)}}\frac{\partial g_\nu^{(ij)}}{\partial t} - \frac{\partial \Xi_{ij}}{\partial t}\Xi_{ij}^{-1}s_i\right\}_{i,j} \tag{6.1.2}$$

is a cocyle. If (6.1.2) is a coboundary, we can write

$$\sum_{\nu=1}^{n} \frac{\partial s_i}{\partial z_\nu^{(i)}}\frac{\partial g_\nu^{(ij)}}{\partial t} - \frac{\partial \Xi_{ij}}{\partial t}\Xi_{ij}^{-1}s_i = \tau_i - \tau_j \qquad \mathrm{mod}\ t^{\ell+1}$$

on $U_i \cap U_j$ for some holomorphic function τ_j on U_j and define

$$\sigma_j = \frac{1}{\ell+1}\left(\tau_j + \frac{\partial s_i}{\partial t}\right)$$

on U_j to get $t^\ell\zeta_{ji} = \sigma_i - \sigma_j$ modulo $t^{\ell+1}$ on $U_i \cap U_j$. Q.E.D.

Remark (6.2). In the notations of Theorem (6.1), let R be the stalk of the first direct image $R^1\pi_*(\mathcal{O}_X(E))$ at the origin of Δ and if q is a positive integer chosen large enough such that the map $t^q R \to t^q R$ defined by multiplication by t is injective, then any element $s \in \Gamma(X_0, mK_{X_0})$ which can be extended to a global holomorphic section of L over the ringed space $\left(X_0, \mathcal{O}_X/\mathcal{O}_X t^{q+1}\right)$ can be extended to a global holomorphic section of E over X. This statement follows from the following argument. Let $\mathcal{F} = \mathcal{O}_X(E)$. The commutative diagram

$$\begin{array}{ccccccccc} 0 & \to & \mathcal{F} & \overset{\psi_{q+1}}{\longrightarrow} & \mathcal{F} & \to & \mathcal{F}/t^{q+1}\mathcal{F} & \to & 0 \\ & & \psi_q \downarrow & & \| & & \downarrow & & \\ 0 & \to & \mathcal{F} & \overset{\psi_1}{\longrightarrow} & \mathcal{F} & \to & \mathcal{F}/t\mathcal{F} & \to & 0, \end{array}$$

where ψ_k is defined by multiplication by t^k, gives rise to the commutative diagram

$$\begin{array}{ccccccc} R^0\pi_*(\mathcal{F}) & \to & R^0\pi_*\left(\mathcal{F}/t^{q+1}\mathcal{F}\right) & \to & R^1\pi_*(\mathcal{F}) & \overset{\psi_{q+1}}{\longrightarrow} & R^1\pi_*(\mathcal{F}) \\ \| & & \downarrow & & \psi_q \downarrow & & \| \\ R^0\pi_*(\mathcal{F}) & \to & R^0\pi_*\left(\mathcal{F}/t\mathcal{F}\right) & \to & R^1\pi_*(\mathcal{F}) & \overset{\psi_1}{\longrightarrow} & R^1\pi_*(\mathcal{F}). \end{array}$$

Since the map from $t^q R$ to itself defined by multiplication by t is injective, it follows that in the commutative diagram above the map $\psi_q : R^1\pi_*(\mathcal{F}) \to R^1\pi_*(\mathcal{F})$ maps the kernel of $\psi_{q+1} : R^1\pi_*(\mathcal{F}) \to R^1\pi_*(\mathcal{F})$ to zero. This implies that in the commutative diagram above the image of $R^0\pi_*\left(\mathcal{F}/t^{q+1}\mathcal{F}\right) \to R^0\pi_*\left(\mathcal{F}/t^q\mathcal{F}\right)$ is contained in the image of $R^0\pi_*(\mathcal{F}) \to R^0\pi_*\left(\mathcal{F}/t\mathcal{F}\right)$.

Let S_k be the kernel of the map from R to itself defined by multiplication by t^k. The ascending chain of submodules $S_k \subset S_{k+1} \subset R$ must stabilizes at some $k = q$ so that $S_q = S_{q+1}$. Thus the integer q with the property that the map $t^q R \to t^q R$ defined by multiplication by t is injective always exist. With such an integer q, Lemma (6.1) gives the condition for the surjectivity of the restriction map $\Gamma(X, E) \to \Gamma(X_0, E)$ when it is applied to $\ell = q + 1$.

Lemma (6.3). Let f be a holomorphic function on an open subset Ω of $\mathbf{C}^n$. Let $\alpha > 1$ be a positive number. Let $z_1, \cdots, z_n$ be the coordinates of $\mathbf{C}^n$ and $1 \le j \le n$. Let $0 < \varepsilon < 1$. Then

$$\left(\frac{\left|\frac{\partial f}{\partial z_j}\right|^2}{|f|^{\frac{2}{\alpha}}}\right)^{\alpha - \varepsilon} \tag{6.3.1}$$

is locally integrable on Ω.

Proof. To verify the local integrability of (6.3.1) at an arbitrarily chosen point $Q \in \Omega$, without loss of generality we can assume that $Q = 0$ and $j = n$. By a linear change of coordinates, we can assume that f is not identically zero on $\{z_k = z_k(Q) \text{ for } 1 \leq k \leq n-1\}$ near Q. For some relatively compact open neighborhood U' of 0 in $\mathbf{C}^{n-1}$ and some relatively compact open neighborhood U'' of 0 in $\mathbf{C}$, we can write $f = uP$ on $U' \times U''$, where u is a nowhere zero holomorphic function on $U' \times U''$ and P is a monic polynomial in z_n of degree q whose coefficients are holomorphic functions on U' such that all q roots in z_n are contained in U'' for $(z_1, \cdots, z_{n-1}) \in U'$. Write

$$P(z_1, \cdots, z_{n-1}, z_n) = \prod_{\nu=1}^{q} (z_n - g_\nu(z_1, \cdots, z_{n-1}))$$

for $(z_1, \cdots, z_{n-1}) \in U'$ so that $g_\nu(z_1, \cdots, z_{n-1}) \in U''$ for $1 \leq \nu \leq q$. Let W be a relatively compact open neighborhood of 0 in $U' \times U''$. Then for $(z_1, \cdots, z_n) \in W$, we have

$$\frac{\left|\frac{\partial f}{\partial z_n}\right|^2}{|f|^{\frac{2}{\alpha}}} = |f|^{2\left(1-\frac{1}{\alpha}\right)} \left|\frac{\partial}{\partial z_n} \log f\right|^2$$

$$\leq |f|^{2\left(1-\frac{1}{\alpha}\right)} \left|\frac{\partial}{\partial z_n} \log u\right|^2 + |u|^{2\left(1-\frac{1}{\alpha}\right)} \left(\prod_{\nu=1}^{q} |z_n - g_\nu|^{2\left(1-\frac{1}{\alpha}\right)}\right) \left|\sum_{\nu=1}^{q} \frac{1}{z_n - g_\nu}\right|^2$$

$$\leq C_1 + C_2 \left(\sum_{\nu=1}^{q} \frac{1}{|z_n - g_\nu|^{\frac{2}{\alpha}}}\right)$$

and

$$\left(\frac{\left|\frac{\partial f}{\partial z_j}\right|^2}{|f|^{\frac{2}{\alpha}}}\right)^{\alpha-\varepsilon} \leq C_3 + C_4 \left(\sum_{\nu=1}^{q} \frac{1}{|z_n - g_\nu|^{2\left(1-\frac{\varepsilon}{\alpha}\right)}}\right),$$

where C_k $(1 \leq j \leq 4)$ are constants. The lemma now follows from the integrability of

$$\sum_{\nu=1}^{q} \frac{1}{|z_n - g_\nu|^{2\left(1-\frac{\varepsilon}{\alpha}\right)}}$$

on $U' \times U''$. Q.E.D.

Theorem (6.4). Let $\pi : X \to \Delta$ be a holomorphic family of compact complex manifolds and L be a holomorphic line bundle with Hermitian metric $e^{-\kappa}$ whose curvature current is bounded from below by some smooth positive $(1,1)$-form. Let $s \in \Gamma\left(X_0, L + mK_{X_0}\right)$. Assume that

$$|s|^{\frac{2\eta m}{1-m\eta'}} e^{-\frac{1}{1-m\eta'}\kappa}$$

is locally integrable for some $\eta < \eta' < \frac{1}{m}$. Then s can be extended to an element of $\Gamma\left(X, \left(\mathcal{O}_X/\mathcal{O}_X t^2\right)\left(L + mK_X\right)\right)$.

Proof. We now apply Lemma (6.1) with $E = (m-1)\,K_X + L$ for the case $\ell = 0$ of the double point at the origin of Δ. We are going to use the vanishing theorem of Kawamata-Viehweg-Nadel [2, 12, 5] to show that the cocycle $\{\Theta_{ji}\}_{i,j}$ from Lemma (6.1) must be a coboundary.

Define the metric

$$e^{-\varphi} = \frac{e^{-\frac{1}{m}\kappa}}{|s|^{\frac{2(m-1)}{m}}}$$

for $E = (m-1)\,K_X + L$ over the reduced fiber X_0. Clearly the curvature current of the metric $e^{-\varphi}$ for whose curvature current is bounded from below by a smooth positive $(1,1)$-form on X_0. By the vanishing theorem of Kawamata-Viehweg-Nadel [2, 12, 5], to show that the cocycle $\{\Theta_{ji}\}_{i,j}$ is a coboundary it suffices to show that the cocycle $\{\Theta_{ji}\}_{i,j}$ locally belongs to the multiplier ideal sheaf $\mathcal{I}_\varphi$ on X_0. Since the term

$$\frac{\partial \Xi_{ij}}{\partial t}\,\Xi_{ij}^{-1} s_i$$

in

$$\Theta_{ji} = \sum_{\nu=1}^{n} \frac{\partial s_i}{\partial z_\nu^{(i)}} \frac{\partial g_\nu^{(ij)}}{\partial t} - \frac{\partial \Xi_{ij}}{\partial t}\,\Xi_{ij}^{-1} s_i$$

is clearly locally L^2 on X_0 with respect to $e^{-\varphi}$, it suffices to show that $\frac{\partial s_i}{\partial z_\nu^{(i)}}$ is also locally L^2 on X_0 with respect to $e^{-\varphi}$ for $1 \leq \nu \leq n$. In order words, the proof of the Theorem is now reduced to verifying the local integrability of

$$\frac{\left|\frac{\partial s_j}{\partial z_\nu^{(j)}}\right|^2 e^{-\frac{1}{m}\kappa}}{|s_j|^{\frac{2(m-1)}{m}}} \tag{6.4.1}$$

on X_0. We rewrite (6.4.1) as

$$\frac{\left|\frac{\partial s_j}{\partial z_\nu^{(j)}}\right|^2}{|s_j|^{2\left(\frac{m-1}{m}+\eta\right)}} \left(|s_j|^{2\eta} e^{-\frac{1}{m}\kappa}\right).$$

We are going to apply Hölder's inequality with $\frac{1}{p} = \frac{m-1}{m} + \eta'$ and $\frac{1}{p'} = \frac{1}{m} - \eta'$ so that $p = \frac{m}{m-1+m\eta'}$ and $p' = \frac{m}{1-m\eta'}$ and $\frac{1}{p} + \frac{1}{p'} = 1$. By Hölder's inequality, for an open subset W of X_0, the integral

$$\int_W \frac{\left|\frac{\partial s_j}{\partial z_\nu^{(j)}}\right|^2}{|s_j|^{2\left(\frac{m-1}{m}+\eta\right)}} \left(|s_j|^{2\eta} e^{-\frac{1}{m}\kappa}\right)$$

is dominated by the product of

$$\left(\int_W \left(\frac{\left|\frac{\partial s_j}{\partial z_\nu^{(j)}}\right|^2}{|s_j|^{2\left(\frac{m-1}{m}+\eta\right)}}\right)^{\frac{m}{m-1+m\eta'}}\right)^{\frac{m-1}{m}+\eta'}$$

and

$$\left(\int_W \left(|s_j|^{2\eta} e^{-\frac{1}{m}\kappa}\right)^{\frac{m}{1-m\eta'}}\right)^{\frac{1}{m}-\eta'}.$$

Since by Lemma (6.3)

$$\left(\frac{\left|\frac{\partial s_j}{\partial z_\nu^{(j)}}\right|^2}{|s_j|^{2\left(\frac{m-1}{m}+\eta\right)}}\right)^{\frac{m}{m-1+m\eta'}}$$

is locally integrable on X_0, it follows that the local integrability of (6.4.1) on X_0 follows from the local integrability of

$$|s_j|^{\frac{2\eta m}{1-m\eta'}} e^{-\frac{1}{1-m\eta'}\kappa}$$

which is assumed in the statement of the Theorem. Q.E.D.

Remark (6.5) The method of proof of Theorem (6.4) applies also to the case when the fibers of $\pi : X \to \Delta$ are of general type and the curvature current of $e^{-\kappa}$ is assumed to be only semipositive instead of being bounded from below by some smooth positive $(1,1)$-form.

Remark (6.6). In Theorem (6.4) we can only handle the case $\ell = 0$ of the double point at the origin of Δ instead of for all nonnegative integral values of ℓ, because there are still no vanishing theorems of Kawamata-Viehweg-Nadel [2, 12, 5] available for the unreduced complex space $\left(X_0, \mathcal{O}_X / \mathcal{O}_X t^{\ell+1}\right)$ when ℓ is a positive integer. For multiplier ideal sheaves for such unreduced complex spaces new formulations have to be introduced. When the metric $e^{-\varphi}$ of the line bundle is defined on an open neighborhood of the unreduced complex subspace of an ambient manifold, one can define a local section s on the unreduced complex subspace to be in the multiplier ideal sheaf if it admits a local extension $\tilde{s}$ to the ambient manifold with $|\tilde{s}|^2 e^{-\varphi}$ locally integrable on the ambient manifold. For algebraic geometric applications the metric used to define the multiplier ideal sheaf on a manifold is constructed from fractional powers of the absolute values of holomorphic sections of line bundles. Such a metric is of the form $\frac{1}{\sum_j |\sigma_j|^{\gamma_j}}$, where σ_j are global sections of line bundles and γ_j are rational positive numbers. For unreduced complex subspaces Y of some ambient complex manifold X we can still use global sections σ_j of line bundles over the unreduced complex space Y and positive rational numbers γ_j. We define a local holomorphic section s over Y to be in the multiplier ideal sheaf if there is a local extension $\tilde{s}$ of s to X and there are local extensions $\tilde{\sigma}_j$ of σ_j to X such that $\frac{|\tilde{s}|^2}{\sum_j |\tilde{\sigma}_j|^{\gamma_j}}$ is locally integrable on X.

One possible approach to vanishing theorems of Kawamata-Viehweg-Nadel [2, 12, 5] for unreduced complex subspaces is to adapt a technique introduced in [10]. The technique uses a holomorphic section s_A of a sufficiently ample line bundle A on the ambient manifold X. After multiplication by s_A we can extend $s_A \sigma_j$ to global holomorphic sections $\hat{\sigma}_j$ of line bundles over X. After multiplication by s_A we can also extend the product $\{s_A \Psi_{ji}\}_{i,j}$ of s_A and the cocycle $\{\Psi_{ji}\}_{i,j}$ on the unreduced complex subspace Y to a cocyle $\left\{\hat{\Psi}_{ji}\right\}_{i,j}$ defined on X. Then one defines a metric for the line bundle on X twisted by A in such a way that, after solving for the cochain $\{\Phi_j\}_j$ whose coboundary is $\left\{\hat{\Psi}_{ji}\right\}_{i,j}$ by the vanishing theorem of Kawamata-Viehweg-Nadel [2, 12, 5] for manifolds, the restriction of the cochain $\{\Phi_j\}_j$ to the unreduced complex subspace Y is divisible by s_A. There are still some technical difficulties with this approach of adapting the techniques of [10].

Remark (6.7). Another approach to apply the argument of Theorem (6.4) to the case of $\ell > 0$ is to differentiate $\ell + 1$ times both sides of (6.1.1) , or equivalently to divide (6.1.1) by $t^{\ell+1}$, instead of just once. This approach has the advantage that one does not have to use vanishing theorems of Kawamata-Viehweg-Nadel [2, 12, 5] for unreduced complex spaces. However, there are other obstacles. The difficulty with this approach is that the resulting cocycle $\{\zeta_{ji}\}_{i,j}$ involves partial derivatives of s_i up to order $\ell + 1$ instead of just the first order. For quotients whose numerator involves partial derivatives of s_i higher than the first order and whose denominator involves s_i, the argument of the proof of

Lemma (6.3) no longer applies and the resulting cocyle $\{\zeta_{ji}\}_{i,j}$ no longer locally belongs the multiplier ideal sheaf $\mathcal{I}_\varphi$. Another alternative is to differentiate (6.1.1) only once and then consider the problem of t^ℓ as a zero-divisor of the quotient $\mathcal{O}_X/\mathcal{I}_\varphi$ formed from the multiplier ideal sheaf $\mathcal{I}_\varphi$. If t^ℓ is not zero-divisor of the quotient $\mathcal{O}_X/\mathcal{I}_\varphi$, then the cocycle $\{\zeta_{ji}\}_{i,j}$ locally belongs to the multiplier ideal sheaf $\mathcal{I}_\varphi$ as a consequence of the fact that $\{t^\ell\zeta_{ji}\}_{i,j}$ locally belongs to $\mathcal{I}_\varphi$. However, there is no good way to handle the problem of t^ℓ as a zero-divisor of $\mathcal{O}_X/\mathcal{I}_\varphi$.

References

[1] J.-P. Demailly, *A numerical criterion for very ample line bundles*, J. Diff. Geom 37 (1993), 323-374.

[2] Y. Kawamata, *A generalization of the Kodaira-Ramanujam's vanishing theorem*, Math. Ann. 261 (1982), 43-46.

[3] J. Kollar, *Higher direct images of dualizing sheaves I.* Ann. of Math. 124 (1986), 11-42. II. Ann. of Math. 124 (1986), 171-202.

[4] M. Levine, *Pluri-canonical divisors on Kähler manifolds*, Invent. math. 74 (1983), 293-903. II. Duke Math. J. 52 (1985), 61-65.

[5] A. Nadel, *Multiplier ideal sheaves and existence of Kähler-Einstein metrics of positive scalar curvature*, Proc. Nat. Acad. Sci. U.S.A. 86 (1989), 7299-7300 and Ann. of Math. 132 (1990), 549-596.

[6] T. Ohsawa, *On the extension of L^2 holomorphic functions*, II, Publ. Res. Inst. Math. Sci. 24 (1988), 265-275. III. Negligible weights, Math. Z. 219 (1995), 215-225. IV, A new density concept, in *Geometry and Analysis on Complex Manifolds*, 157–170, World Sci. Publishing, 1994. V. Effects of generalization, Nagoya Math. J. 161 (2001), 1-21.

[7] T. Ohsawa and K. Takegoshi, *On the extension of L^2 holomorphic functions*, Math. Z. 195 (1987), 197-204.

[8] Y.-T. Siu, *Invariance of plurigenera*. Invent. Math. 134 (1998), 661–673.

[9] Y.-T. Siu, *Extension of twisted pluricanonical sections with plurisubharmonic weight and invariance of semipositively twisted plurigenera for manifolds not necessarily of general type*. In: *Complex Geometry*, ed. I. Bauer et al, Springer-Verlag 2002, pp. 223–277.

[10] Y.-T. Siu, *A New Bound for the Effective Matsusaka Big Theorem*, Houston J. of Mathematics, Issue for 90th Birthday of S. S. Chern, Volume 28 (2002), 389-409.

[11] H. Tsuji, *Deformational invariance of plurigenera*, Preprint 2001, math. AG/0011257; (earlier version: Invariance of plurigenera of varieties with nonnegative Kodaira dimensions, math.AG/0012225).

[12] E. Viehweg, *Vanishing theorems*, J. Reine Angew. Math. 335 (1982), 1-8.

5

MAXWELL'S EQUATIONS IN CLIFFORD CALCULUS FRAMEWORK – AN OVERVIEW ON THE DEVELOPMENT

Wolfgang Sprößig
Freiberg University of Mining and Technology, Germany
E-mail: sproessig@math.tu-freiberg.de

Abstract: Quaternionic and Clifford calculus is applied on wide classes of problems in very different fields of science. The aim of our overview article is to line out the historical guidelines of electromagnetism in Clifford algebra framework. We want to introduce into the structure of low dimensional Clifford algebras and show the possibilities for the description of well-known terms in electromagnetics. Using Teodorescu transform we present an iteration procedure for solving Maxwell' s equations in isotropic homogeneous media. Assuming the so-called Drude-Born-Feodorov constitutive laws Maxwell' s equations in chiral media are formulated. Time-dependent problems are reduced to the consideration of Weyl operators and their right inverses the Teodorescu transforms. New tendencies are discussed.

1. Introduction

The discovery of electromagnetic field equations is one of the consequences of the industrial revolution in the second half of the 19 th century. For about 150 years these equations have attracted and fascinated generations of mathematicians and physicists. The study and the treatment of the so-called Maxwell' s equations can be done in two manners: by using a suitable algebraical structure or in a more "pedestrian" way. The aim of this paper is to show the close connection between electromagnetic field theories and Clifford algebras as well as to develop a solution theory under some conditions. All the explanations are given in a historical context.

One of the basic structures needed here are the algebras of real and complex quaternions. The famous Irish physicist W.R. Hamilton [13] discovered the skew-field of real quaternions in 1843. In his early papers already J.C. Maxwell used quaternions for the description of electromagnetic equations. The first comprehensive discussion of these topics is found in the book of J. C. Maxwell titled " A Treatise on Electricity and Magnetism" in 1873. On the basis of H. Grassmann' s "Ausdehnungslehre" and W. R. Hamilton' s quaternions W. K. Clifford discovered [5] geometric algebras, nowadays called Clifford algebras, in 1978. His fundamental work has the modest title: "Applications of Grassmann' s extensive algebra".

2. Clifford algebras

Let us start with the description of this algebra in expressing the main ideas.

Definition 1. (Real Clifford algebra). *An associative algebra over* $\mathbb{R}$ *with unit* $e_0 := 1$ *which is freely generated by* n *basis elements* $e_1, ..., e_n$ *together with the defining relations*

$$e_i e_j + e_j e_i = -2\delta_{ij} \quad (i, j = 1, ..., n). \tag{1}$$

is called real Clifford algebra (denoted by: $C\ell_{0,n}$*). Here* δ_{ij} *denotes the Kronecker symbol. This means*

$$e_i e_j + e_j e_i = 0 \quad \textit{iff} \quad i \neq j, \quad e_i^2 = -1 \quad (i, j = 1, ..., n).$$

Usually the elements $e_1, e_2, ..., e_n$ belong to the vector space $\mathbb{R}^n$ and form there a basis. After the canonical embedding we have identified the "initial" vectors $e_1, e_2, ..., e_n$ in $\mathbb{R}^n$ with corresponding elements in $C\ell_{0,n}$. Furthermore, we identify the unit element in $C\ell_{0,n}$ with $1 \in \mathbb{R}$.

Definition 2. (Clifford algebra of the (p,q)-type). *Let* $e_1, ..., e_q, ..., e_{p+q}$ *form a basis of* $\mathbb{R}^{p+q}$. *We introduce the denotation*

$$\dot{e}_j := \mathbf{i}\ e_{q+j}.$$

Following an idea of F. Sommen the Clifford algebra of the (p,q)-type (denoted by: $C\ell_{p,q}$*) is defined by the real associative algebra with unit* $e_0 := 1$ *freely generated by elements of the form*

$$\dot{e}_1, ..., \dot{e}_p, e_1, ..., e_q,$$

which fulfil the defining relations

$$\dot{e}_j\, \dot{e}_k + \dot{e}_k\, \dot{e}_j = 2\delta_{jk},$$

$$e_j\, e_k + e_k\, e_j = -2\delta_{jk},$$

$$e_j\, \dot{e}_k + \dot{e}_k\, e_j = 0 \quad .$$

Definition 3. (Complex Clifford algebra) *The Clifford algebra $C\ell_{n,0}$ is called complex Clifford algebra can be seen as complexification of the real Clifford algebra $C\ell_{0,n}$ i.e. $C\ell_{n,0} = \mathbb{C} \otimes C\ell_{0,n}$.*

Using Pauli matrices we get for low-dimensional Clifford algebras the following representations. Firstly, we remember the definition of Pauli matrices:

$$\sigma_0 = \begin{pmatrix} 1 & 0 \\ 0 & 1 \end{pmatrix}, \quad \sigma_1 = \begin{pmatrix} 1 & 0 \\ 0 & -1 \end{pmatrix}, \quad \sigma_2 = \begin{pmatrix} 0 & -i \\ i & 0 \end{pmatrix}, \quad \sigma_3 = \begin{pmatrix} 0 & 1 \\ 1 & 0 \end{pmatrix}.$$

$\mathbf{C\ell_{0,1}}$: This algebra is generated by σ_0 and $i\sigma_2$ with $\sigma_2^2 = -\sigma_0$, therefore $C\ell_{0,1} = \mathbb{C}$. This is clear by the representation $x = x_0\sigma_0 + x_1 i\sigma_2$.

$\mathbf{C\ell_{0,2}}$: The matrices $\sigma_0, -i\sigma_1, -i\sigma_2, \sigma_1\sigma_2$ $(= -i\sigma_3)$ form a basis in $C\ell_{0,2}$. For any element x of this algebra we obtain the representation

$$x = \begin{matrix} x_0 - ix_1 & -x_2 + ix_3 \\ x_2 + ix_3 & x_0 + ix_1 \end{matrix}$$

with $x_i \in \mathbb{R} \quad (i = 0, 1, 2, 3)$.

Later we will see that in this way can also be described a matrix realization of the *algebra of real quaternions* Δ. Hence, $C\ell_{0,2} = \Delta$.

$\mathbf{C\ell_{3,0}}$: The matrices $\sigma_0, \sigma_1, \sigma_2, \sigma_3, \sigma_1\sigma_2, \sigma_2\sigma_3, \sigma_3\sigma_1, \sigma_1\sigma_2\sigma_3$ form a basis in the algebra $C\ell_{3,0}$. We find that $\sigma_1\sigma_2 = -i\sigma_3, \sigma_2\sigma_3 = -i\sigma_1, \sigma_3\sigma_1 = -i\sigma_2$ and $\sigma_1\sigma_2\sigma_3 = -i\sigma_0$. An arbitrary element x of this algebra can be represented by

$$x = (x_0 - x_7 i)\sigma_0 + (x_1 - x_5 i)\sigma_1 + (x_2 - x_6 i)\sigma_2 + (x_3 - x_4 i)\sigma_3.$$

Then it is easy to get $C\ell_{3,0} = \mathbb{C} \bigotimes \Delta$. This algebra coincides with the *Pauli algebra.* The denotation by P is made in honour of Wolfgang Pauli.

The algebra $C\ell_{p,q}$ is isomorphic to a matrix algebra with matrices of the order $(p-q)$mod 8. This result was obtained in 1908 by E. Cartan. H. Grassmann introduced another algebra $\mathcal{A}_G$ over the a real vector space V, which is defined by the so-called exterior multiplication $x \wedge x =: x^2$ where $x \in V$. This algebra is nowadays called *Grassmann algebra.* As linear spaces the Clifford algebra $C\ell_{p,q}$ and $\mathcal{A}_G$ are isomorphic.

In the nice book "Clifford algebras and Dirac operators in harmonic analysis" J. Gilbert and A. Murray [17] line out by introduction of a creation and an annihilation operator the transformation of the Grassmann algebra into a Clifford algebra $C\ell_{p,q}$. This can be done as follows:

Let us write for brevity $B(x,\xi) =:< x, \xi >$. This symmetric bilinear form can be extended to simple p-vectors in $\bigwedge^p X$ by the determination

$$< x_1 \wedge ... \wedge x_p, \xi_1 \wedge ... \wedge \xi_p >:= \det(< x_i, \xi_j >)_{i,j=1}^p.$$

Then we extend to all p-vectors by linearity and to all elements in $\bigwedge X$ by orthogonality.

Let $x^{(p)}$ be a simple p-vector and $x \in X$, then the operator of the *creation operator* E_x is defined by the relations

$$E_x(1) := x \, , \; E_x(x^{(p)}) := x \wedge x^{(p)}$$

and by the linear extension onto $\bigwedge X$. The *annihilation operator* J_x is defined by

$$J_x(1) := 0 \, , \; J_x(x^{(p)}) := \sum_{i=1}^{p} (-1)^{(i-1)} B(x, x_i) x_1 \wedge ... \wedge \hat{x}_i \wedge ... \wedge x_p$$

and by linear extension onto the whole $\bigwedge X$, where $\hat{x}_i$ means that this factor is omitted in this product.

Let $u \in \bigwedge^p X$ then $E_x(u) \in \bigwedge^{p+1} X$ and $J_x(u) \in \bigwedge^{p-1} X$. In addition, in the case of $\xi \in X$ holds, $J_x(\xi) =< x, \xi >= B(x, \xi)$. We have for $x \in X, \; u, v \in \bigwedge X$ the relation

$$< E_x(u), v >=< u, J_x(v) >,$$

which means J_x is the adjoint operator to E_x. In tensor calculus notation the annihilation operator is called *contraction.* In physical literature it is usually used for the denotation: $x \rfloor v, x \in X, \; v \in \bigwedge X$. In [20] this definition is generalized to $x \in \bigwedge X$. It is easy to prove that

$$E_x^2 \; = \; J_x^2 \; = 0$$

Introducing the CLIFFORD product of $x \in X$ and $u \in \bigwedge X$ by

$$xu := E_x(u) - J_x(u)$$

and extending this by linearity and associativity to all $\bigwedge X$, then $\bigwedge X$ is isomorphic to $C\ell_{p,q}$ as a linear vector space. By the way C. Chevalley [4] proved in 1954 that each p-vector corresponds with a skew-symmetric tensor as element of the tensor algebra.

3. Algebraisation conceptions

One of the main goals from the very beginning was to write Maxwell' s equations in a more simply form maybe

$$Du + Au = F$$

with a suitable derivation D and a potential operator A, which allows an advandageously mathematical treatment. So already W.R. Hamilton introduced his well-known nabla-operator :

$$\nabla := i\partial_1 + j\partial_2 + k\partial_3$$

which is nothing else as a vector derivative. By the way the denotation "vector" was introduced by W.R. Hamilton as a pure quaternion. Important books followed by L. Silberstein ("The theory of relativity", 1914) and H. Weyl ("Raum-Zeit-Materie", 1921) In 1929 appear in a paper by C. Lanczos the notion of a "spinor" as an element of a spinor space. Later was shown that these spaces can be generated by primitive idempotent elements of a Clifford algebra. A. Proca at the first time used in his physical work Clifford numbers. Some progress brought the usage of the so-called semi-vectors in a paper by A. Einstein and A. Mayer in 1932. The first complete representation of Maxwell's equations with Clifford numbers one can find in the theses of M. Mercier (Geneva) titled: "Expression des quations de lectromagnetisme au moyen des nombres de Clifford". Maybe caused by the second world war there is a gap of more than 20 years in this delvelopment. Intersting new ideas came from F. Bolinder in 1957 who used complex 4-d forms in order to describe electromagnetic equations. One year later M. Riesz developed his theory of spinors and considered general field equations in Clifford algebra framework. G. Kron used in 1959 skew-symmetric tensors for the representation of Maxwell's equations. It is the great merit of D. Hestenes to outline the advandage of the approach to electromagnetism by the help of real Clifford algebras ([15]). This book can be seen as a basis for further progression. In order to show the usefulness of working in Clifford structures we will demonstrate here the following example:

Example 1. *We consider the algebra $C\ell_{3,1}$ (Majorana algebra). The elements*

$$e_0 = \begin{pmatrix} i & 0 \\ 0 & i \end{pmatrix}, \quad e_j = \begin{pmatrix} 0 & \sigma_j \\ \sigma_j & 0 \end{pmatrix} \quad (j = 1, 2, 3)$$

form a basis in $C\ell_{3,1}$. Obviously, it holds

$$e_i e_j + e_j e_i = 2\delta_{ij} \quad (i, j = 1, 2, 3).$$

Introducing now the *electric field*

$$E := E_1 e_1 e_0 + E_2 e_2 e_0 + E_3 e_3 e_0$$

the *magnetic flux density*

$$B := B_1 e_1 + B_2 e_2 + B_3 e_3$$

as well as the *electromagnetic field*

$$F := E e_0 + (-B) e_{123} \quad \text{with} \quad e_{123}^2 = (e_1 e_2 e_3)^2 = -1.$$

Using the ideas formulated in Hestenes' books, one obtains

$$\partial F = J = \rho + I$$

where

$$\partial = -e_0 \partial_0 + D, \quad D = \sum_{k=1}^{3} e_k \partial_k.$$

ρ is the *charge density* and I the *current vector*. Because of

$$\partial F = \partial \cdot F + \partial \wedge F$$

we get

$$\partial \cdot F = \rho \quad \text{and} \quad \partial \wedge F = I.$$

Here $\partial\cdot$ is the annihilation operator and $\partial\wedge$ is the creation operator. By introduction of the so-called *four potential* $A = A_0 + A_1$ with a scalar part A_0 and a vector part A_1 it follows that

$$\partial A = F$$

is equivalent to the pair of relations $\partial \wedge A = -F$ and $\partial \cdot A = 0$. The latter one is called *Lorenz condition* and also written as

$$\partial_0 A_0 + \nabla \cdot A_1 = 0$$

It can be proved that

$$E = -\partial_0 A_1 - \nabla A_0$$
$$B = \nabla \wedge A_1.$$

The generation of *Maxwell's stress tensor* makes use of the Clifford number valued fields

$$T_k = -\frac{1}{2} F e_k F \quad (k = 0, 1, 2, 3).$$

For $k = 0$ we obtain

$$T_0 = \frac{1}{2}\left(E^2 + B^2\right) e_0 - e_{123} E \cdot B.$$

The scalar part of T_0 is just the energy density and $-e_{123}E \cdot B$ is the *Poynting vector.* Finally, we get for Maxwell' s stress tensor the representation

$$T = (T_k \cdot e_\ell)$$

(cf. M. Riesz [25]).

4. Maxwell' s equations

4.1 Theories with influence on our treatment of Maxwell' s equations

It was the work of the famous Swiss mathematician R. Fueter and his students like Bosshard, Nef and Haefeli which developed basis ideas of a generalized function theory of quaternions in the period 1928 – 1949 . All the time they looked for connections between his theory and the theory of several complex variables (cf. [6],[7]). In a consequent continuation of Fueter' s ideas the Belgium research group in Clifford analysis under the leadership of R. Delanghe developed a function theory in real Clifford algebras. The book "Clifford analysis" [9] by F. Brackx, R. Delanghe and F. Sommen in 1982 was the first comprehensive description of this new field of research. Many impulses and motivations came from books by D. Hestenes and his coworkers (cf. [14],[15],[16]). At the same time a function theoretic approach to the treatment of systems of first order differential equations appeared by R.P. Gilbert and J. Buchanan (cf. [8]). The idea was to study functions defined on the complex plane with values in a so-called Douglis algebra, develop a corresponding function theory and use this in the investigation of partial differential equations.

Under the influence of a general operator theory (D. Przeworska-Rolewicz, M. Tasche), the so-called *Vekua theory* by I.N. Vekua appears in 1959 [28]. On the other hand the theory of multidimensional singular integral equations was developed by S. Michlin and S. Prößdorf [23]. So new assumptions for a successful treatment of boundary value problems have been given. In the sequence our book (with K. Gürlebeck) "Quaternionic analysis and elliptic boundary value problems" [18] appeared. Later these ideas were transmitted to further classes of partial differential equations [19].

4.2 Formulation of Maxwell' s equations

The four partial differential equations first appear in fully developed form in the above mentioned work "A Treatise on Electricity and Magnetism" by J. C. Maxwell. Let G be a bounded domain with sufficient smooth boundary Γ that is filled out with an isotropic homogeneous material.

Using Gauss units Maxwell' s equations read as follows:

$$\begin{aligned} c \operatorname{rot} H &= 4\pi J + \partial_t D && \text{(Biot-Savart-Ampere's law)},\\ c \operatorname{rot} E &= -\partial_t B && \text{(Faraday's law)},\\ \operatorname{div} D &= 4\pi\rho && \text{(Coulomb's law)},\\ \operatorname{div} B &= 0 && \text{(no free magnetic charge)}. \end{aligned}$$

Furthermore, the continuity condition has to be fullfilled:

$$\operatorname{div} J = -\partial_t \rho,$$

where $E = E(t, x)$ *electric field*, $H = H(t, x)$ *magnetic field*, $J = J(t, x)$ *electric current density*, $D = D(t, x)$ *electric flux density* , $B = B(t, x)$ *magnetic flux density* , $\rho = \rho(t, x)$ *charge density* and c *speed of light in vacuum*.

The relations between flux densities and the electric and magnetic fields are depending on the material. It is well-known that for instance all organic materials contain carbon and realize in this way some kind of optical activity. Therefore, Lord Kelvin introduced the notion of the *chirality measure* of a medium. This coefficient expresses the optical activity of the underlying material. The correspondent constitutive laws are the following:

$$\begin{aligned} D &= \varepsilon\, E + \varepsilon\, \beta \operatorname{rot} E && \text{(Drude-Born-Feodorov laws)},\\ B &= \mu\, H + \mu\, \beta \operatorname{rot} H, \end{aligned}$$

where $\varepsilon = \varepsilon(t, x)$ is the *electric permittivity*, $\mu = \mu(t, x)$ is the *magnetic permeability* and the coefficient β describes the *chirality measure* of the material. In absorbing media electric permittivity is changing in time. A good reference to chiral media is the article [1]. In [17] constitutive laws with memory effects are described. Moreover, non-linear constitutive laws are studied (cf. [29],[24]) which are important for ferromagnetic materials for instance. In this case we have $B = f(H)$ with a monotonic function $f : \mathbb{R}^3 \to \mathbb{R}^3$, which reflects the behaviour of soft iron for high field saturation. The connection between the electric field E and current density J is given by

$$J = \sigma\, E + \sigma\, g,$$

where σ is the *electric conductivity* and g a given electric source.

Starting with $\beta = 0$ and replacing D and B by $D = \varepsilon\, E$ and $B = \mu\, H$ we get in the case of $\varepsilon = \varepsilon(x)$, $\mu = \mu(x)$

$$-\varepsilon\partial_t E + c \operatorname{rot} H = 4\pi\, J\,, \tag{2}$$

$$\mu\partial_t H + c \operatorname{rot} E = 0\,, \tag{3}$$

$$\varepsilon \operatorname{div} E = 4\pi\rho - (\nabla\varepsilon \cdot E)\,, \tag{4}$$

$$\mu \operatorname{div} H = -(\nabla\mu \cdot H)\,. \tag{5}$$

After summing (4.2) and (4.5) as well as (4.3) and (4.4) we obtain

$$-\varepsilon\partial_t E + c \operatorname{rot} H + \mu \operatorname{div} H = -(\nabla\mu \cdot H) + 4\pi\, J, \tag{6}$$

$$\mu\partial_t H + c \operatorname{rot} E + \varepsilon \operatorname{div} E = -(\nabla\varepsilon \cdot H) + 4\pi\, \rho. \tag{7}$$

In the case of ε, μ being constants we can introduce the new functions $\tilde{E}$, $\tilde{H}$ which are defined on a homogeneous space with a first coordinate x_0 and the other coordinates $\tilde{x} = (\tilde{x}_1, \tilde{x}_2, \tilde{x}_3)$. We obtain:

$$E(t,x) =: \tilde{E}\left(-\frac{1}{\varepsilon}\, t, \frac{1}{c}\, c\right),$$

$$H(t,x) =: \tilde{H}\left(\frac{1}{\mu}\, t, \frac{1}{c}\, c\right).$$

The equations (4.6) - (4.7) transform into

$$\partial_1 \tilde{E} + \operatorname{rot} \tilde{H} + \mu\, c \operatorname{div} \tilde{H} = 4\pi\, J,$$

$$\partial_1 \tilde{H} + \operatorname{rot} \tilde{E} + \varepsilon\, c \operatorname{div} \tilde{E} = 4\pi\, \rho.$$

4.3 A solution theory

Let G be a bounded domain in $\mathbb{R}^3$ and a a positive constant. We consider in $\mathbb{R}^4$ the cylinder $Z = G \times [-a, a]$. A right inverse to the Weyl operator $\partial := \partial_0 + D$ (cf.[27]) is the so-called *Teodorescu transform*:

$$(T_Z u)(x) = \frac{-1}{\sigma_3} \int\limits_Z e(x-y)u(y)dy \ , \quad Z = G \times [-a,a]$$

with $e(x) = \frac{\overline{x}}{|x|^4}$, $\sigma_3 = \frac{2\pi^{3/2}}{\Gamma(\frac{3}{2})}$. We obtain in a straightforward manner

$$\partial\, T_Z\, u = \begin{cases} u & \text{in} & Z \\ 0 & \text{in} & \mathbb{R}^4 \setminus \overline{Z} \end{cases}$$

and

$$T_Z\, \partial\, u + \phi_Z = \begin{cases} u & \text{in} & Z \\ 0 & \text{in} & \mathbb{R}^4 \setminus \overline{Z} \end{cases}$$

with $\phi_Z \in \ker \partial$. In complete analogy is introduced a conjugate Teodorescu transform T_Z^*. We just have to replace $e(x)$ by its conjugate. In our paper [27] we obtained the following result:

Theorem 1. *[27] If h, h^* are known, then under smallness conditions the iteration procedure:*

$$\begin{aligned} v_n &= 4\pi\sigma T_Z(v_{n-1} + w_{n-1}) + 4\pi T_Z \left(\sigma g + \frac{\rho}{\varepsilon c}\right) + h \\ w_n &= 4\pi\sigma T_Z^*(v_{n-1} + w_{n-1}) + 4\pi T_Z^* \left(\sigma g - \frac{\rho}{\varepsilon c}\right) + h^* \end{aligned}$$

with ($v_0 = w_0 = 0$) will converge in suitable Banach spaces. The connection to E and H is given by: $\tilde{E} + \tilde{H} =: v$ and $\tilde{E} - \tilde{H} =: w$. v_n and w_n are approximations of v and w.

If $\beta \neq 0$ io than we get the following modified version of Maxwell' s equations

$$\begin{aligned} -\text{div}\, E &= f \\ \text{rot}\, E &= \gamma^2 \beta E + i\omega\mu \left(\frac{\gamma}{k}\right)^2 H \\ \text{rot}\, H &= \gamma^2 \beta H - i\omega\varepsilon \left(\frac{\gamma}{k}\right)^2 E \\ -\text{div}\, H &= 0 \end{aligned}$$

where $k := \omega\sqrt{\varepsilon\mu}$ and $\gamma^2 := k^2(1 - k^2\beta^2)^{-1}$. Further, in a shorthand notation we simply will write $\alpha_1 := \gamma^2\beta$ and $\alpha_2 := i\omega\varepsilon \left(\frac{\gamma}{k}\right)^2$. Then we obtain the following representation with the Dirac operator D

$$-(D - \alpha_1)\frac{1}{\alpha_2}(D - \alpha_1)H = \alpha_2 H + f.$$

In case of constant electric permittivity and magnetic permeability we can write

$$(d-\alpha_1)^2 H = -\alpha_2 H + f.$$

Using modified Teodorescu transformation it is possible to get

$$H = -\alpha_2^2 T_{\alpha_1}^2 H + T_{\alpha_1}^2 \alpha_2 f + T_{\alpha_1}\Phi_1 + \Phi_2.$$

Here $\Phi_1, \Phi_2 \in \ker(D-\alpha_1)$. This modified Teodorescu transformation are well studied in [2].

4.4 Maxwell' s equations with magnetic monopols

In [21] are considered magnetic monopols. Then we have the following model:

In vectorial form we obtain:

$$\nabla \times E + \frac{1}{c}\partial_t\, B = -\frac{4\pi}{c}J',$$

$$\nabla \times B - \frac{1}{c}\partial_t\, E = \frac{4\pi}{c}J,$$

$$\nabla \cdot E = 4\pi\rho,$$

$$\nabla \cdot B = 4\pi\rho'.$$

Here J, J'' are the *electric and the monopole current density.*

With the help of the *Faraday vector* $W = E + \mathbf{i}B$ we get:

$$\nabla \times -\frac{\mathbf{i}}{\mathbf{c}}\partial_t\, W = \frac{4\pi\mathbf{i}}{c}(J' + \mathbf{i}J),$$

$$\nabla \cdot W = 4\pi(\rho + \mathbf{i}\rho').$$

Remark 1. *Theorem 1 is also true in the case of the appearance of magnetic monopols.*

4.5 Fluid flow with magnetic induction

In hydromechanics the following problem is interesting. As before let u be the velocity and B the field induction. A fluid is flowing under the influence of a magnetic field. Then we have the equations:

$$\begin{aligned}
-\Delta \underline{u} + \frac{\rho}{\eta}(\underline{u} \cdot \nabla)\underline{u} + \frac{1}{\eta}\nabla p - \frac{1}{\mu\eta}\underline{B} \times \operatorname{rot} \underline{B} &= \frac{\rho}{\eta} f \quad \text{in} \quad G \\
\Delta \underline{B} + \mu\sigma \operatorname{rot}(\underline{B} \times \underline{u}) &= \underline{g} \quad \text{in} \quad G \\
\operatorname{div} \underline{B} &= 0 \quad \text{in} \quad G \\
\operatorname{div} \underline{u} &= 0 \quad \text{in} \quad G \\
\underline{B} = \underline{u} &= 0 \quad \text{on} \quad \Gamma,
\end{aligned}$$

where μ is the permeability and σ is a measure of the electric charge.

In quaternionic denotation this problem can be reformulated as follows:

$$\begin{aligned}
DDu + \frac{1}{\eta} D\, p + M(u) &= \frac{1}{\mu\eta} B \times (D \times B) \quad \text{in} \quad G \\
DDB + \mu\sigma[u, B] &= g \quad \text{in} \quad G \\
Sc\, D\, u &= 0 \quad \text{in} \quad G \\
Sc\, DB &= 0 \quad \text{in} \quad G \\
B = u &= 0 \quad \text{on} \quad \Gamma
\end{aligned}$$

where $D = \partial_1 e_1 + \partial_2 e_2 + \partial_3 e_3$. Only for simplicity we will add the following trivial problems:

$$\begin{aligned}
\Delta u_0 &= 0, \quad \Delta B_0 = 0 \quad \text{in} \quad G, \\
u_0 &= 0, \quad B_0 = 0 \quad \text{on} \quad \Gamma.
\end{aligned}$$

Furthermore, we abbreviate:

$$\begin{aligned}
M(u) &:= \frac{\rho}{\eta}[M^*(u) - f], \; M^*(u) := (u \cdot D)u \\
[u, B] &:= D \times (B \times u) = (u \cdot D)B - (B \cdot D)u.
\end{aligned}$$

Notice that $[u, B]$ is just a Lie bracket, $f = f$ and $g = g$. Finally, we get

$$\begin{aligned}
DDu + \frac{1}{\eta} Dp + M(u) &= \frac{1}{\mu\eta} B \times (D \times B) \quad \text{in} \quad G \\
DD\,B + \mu\sigma[u, B] &= g \quad \text{in} \quad G \\
Sc\, Du &= 0 \quad \text{in} \quad G \\
Sc\, DB &= 0 \quad \text{in} \quad G \\
u &= 0 \quad \text{on} \quad \Gamma \\
B &= 0 \quad \text{on} \quad \Gamma\ .
\end{aligned}$$

Proposition 1. *[19] Let $G \subset \mathbb{R}^3$ be a bounded domain with a piecewise smooth boundary. Then the right–linear set*

$$\mathcal{L} := L_2(G,\Delta) \cap \ker D$$

is a subspace in $L_2(G,\Delta)$.

It is important to have for our calculus the following Hodge-type decomposition of the Hilbert space $L_2(G,\Delta)$, namely

$$L_2(G,\Delta) = \ker D \cap L_2(G,\Delta) \oplus D \overset{\circ}{W}{}_2^1 (G,\Delta).$$

The corresponding projections P (onto $\ker D \cap L_2(G,\Delta)$) and Q onto $D \overset{\circ}{W}{}_2^1 (G,\Delta)$ are orthoprojections and P is just the well-known *Bergman projection.* After transformation into an operator integral equation which can be done by the help of the preliminary results we obtain

$$u = -\frac{\rho}{\eta} T_G \mathcal{Q} T_G\, M(u) - \frac{1}{\eta} T_G \mathcal{Q} p + \frac{1}{\mu\eta} T_G \mathcal{Q} T_G (B \times (D \times B) \qquad (8)$$

$$B = T_G \mathcal{Q} T_G\, g - \mu\sigma\, T_G \mathcal{Q} T_G [u, B] \qquad (9)$$

$$0 = Sc\, \rho\, \mathcal{Q} T_G M(u) + Sc\, \mathcal{Q} p - \frac{1}{\mu} Sc\, \mathcal{Q} T_G (B \times (D \times B) \qquad (10)$$

$$0 = Sc\, \mathcal{Q} T_G g - \mu\sigma\, Sc\, \mathcal{Q} T_G [u, B] \ . \qquad (11)$$

Note that the boundary conditions are fulfilled automatically. It is sufficient to show this fact for the velocity. Indeed, we know that any element $\mathcal{Q}u$ of im $\mathcal{Q}$ has the representation $\mathcal{Q}u = Dv$ with $v \in \overset{\circ}{W}{}_2^1 (G,\Delta)$. Then it follows

$$tr_\Gamma T_G \mathcal{Q} u = tr_\Gamma T_G D v = tr_\Gamma v - tr_\Gamma F_\Gamma\, tr\, v = 0 \ .$$

Remark 2. *Let $g = 0$. The solution $\{u, p, 0\}$ of (4.8) - (4.11) solves Navier–Stokes equations. We look for solutions with $B \not\equiv 0$.*

4.6 An iteration procedure

We will now consider the following iteration method. For $n = 0, 1, 2, ...$ we get

$$\begin{aligned}
u_n &= -T_G \mathcal{Q} T_G [\frac{\rho}{\eta} M(u_{n-1}) - \frac{1}{\mu\eta} B_{n-1} \times (D \times B_{n-1})] - \frac{1}{\eta} T_G \mathcal{Q} p_n \\
0 &= Sc[\rho \mathcal{Q} T_G M(u_{n-1}) + \mathcal{Q} p_n - \frac{1}{\mu} \mathcal{Q} T_G [B_{n-1} \times (D \times B_{n-1})] \\
B_n^{(j)} &= -\mu\sigma T_G \mathcal{Q} T_G [u_n, B_n^{(j-1)}] + T_G \mathcal{Q} T_G g \qquad (j = 1, 2, ...) \\
0 &= -\mu\sigma Sc\, \mathcal{Q} T_G [u_n, B_n^{(j-1)}] + Sc\, \mathcal{Q} T_G g.
\end{aligned}$$

Here B_n will be computed by using of an "inner" iteration The convergence of the sequence $\{u_n, w_n, p_n\}$ is proved in [26].

5. Conclusions

In [18] by using complexified Clifford algebras connections between elliptic and hyperbolic Dirac operators could be obtained . There is a natural application to the time-dependent inhomogeneous Maxwell equations. In the time-harmonic case in [20] is found a further approach by the so-called quaternionic MAXWELL operator. For MAXWELL equations in the vacuum

$$\operatorname{rot} \mathbf{H} = \sigma \mathbf{E}, \quad \operatorname{rot} \mathbf{E} = i\omega\mu \mathbf{H}$$

$$\operatorname{div} \mathbf{E} = 0, \quad \operatorname{div} \mathbf{E} = 0$$

with the *complex electrical conductivity* $\sigma := \rho^{-1} - i\omega\varepsilon$ one obtains the Helmholtz equations

$$\Delta \mathbf{E} + \lambda \mathbf{E} = 0 \quad \text{and} \quad \Delta \mathbf{H} + \lambda \mathbf{H} = 0$$

with $\lambda := i\omega\mu\rho^{-1} + \omega^2\mu\varepsilon$. Here $\sqrt{\lambda} = \alpha$ is called *medium wave number.* The *Maxwell operator* is defined by

$$M := \begin{pmatrix} \sigma & \text{div}-\text{rot} \\ -\text{div}+\text{rot} & -i\omega\mu \end{pmatrix} .$$

V.V Kravchenko and M.V. Shapiro worked out the following connection between Maxwell' s equations and the Dirac equation. It reads as follows:

$$AMB = \begin{pmatrix} \overline{D-\alpha} & 0 \\ 0 & D-\alpha \end{pmatrix}$$

with

$$A := \begin{pmatrix} \alpha & -\sigma \\ -\alpha & -\sigma \end{pmatrix}, \quad B := \frac{1}{2} \begin{pmatrix} \sigma^{-1} & -\sigma^{-1} \\ \alpha^{-1} & \alpha^{-1} \end{pmatrix} .$$

In [22] we find an approach for anisotropic and inhomogeneous media with different kinds of boundary conditions.

References

[1] Athanasiadis C., Costakis G. and Stratis I.(2000) *Electromagnetic scattering by homogeneous chiral abstacle in a chiral enviroment,* IMA Journal of Applied Mathematics 64, 245–258.

[2] H. Bahmann, K. Gürlebeck , M. Shapiro and W. Sprößig, *On a modified Teodorescu transform* ,Integral Transforms and Special Functions Vol. 12 (2001), Number 3, 213-226.

[3] Brackx F., Delanghe R. and Sommen F. (1982) *Clifford analysis.* Pitman Research Notes in Math., Boston, London, Melbourne.

[4] Chevalley Cl. (1954) *The algebraic theory of spinors.* Columbia U.P., New York.

[5] Clifford W.K. (1978) *Applications of Grassmann's extensive algebra.* Americ. J. of Math. Pure and Appl. 1: 350–358.

[6] Fueter R. (1937) *Die Theorie der regulren Funktionen einer Quaternionenvariablen*, Comp. Ren. Congrs Intern. des Mathmaticiens, Oslo 1936, 75 – 91.

[7] Fueter R. (1949) *Funktionentheorie im Hyperkomplexen*, Vorlesung Wintersemester 1948/49. Ausgearbeitet und erweitert von Erwin Barneis, 318 pp. Mathematisches Institut der Universitt Zürich

[8] Gilbert R.P. and Buchanan J. (1983) *First-order ellipitic systems: A function theoretic approach.* Academic Press, San Diego.

[9] Gilbert J.E. and Murray M.A.M. (1990) *Clifford algebras and Dirac operators in harmonic analysis.* Cambr. Studies in advanced mathematics, Cambridge 26.

[10] Grassmann H. (1877) *Der Ort der Hamiltonschen Quaternionen in der Ausdehnungslehre.* Math. Ann. 12: 375–386.

[11] Gürlebeck K. and Sprößig W. (1990) *Quaternionic analysis and elliptic boundary value Problems.* Birkhäuser, Basel.

[12] Gürlebeck K. and Sprößig W. *Quaternionic and Clifford calculus for physicists and engineers*, John Wiley & Sons, Series: Mathematical Methods in Practice, Vol.1, 1997.

[13] Hamilton W.R. (1866, 1969) *Elements of quaternions.* Chelsea (reprint).

[14] Hestenes D. (1966) *Space-time algebra.* Gordon and Breach, New York.

[15] Hestenes D. (1968) *Multivector calculus.* J. Math. Anal. Appl. 24: 313–325.

[16] Hestenes D. (1985) *New foundations for classical mechanics.* Reidel, Dordrecht, Boston.

[17] Hillion P. (1995) *Constitutive relations and Clifford algebra in electromagnetism.* Adv. in Appl. Clifford Algebras 5, Nr. 2: 141–158.

[18] Hillion P. *Complex-Distance Potential theory and hyperbolic equations, in Clifford Algebras and their Applications in Mathematical*

Physics, Volume 2: *Clifford Analysis*, edited by J. Ryan and W. Sprößig, Birkhauser-Boston, July, 2000.

[19] Kravchenko V.V. and Shapiro M.V. (1996) *Integral representations for spatial models of mathematical physics*. Pitman Research Notes in Math. Series 351.

[20] Lounesto P. (1993) *Clifford algebras and Hestenes spinors*. Foundation of Physics, Special issue dedicated to D. Hestenes 23, No. 9: 1203–1237.

[21] Majernik V. (1999) *Quaternionic formulation of the classical fields*, Advances in Applied Clifford Algebras, 9 No. 1, 119-130.

[22] Meister E. and Meister L. *Quaternion Boundary Value Problems for Canonical Objects in Continuum Physics* In: Int. Conference on Clifford Analysis Its Appl. and Related Topics, Beijing, August 1–6, 2000. Advances in Applied Clifford Algebras 11(S2), 231 - 246 (2001).

[23] Michlin S.G. and Prößdorf S. (1986) *Singular integral operators*, Akademie-Verlag Berlin, Berlin.

[24] Picard R. and Zajaczkowski W.M. (1995) *Local existence of solutions of impedance initial-boundary value problem for non-linear Maxwell equations*,MMAS 18, 169–199.

[25] Riesz M. (1958) *Clifford numbers and spinors*. Lecture Series, Institute of Physical Science and Technology, Nr. 38, Maryland.

[26] Sprößig W. (2001) *Quaternionic Analysis in Fluid Mechanics*, In: J. Ryan and W. Sprößig (Eds.) Clifford Algebras and their Applications in Mathematical Physics Volume 2, Clifford Analysis, Series Progress in Physics, Birkhuser Boston, 38–53.

[27] Sprößig W. and Venturino E. (2001) *An alternative approach for solving Maxwell equations*, Proceedings of the A4A4 Conference held in Huddersfield 2001, accepted for publication.

[28] Vekua I.N. (1959) *Generalized analytic functions*. Nauka, Moskow (Russian).

[29] Visintin A. (1996) *Models of Phase Transitions*. Serie: Progress in Nonlinear Differential Equations, Birkhäuser Basel.

6

GENERALIZED ANALYTIC FUNCTIONS AND THEIR CONTRIBUTIONS TO THE DEVELOPMENT OF MATHEMATICAL ANALYSIS

W. Tutschke
Graz University of Technology
Department of Mathematics D
Steyrergasse 30/3, A-8010 Graz
Austria
E-mail: tutschke@matd.tu-graz.ac.at

Abstract: The lecture tries to show that the field of Generalized Analytic Functions (founded by the Georgian I. N. Vekua and the American L. Bers) changed the structure not only of Complex Analysis but also of Mathematical Analysis as a whole. This will be outlined in connexion with integral representations and their applications to boundary value problems and to initial value problems as well. Also some open problems will be formulated. At the end of the lecture it will be discussed how the theory of Generalized Analytic Functions can help to unify Mathematical Analysis, and how this could be reflected in teaching Mathematical Analysis.

1. Generalized analytic functions as a branch of today's Complex Analysis

An important method for solving partial differential equations is the transformation of solutions of a simplified equation into solution of the equation under consideration. An overview on such transformations can be found in H. Begehr's and R. P. Gilbert's books [1] entitled "Transformations, transmutations and reproducing kernels". In many cases such transformations are generated by a fundamental solution of the simplified equation. Unfortunately, fundamental solutions are explicitly known only in special cases.

Using the classical Cauchy kernel $\dfrac{1}{z-\zeta}$, the lecture will outline that the theory of generalized analytic function is in a position to reduce general fully non-linear (elliptic) first order systems in the x, y-plane for two desired real-valued functions to the Cauchy-Riemann system

$$\frac{\partial w}{\partial \overline{z}} = 0.$$

At the first stage the theory of generalized analytic functions considered the equation

$$\frac{\partial w}{\partial \overline{z}} = A(z)w + B(z)\overline{w} \tag{1}$$

where $w = u + iv$ is the desired complex-valued function, and $A(z)$ and $B(z)$ are given coefficients, $z = x + iy$ (see [3, 18]). The system (6) is the canonical form of linear and uniformly elliptic systems of type

$$\begin{aligned} a_{11}\frac{\partial u}{\partial x} + a_{12}\frac{\partial u}{\partial y} + b_{11}\frac{\partial v}{\partial x} + b_{12}\frac{\partial v}{\partial y} + c_1 u + d_1 v = f_1 \\ a_{21}\frac{\partial u}{\partial x} + a_{22}\frac{\partial u}{\partial y} + b_{21}\frac{\partial v}{\partial x} + b_{22}\frac{\partial v}{\partial y} + c_2 u + d_2 v = f_2 \end{aligned}$$

whose coeffifients $a_{11}, a_{12}, \ldots$ may depend on (x, y). Later on linear uniformly elliptic systems for $2n$ desired real-valued functions, too, has been included in the theory. This lead to the theory of generalized analytic vectors (see Bojarski [4]) and to the theory of generalized hyperanalytic functions (see R. P. Gilbert [8] where further references can be found). Finally, also uniformly elliptic fully non-linear first order systems

$$H_j\left(x, y, u_1, v_1, \ldots, u_n, v_n, \frac{\partial u_1}{\partial x}, \ldots, \frac{\partial v_n}{\partial y}\right) = 0, \quad j = 1, 2, \ldots, 2n, \tag{2}$$

for $2n$ desired real-valued functions $u_1, v_1, \ldots, u_n, v_n$ can be solved using a complex rewriting: Solving this system (2) for suitably chosen variables, one gets the complex normal form

$$\frac{\partial w}{\partial \overline{z}} = F\left(z, w, \frac{\partial w}{\partial z}\right) \tag{3}$$

for a desired vector $w = (w_1, \ldots, w_n)$ with complex-valued components $w_j = u_j + iv_j$, $j = 1, \ldots, n$ (cf. [14]).

Using the well-known two-dimensional T_Ω-operator,

$$(T_\Omega h)[\zeta] = -\frac{1}{\pi} \iint_\Omega \frac{h(z)}{z-\zeta} dxdy,$$

the system (3) is equivalent to the operator equation

$$w = \Phi + T_\Omega F\left(\cdot, w, \frac{\partial w}{\partial z}\right) \tag{4}$$

where Φ is a holomorphic function in view of Weyl's Lemma. That way the construction of solutions of the non-linear system (3) can be reduced to the construction of holomorphic functions.

2. Applications to boundary value problems

Starting from (4), define an operator by

$$W = \Phi_{(w)} + T_\Omega F\left(\cdot, w, \frac{\partial w}{\partial z}\right) \tag{5}$$

where $\Phi_{(w)}$ is a holomorphic solution of the boundary value problem

$$\mathcal{B}\left(\Phi_{(w)} + T_\Omega F\left(\cdot, w, \frac{\partial w}{\partial z}\right)\right) = g,$$

$\mathcal{B}$ is an operator acting on the boundary, and g are boundary data given on the boundary. Then each image W satisfies the boundary condition

$$\mathcal{B}W = g$$

and, therefore, a possibly existing fixed element satisfies not only the differential equation (3) but also the given boundary condition

$$\mathcal{B}w = g. \tag{6}$$

Since the definition of the operator (4) contains also the derivative $\partial w/\partial z$, this operator must be investigated in a space whose elements possess derivatives with respet to z in the distributional sense, at least. Note further that the operator (4) has to map this space into itself. Consequently, the derivative

$$\frac{\partial W}{\partial z} = \Phi'_{(w)} + \Pi_\Omega F\left(\cdot, w, \frac{\partial w}{\partial z}\right)$$

has to exist where Π_Ω is the strongly singular operator

$$(\Pi_\Omega h)[\zeta] = -\frac{1}{\pi}\iint\limits_{\Omega} \frac{h(z)}{(z-\zeta)^2}dxdy.$$

To sum up, we have proved the following statement:

Using the T_Ω- and the Π_Ω-operators with the explicitly given kernels $1/(z-\zeta)$ and $1/(z-\zeta)^2$ resp., the boundary value problem (3), (5) can be reduced to an analogous boundary value problem for holomorphic functions.

Note that the Π_Ω-operator maps $L_p(\Omega)$ into itself, whereas T_Ω maps $L_p(\Omega)$ into $C^\beta(\Omega)$ where $\beta = \dfrac{p-2}{p}$. Therefore, a suitable space consists of Hölder-continuous functions (with exponent β) whose derivative with respect to z belongs to $L_p(\Omega)$. If the boundary values of a holomorphic function (in a domain whose boundary is sufficiently smooth) are Hölder continuous with exponent α, then its derivative belongs to $L_p(\Omega)$ provided $p < \dfrac{1}{1-\alpha}$. Therefore, we assume $\alpha > \dfrac{1}{2}$ and choose p in the interval $2 < p < \dfrac{1}{1-\alpha}$.

Another important class of boundary values is given by functions belonging to a fractional order space. Since solutions of the Laplace equation with boundary data from $\mathcal{W}_p^s(\partial\Omega)$ with $s = 1 - \dfrac{1}{p}$ belong to $\mathcal{W}_p^1(\Omega)$, in this case one gets solutions of (3) also belonging to $\mathcal{W}_p^1(\Omega)$.

A still more general class of boundary data are those piecewise belonging to a fractional order space, i.e., one can decompose the given boundary data in three parts $g_1 + g_2 + g_3$ where g_1 belongs to a fractional order space, g_2 is Hölder continuous, and g_3 is piecewise linear (in the arc length parameter of the boundary). A piecewise Hölder continuous boundary data belongs thus to this class. If the boundary data have a jump discontinuity at c_j, then the holomorphic solution has a logarithmic singularity $\log(z - c_j)$. Its derivative has thus a singularity of type

$$\frac{1}{z - c_j}$$

not belonging to $L_p(\Omega)$ with $p > 2$. However, this singularity belongs to a weighted Lebesgue space with weight

$$|z - c_j|^\gamma$$

provided $p-\gamma<2$. If $\rho=\Pi_j|z-c_j|$ where j runs over all jump discontinuities, then the T_Ω-operator maps $L_p(\Omega,\rho^\gamma)$ into $C^\lambda(\Omega,\rho^\mu)$ where

$$\lambda=\min\left(\frac{p-2}{p},\frac{\gamma}{p}-\varepsilon\right),\qquad \mu=\frac{\gamma}{p},$$

and $0<\varepsilon<1$ can be chosen arbitrarily. Choosing $\gamma=(p-2)+\varepsilon\cdot p$, we get as suitable space now $C^\beta(\Omega,\rho^\mu)$, while the derivative has to belong to $L_p(\Omega,\rho^\gamma)$.

Using weighted function spaces, the usual boundary value problems can also be solved [**?**, 16] for boundary data piecewise belonging to fractional order spaces.

Note that piecewise Hölder continuous boundary data are a special case of boudary data which belong piecewise to fractional order spaces. Notice, further, that similar methods using weights can be applied for domains with corners.

3. Applications to initial value problems

Let t be a real variable (=time) and $x=(x_1,...,x_n))$ a spacelike variable. We look for a function $u=u(t,x)$ which is a solution of the following initial value problem

$$\frac{\partial u}{\partial t} = \mathcal{F}\left(t,x,u,\frac{\partial u}{\partial x_j}\right),\quad j=1,...,n \tag{7}$$

$$u(0,x) = \varphi(x). \tag{8}$$

Clearly, this first order initial value problem is equivalent to the following integro-differential equation:

$$u(t,x)=\varphi(x)+\int_0^t \mathcal{F}\left(\tau,x,u,\frac{\partial u}{\partial x_j}\right)d\tau. \tag{9}$$

If the right-hand side is a power series in x, u, and the derivatives $\partial u/\partial x_j$, and if the initial function φ is a power series in x, then the classical Cauchy-Kovalevskaya Theorem states that the initial value problem (7), (8) is solvable by a uniquely determined power series in x. Moreover, the classical Holmgren Theorem states, that there are no other C^1-solutions than power series. Consequently, the initial value problem (7), (8) is uniquely solvable under the hypotheses of the classical Cauchy-Kovalevskaya Theorem. Already in the fourtieth the Japanese mathematician M. Nagumo used the integral rewriting (9) of the initial value problem (7), (8) in order to give an functional-analytic proof of the

Cauchy-Kovalevskaya theorem. His proof is based on Schauder's fixed point theorem.

It turned out to be quite difficult to solve the initial value problem (7), (8) under weaker conditions. This is not surprising because in the fiftieth H. Lewy [10] contructed an example of a first order differential equation of the form (7) with C^∞-coefficients not having any solution. The H. Lewy example is the differential equation

$$\frac{\partial w}{\partial t} = -\frac{i}{z}\frac{\partial w}{\partial \bar{z}} + f(t,z)$$

for a desired complex valued function $w = w(t,z)$.

A certain progress has made by solving the initial value problem in Gevrey classes. These are classes of C^∞-functions whose derivatives satisfy certain estimates. In the present talk another approach for solving the initial value problem (7), (8) under weaker assumptions will be introduced.

This approach generalizes W. Walter's paper [19] in which the classical Cauchy-Kovalevskaya Theorem is proved by the contraction-mapping principle in a Banach space of (complex-valued) functions $w = w(t,z)$ depending holomorphically on the spacelike variable z, while they are continuous in (t,z). The basic idea of W. Walter's proof is the following:

Consider the initial value problem

$$\begin{aligned} \frac{\partial w}{\partial t} &= \mathcal{F}\left(t,z,w,\frac{\partial w}{\partial z}\right), \quad j=1,...,n \\ w(0,z) &= \varphi(z). \end{aligned}$$

Its integral rewriting is

$$w(t,z) = \varphi(z) + \int_0^t \mathcal{F}\left(\tau,z,u,\frac{\partial u}{\partial z}\right) d\tau.$$

Suppose the right-hand side of the differential equation is given in a bounded domain Ω of the z-plane. Denote the distance of a point $z \in \Omega$ from the boundary $\partial\Omega$ by $d(z)$. Then introduce a conical domain M in the t,z-space by

$$M = \left\{(t,z) : z \in \Omega, 0 \leq t < \eta d(z)\right\}$$

where the height of the conical domain is given by $\eta \cdot \sup_{\Omega} d(z)$. Then

$$d(t,z) = d(z) - \frac{t}{\eta}$$

measures the distance of a point (t, z) of M from the lateral surface. Introduce the Banach space $\mathcal{H}_*(M)$ of functions $w = w(t, z)$ having the following properties:

1 $w(t, z)$ is continuous in (t, z)

2 $w(t, z)$ is holomorphic in z

3 the functional

$$\|w\|_* = \sup_M |w(t, z)| d^\sigma(t, z)$$

is finite where $\sigma > 0$ is fixedly chosen.

In interior points (t, z) of M the absolute value of a function $w(t, z)$ belonging to $\mathcal{H}_*(M)$ can be estimated by

$$|w(t, z)| \leq \frac{\|w\|_*}{d^\sigma(t, z)}.$$

Since the right-hand side of the differential equation under consideration depends holomorphically on z, the images

$$W(t, z) = \varphi(z) + \int 0^t \mathcal{F}\left(\tau, z, w, \frac{\partial w}{\partial z}\right) d\tau. \tag{10}$$

of elements of $\mathcal{H}_*(M)$ have the same property.

In order to estimate the corresponding integro-differential operator (10), it is also necessary, however, to estimate of the absolute value of the derivative of functions $w(t, z)$ belonging to $\mathcal{H}_*(M)$. Since these functions are holomorphic in z, one can apply Cauchy's Integral Formula yielding the following statement:

The supremum norm of the derivative of a holomorphic function in a smaller domain can be estimated by the supremum norm in a larger domain divided by the distance of the smaller domain from the boundary of the larger domain.

This estimate leads to the following estimate of the distance of the images W_1 and W_2 of two elements $w_1, w_2 \in \mathcal{H}_*(M)$:

$$\|W_1 - W_2\|_* \leq \eta K \|w_1 - w_2\|_*$$

where the constant K depends on the coefficients of the differential equation.

To sum up, one gets the following statement:

In case $\eta < \frac{1}{K}$, the corresponding integro-differential operator is contractive, and thus the initial value problem can be solved by the contraction-mapping principle.

Modifying these construction, one can solve more general initial value problems of the form (7), (8). This generalization works as follows:

1 The Cauchy-Riemann system $\partial w/\partial\overline{z} = 0$ is to be replaced by a more general elliptic differential equation $\mathcal{G}u = 0$ whose coefficients do not depend on t.

2 The right-hand side $\mathcal{F}$ of the differential equation (7) and this elliptic operator $\mathcal{G}$ form (for each t) a so-called associated pair, i.e., $\mathcal{F}$ transforms the space of all solutions of $\mathcal{G}u = 0$ into itself.

In order to estimate the corresponding integro-differential operator (defined by the right-hand side of (9)), one needs an interior estimate for the first order derivatives which is analogous to that for the first order derivative of a holomorphic function. However, such an interior estimate is not always available for the supremum norm. Therefore, the definition of the *-norm, too, is to be modified. This can be done in the following way:

One considers an exhaustion of the given domain Ω by a family of subdomains Ω_s, $0 < s < s_0$, satisfying the following two conditions:

1 If $s' < s''$, then dist $(\Omega_{s'}, \partial\Omega_{s''}) \geq c_0(s'' - s')$.

2 Each point $x \in \Omega$ (with the exception of one point x_0) belongs to the boundary of a uniqely determined domain $\Omega_{s(x)}$ of the exhaustion.

The conical domain M is now to be defined by

$$M = \Big\{(t,x) : \ x \in \Omega, \ 0 \leq t < \eta(s_0 - s(x))\Big\}$$

and, correspondingly, the distance of a point (t,x) of M is to be measured by

$$d(t,x) = s_0 - s(x) - \frac{t}{\eta}.$$

Then the corresponding *-norm can be defined by

$$\|u\|_* = \sup_M \|u\|_{s(x)} \, d^\sigma(t,x)$$

where $\|u\|_{s(x)}$ means the norm of the restriction of $u(t,\cdot)$ to $\Omega_{s(x)}$.

Using this modified *-norm, analogous considerations lead to the following result:

Suppose $\mathcal{F}$ and $\mathcal{G}$ are an associated pair where $\mathcal{G}$ is an elliptic operator with an interior estimate for the first order derivatives. Suppose, further, *thatthe constant in the interior estimate is proportional to* $1/\delta$ *where* δ *is the distance of a subdomain from the boundary of the larger domain. Provided the initial function* φ *satisfies the side condition* $\mathcal{G}\varphi = 0$, *the initial value problem*

$$\begin{aligned} \frac{\partial u}{\partial t} &= \mathcal{F}\left(t, x, u, \frac{\partial u}{\partial x_j}\right), \quad j = 1, ..., n \\ u(0, x) &= \varphi(x). \end{aligned}$$

is solvable by a function $u = u(t, x)$ *satisfying the side condition* $\mathcal{G}u(t, \cdot) = 0$ *for each* t. *The solution exists in the interval* $0 \leq t < \eta \cdot (s_0 - s(x))$ *for a sufficiently small* η.

Te proof is based on the fact that one has again an estimate of the form

$$\|U_1 - U_2\|_* \leq \eta K \|u_1 - u_2\|_*$$

for the corresponding integro-differential operator and, therefore, η has to satisfy the condition $\eta K < 1$. Then the solution can also be calculated by successive approximations.

As an example we consider initial value problems with generalized analytic functions as initial functions, i.e., the initial function satisfies a differential equation of the form (6). In order to get the necessary interior estimates, define

$$\Phi = w - T_\Omega\left(Aw + B\overline{w}\right).$$

In view of Weyl's Lemma the function Φ turns out to be holomorphic. Consequently,

$$\frac{\partial w}{\partial z} = \Phi' + \Pi_\Omega\left(Aw + B\overline{w}\right)$$

where Φ' is the ordinary complex derivative of the holomorphic function Φ. Taking into account the boundedness of the T_Ω- and the Π_Ω-operators, the desired interior estimate (in the space of Hölder-continuous functions, for instance) follows from an interior estimate for holomorphic functions.

4. Present trends and open problems

4.1 Interactions between different generalizations of classical Complex Analysis

There are various generalizations of classical Complex Analysis dealing with several complex variables, monogenic functions, hyperanalytic functions and generalized analytic functions. A main feature of present trends in Complex Analysis is the combination of methods developed in different branches. For instance, in the theory of multi-monogenic functions one can apply ideas coming from the theory of holomorphic functions in several complex variables such as methods for proving extension theorems.

In my opinion interesting questions are the following

- Characterization of the set of all zeros of a solution, including factorization theorems.
- Characterization and classification of the singularities of solutions.
- Solution of overdetermined systems, including systems with singular coefficients.

4.2 Boundary value problems

Using complex methods, boundary value problems can be solved not only for first order systems but also for equations and systems of higher order (see the book [20] written by Wen Guo Chun and H. Begehr).

Boundary value problems can also be solved for poly-analytic functions. Poly-analytic functions are polynomials in $\overline{z}$ with holomorphic coefficients, i.e., they are solutions of the differential equation

$$\frac{\partial^n w}{\partial \overline{z}^n} = 0.$$

Recently a nice generalization of the theory of poly-analytic functions has been obtained by P. Berglez who considered the differential equation

$$D^n w = 0$$

where $Dw = \dfrac{\partial w}{\partial \overline{z}} - A(z)w - B(z)\overline{w}$. Details can be found in P. Berglez' paper [2] in the Proceedings [7] of the Graz Workshop which took place in February this year.

The construction of Section 2 can also be carried out in the following situation:

Consider the boundary value problem

$$\begin{aligned} \mathcal{L}u &= \mathcal{F}(.,u) \quad \text{in} \quad \Omega \\ \mathcal{B}u &= g \quad \text{on} \quad \partial\Omega \end{aligned} \tag{11}$$

where Ω is a domain in $\mathbf{R}^n$ and the right-hand side $\mathcal{F}(.,u)$ of the differential equation (6) may also depend on (lower order) derivatives of the desired function u. Knowing a fundamental solution $E(x,\xi)$ of the (linear) operator $\mathcal{L}$ on the left-hand side, this boundary value problem leads to the operator

$$U = u_0 + \int_{\Omega} E(\cdot,\xi)\mathcal{F}(\xi,u(\xi))d\xi$$

where u_0 is a solution of the simplified equation $\mathcal{L}u_0 = 0$ such that the image U of u satisfies the given boundary condition $\mathcal{B}U = g$. Then fixed points of this operator are solutions of the boundary value problem under consideration.

In connexion with this general approach to boundary value problems, the following two problems have to be answered:

- Find a fundamental solution of the operator $\mathcal{L}$.
- Which boundary value problem is well-posed for $\mathcal{L}u = 0$?

Generally speaking, a well-posed boundary value problem for $\mathcal{L}u = 0$ is well-posed for (6), too. Note, finally, that fundamental solutions can also be used for getting interior estimates (see, for instance, [17]).

4.3 Initial value problems

Concerning the application of Complex Analysis to initial value problems, mainly the following questions are to be answered:

- Proof of interior estimates, especially it is to be investigated how the constant in the estimate depends on the distance of a compact subset from the distance from the boundary.
- Construction of associated differential operators. Example: Which differential operators map monogenic functions into themthelves? Note that in connexion with initial value problems associated spaces lead to *conservation laws.*
- Formulation and proof of uniqueness theorems for general initial value problems (generalized Holmgren type theorems).
- Investigation of the behaviour of the integral operators of Complex Analysis in weighted function spaces.

4.4 Infinite dimensions

Concerning infinite dimensions the two following situations are to be investigated:

- Behaviour of complex-valued functions of infinitely many complex variables.
- Banach-space-valued functions of one or several complex variables.

Concerning the first problem one can solve, for instance, the inhomogeneous Cauchy-Riemann equation in infinitely many complex variables:

$$\frac{\partial w}{\partial \overline{z}_j} = f_j, \quad j = 1, 2, \ldots$$

Provided the compatibility conditions

$$\frac{\partial f_j}{\partial \overline{z}_k} = \frac{\partial f_k}{\partial \overline{z}_j}$$

are satified, this system is solvable in a polydisk

$$\Omega_1 \times \Omega_2 \times \cdots$$

whose measure is small enough (see [13]).

Concerning the second item, A. L. Bukhgeim and S. G. Kazantsev [5] developed a theory of so-called A-analytic functions. These are functions $u = u(z)$ depending on a complex variable z, having values in a complex Banach space, and satisfying a Beltrami type equation $u_{\overline{z}} - Au_z = 0$. Two-dimensional problems in tomography can be reduced to boundary value problems for this equation. The theory includes not only the case $\|A\| < 1$ but also the case $\|A\| \leq 1$. S. G. Kasantsev [9] obtains also a Cauchy Integral Formula for A-analytic functions.

5. A unified approach to Mathematical Analysis

Calculus should be thaught within the framework of Banach spaces (simultaneous investigation of real-valued functions of one real variable, systems of real-valued functions depending on several real variables, complex-valued functions of one complex variable, mappings between Banach spaces such as integral operators mapping the space of continuous functions into itself).

Of course, concerning Complex Analysis, Generalized Analytic Function cannot be a compulsory subject for all students of Mathematics. However, the spirit of that theory should be used when teaching Complex Analysis. That means, especially,

- Complex differentiation has to be introduced in the framework of partial complex differentiation.

- Instead of starting from the classical Cauchy Integral Formula, the starting point should be the generalized Cauchy Integral Formula

$$f(\zeta) = \frac{1}{2\pi i} \int\limits_{\partial\Omega} \frac{f(z)}{z-\zeta} dz - \frac{1}{\pi} \iint\limits_{\Omega} \frac{1}{z-\zeta} \frac{\partial f}{\partial \overline{z}}(z) dx dy$$

(Cauchy-Pompeiu Integral Representation).

- Weak (distributional) derivatives in connexion with the concept of distributional solutions of Partial Differential Equations should be compulsory for each student of Mathematics.

As far as possible, existence theorems should be proved using methods of high generality such as the contraction-mapping principle. E.g., in analogy to the integral rewriting of ordinary differential equations, partial differential equations with a linear elliptic principal part shoud be reduced to fixed-point problems in suitable function spaces (example: W. Walter's proof of the Cauchy-Kovalevskaya Theorem by the contraction-mapping principle).

References

[1] H. Begehr, R. P. Gilbert, *Transformations, transmutations and reproducing kernels*, Vol. **I** and Vol. **II**, Pitman-Longman, 1993 and 1994.

[2] P. Berglez, *On the solutions of the iterated Bers-Vekua equation*. Contained in [7], pp.266-275.

[3] L. Bers, *Theory of pseudo-analytic functions*. New York 1953.

[4] B. Bojarski, *Theory of generalized analytic vectors* (in Russian). Ann. Polon. Math., vol. **17**, 281-320, 1966.

[5] A. L. Bukhgeim and S. G. Kazantsev, *Elliptic systems of Beltrami type and tomography problems* (in Russian). Dokl. Akad. Nauk **42**, no. 3, pp. 704-707, 1991.

[6] A. Dzhuraev, *On a class of elliptic systems in the halfplane which are singular at the boundary*. Contained in [7], pp. 341-377.

[7] H. Florian, N. Ortner, F. J. Schnitzer and W. Tutschke (eds.), *Functional-analytic and complex methods, their interactions, and applications to Partial Differential Equations*. Proceedings of the Workshop held at Graz University of Technology, February 12-16, 2001. World Scientific Publishing Company 2001.

[8] R. P. Gilbert, *Constructive methods for elliptic equations*. Lecture Notes in Math. **365**, Springer-Verlag Berlin-Heidelberg-New York, 1974.

[9] S. G. Kazantsev, *Integral formulas of Cauchy type for generalized A-analytic functions* (in Russian). Group and metric properties of mappings, pp. 126-141, Novosibirsk Gos. Univ., Novosibirsk 1995.

[10] H. Lewy, *An example of a smooth linear partial differential equation without solution.* Ann. of Math., vol. **66**, 155-158, 1957.

[11] G. F. Manjavidze and W. Tutschke, *Some boundary value problems for non-linear first order systems in the plane* (in Russian). Contained in the book "Boundary value problems in the theory of generalized analytic functions and their applications", Trudy Inst. Appl. Mathem. of University Tbilisi, pp. 79-128, Tbilisi 1983.

[12] M. Nagumo, *Über das Anfangswertproblem partieller Differentialgleichungen.* Japan. Journ. Math., vol. 18, 41-47, 1941.

[13] W. Tutschke, *The inhomogeneous Cauchy-Riemann differential equation in infinitely many complex variables* (in German). Beiträge zur Analysis, vol. **7**, 7-10, 1975.

[14] — , *Solution of initial value problems in classes of generalized analytic functions.* Teubner Leipzig and Springer-Verlag 1989.

[15] —, *Solution of initial value problems for first order systems using matrix notation.* Appl. Anal., vol. 73, pp. 271-280, 1999.

[16] —, *The method of weighted function spaces for solving initial value and boundary value problems.* Contained in [7], pp. 75-90.

[17] W. Tutschke and U. Yüksel, *Interior L_p-Estimates for Functions with Integral Representations.* Appl. Anal., vol. 73, pp. 281-294, 1999.

[18] I. N. Vekua, *Generalized analytic functions.* 2nd edit. Moscow 1988 (in Russian; Engl. transl. Pergamon Press 1962).

[19] W. Walter, *An elementary proof of the Cauchy-Kowalevsky theorem.* Amer. Math. Monthly, vol. **92**, 115-125, 1985.

[20] Wen Guo Chun and H. Begehr, *Boundary Value Problems for Elliptic Equations and Systems.* Pitman Monographs and Surveys in Pure and Applied Mathematics, vol. **46**, 1990.

II. COMPLEX METHODS IN THE PLANE

7

ON Q_p SEQUENCES

Rauno Aulaskari
Department of Mathematics, University of Joensuu
P. O. Box 111, FIN-80101, Joensuu, Finland
E-mail: rauno.aulaskari@joensuu.fi

Shamil Makhmutov
Department of Mathematics
Ufa State Aviation Technical University Ufa, Russia

Hasi Wulan
Department of Mathematics
Shantou University Shantou, Guangdong 515063, P. R. China
E-mail: wulan@stu.edu.cn

Abstract: For a non-normal function f the sequences of points $\{a_n\}$ and $\{b_n\}$ for which $\lim_{n\to\infty}(1-|a_n|^2)f^{\#}(a_n) = +\infty$ and $\lim_{n\to\infty}\iint_{\Delta}(f^{\#}(z))^2 g^p(z,a_n)\,dA(z) = +\infty$ are considered and compared with each other.

AMS-classification: 30D45

1. Introduction

Let Δ be the unit disk in the complex plane and let $dA(z)$ be the Euclidean area element on Δ. Let $M(\Delta)$ denote the class of functions meromorphic in Δ. Green's function in Δ with pole at $a \in \Delta$ is given by $g(z,a) = \log\frac{1}{|\varphi_a(z)|}$, where $\varphi_a(z) = \frac{a-z}{1-\bar{a}z}$ is a Möbius transformation of Δ and $\sigma(z,w) = |\varphi_a(z)|$ is the pseudohyperbolic distance between z

and w. For $0 < r < 1$, let $\Delta(a,r) = \{z \in \Delta : \sigma(z,a) < r\}$ be the pseudohyperbolic disk with center $a \in \Delta$ and radius r.

For $0 < p < \infty$, we define classes $Q_p^\#$ (cf. [8] and [6])

$$Q_p^\# = \{f \in M(\Delta) : \sup_{a\in\Delta} \iint_\Delta (f^\#(z))^2 g^p(z,a)\, dA(z) < \infty\},$$

where $f^\#(z) = \frac{|f'(z)|}{1+|f(z)|^2}$ is the spherical derivative of f. We know that $Q_1^\# = UBC$, the class of all meromorphic functions of uniformly bounded characteristic in Δ (cf. [9]), and for each $p \in (1,\infty)$ the class $Q_p^\#$ is the class of normal functions $\mathcal{N}$ (cf. [6]), which is defined as follows:

$$\mathcal{N} = \{f \in M(\Delta) : \|f\|_\mathcal{N} = \sup_{z\in\Delta}(1-|z|^2)f^\#(z) < \infty\}.$$

Definition 1. *Let f be a meromorphic function in Δ. A sequence of points $\{a_n\}$ ($|a_n| \to 1$) in Δ is called a $q_\mathcal{N}$-sequence if*

$$\lim_{n\to\infty} f^\#(a_n)(1-|a_n|^2) = +\infty.$$

Definition 2. *Let f be a meromorphic function in Δ and $0 < p < \infty$. A sequence of points $\{a_n\}$ ($|a_n| \to 1$) in Δ is called a q_p-sequence if*

$$\lim_{n\to\infty} \iint_\Delta (f^\#(z))^2 g^p(z,a_n)dA(z) = +\infty.$$

2. q_p and $q_\mathcal{N}$ sequences

Theorem 1. *Let f be a meromorphic function in Δ. If $\{a_n\}$ is a $q_\mathcal{N}$-sequence, then any sequence of points $\{b_n\}$ in Δ for which $\sigma(a_n,b_n) \to 0$ is a q_p-sequence for all p, $0 < p < \infty$.*

Proof. By [[4], Theorem 7.2] there exist a sequence of points $\{c_n\}$ in Δ and a sequence of positive numbers $\{p_n\}$ satisfying for the pseudohyperbolic distance $\sigma(a_n,c_n) \to 0$ and $\frac{p_n}{1-|c_n|} \to 0$ such that the sequence of functions $\{f_n(t)\} = \{f(c_n + p_n t)\}$ converges uniformly on each compact subset of the complex plane to a nonconstant meromorphic function $g(t)$.

For a fixed $R > 0$ set $\Delta_n = \{z : z = c_n + p_n t, |t| < R\}$. For any sequence of points $\{b_n\}$ in Δ for which $\sigma(a_n,b_n) \to 0$, we have $\sigma(b_n,c_n) \to 0$ since $\sigma(a_n,c_n) \to 0$. Thus, for n sufficiently large, we obtain that

$$\Delta_n = \{z : z = c_n + p_n t, |t| < R\} \subset \Omega_n = \{z : |\varphi_{b_n}(z)| < 1/e\}.$$

Therefore, we get by change of variables

$$\begin{aligned}&\iint_{\Omega_n} (f^{\#}(z))^2 g^p(z, b_n) dA(z)\\ &\geq \iint_{\Delta_n} (f^{\#}(z))^2 g^p(z, b_n) dA(z)\\ &= \iint_{|t|<R} (f_n^{\#}(t))^2 g^p(c_n + p_n t, b_n) dA(t).\end{aligned}$$

By the uniformly convergence we have

$$\iint_{|t|<R} (f_n^{\#}(t))^2 dA(t) \to \iint_{|t|<R} (g^{\#}(t))^2 dA(t),$$

and this last integral is finite and nonzero, because $g(t)$ is a nonconstant meromorphic function and the integral is over a fixed bounded set. However, $g(c_n + p_n t, b_n) \to +\infty$ as $n \to \infty$ uniformly for $|t| < R$, and it follows for p, $0 < p < \infty$, that

$$\iint_{|t|<R} (f_n^{\#}(t))^2 g^p(c_n + p_n t, b_n) dA(t) \to +\infty.$$

In fact,

$$g(c_n + p_n t, b_n) = \log \left| \frac{1 - \overline{b_n}(c_n + p_n t)}{c_n + p_n t - b_n} \right|,$$

and for $|t| < R$

$$\left| \frac{c_n + p_n t - b_n}{1 - \overline{b_n}(c_n + p_n t)} \right| \leq \frac{|c_n - b_n| + p_n |t|}{|1 - \overline{b_n} c_n| - p_n |b_n t|} \leq \frac{\left| \frac{c_n - b_n}{1 - \overline{b_n} c_n} \right| + \frac{p_n |t|}{|1 - \overline{b_n} c_n|}}{1 - \frac{p_n |b_n t|}{|1 - \overline{b_n} c_n|}} \to 0.$$

Thus the proof of Therem 1 is complete. □

Corollary 1. *Let f be a meromorphic function in Δ. Suppose $\{a_n\}$ is a q_N-sequence, then $\{a_n\}$ is a q_p-sequence for all p, $0 < p < \infty$.*

In view of Corollary 1, it is natural to ask whether the converse of Corollary 1 is also true. However, the following result shows that no such assertion can be true.

Theorem 2. *There exist a non-normal function f and a sequence $\{a_n\}$ in Δ which is a q_p-sequence for all p, $0 < p < \infty$, but $\{a_n\}$ is not a q_N-sequence.*

Proof. Consider function $f(z) = \exp \frac{i}{1-z}$. It is easy to see that f is not normal. Choose sequence $\{b_n\} = \{\frac{n^2}{1+n^2}\}$ and by a simple computation we obtain that

$$\lim_{n\to\infty} (1 - |b_n|^2) f^{\#}(b_n) = +\infty.$$

By Theorem 1 for any sequence of points $\{a_n\}$ in Δ for which $\sigma(a_n, b_n) \to 0$,

$$\lim_{n\to\infty} \iint_{\Delta} (f^{\#}(z))^2 g^p(z, a_n) dA(z) = +\infty$$

for all p, $0 < p < \infty$. Now we choose $\{a_n\} = \{\frac{n^2}{1+n^2} - \frac{i}{n+n^3}\}$ and notice that $\sigma(a_n, b_n) \to 0$. But

$$\lim_{n\to\infty} (1 - |a_n|^2) f^{\#}(a_n) = 0.$$

Thus $\{a_n\}$ is just one we needed. □

Remark 1. There exist a non-normal function f and two sequences of points $\{a_n\}$ and $\{b_n\}$ in Δ for which $\sigma(a_n, b_n) \to 0$ such that

$$\lim_{n\to\infty} (1 - |a_n|^2) f^{\#}(a_n) = +\infty$$

and

$$\lim_{n\to\infty} (1 - |b_n|^2) f^{\#}(b_n) < +\infty.$$

Theorem 3. *Let f be a meromorphic function in Δ and let $0 < p' < p < \infty$. If, for a sequence of points $\{a_n\} \subset \Delta$,*

$$\lim_{n\to\infty} \iint_{\Delta} (f^{\#}(z))^2 g^p(z, a_n) dA(z) = +\infty, \tag{1}$$

then

$$\lim_{n\to\infty} \iint_{\Delta} (f^{\#}(z))^2 g^{p'}(z, a_n) dA(z) = +\infty. \tag{2}$$

Remark 2. By assumption (1) we know that $f \notin Q_p^{\#}$. Since the function classes $Q_p^{\#}$ have a nesting property, $f \notin Q_{p'}^{\#}$ for $p' < p$. However, Theorem 3 gives more information about this situation showing that the same sequenec $\{a_n\}$, which breaks the $Q_p^{\#}$-condition, also breaks the $Q_{p'}^{\#}$-condition.

Proof of Theorem 3. Let $U_1^n = \{z : g(z, a_n) > 1\}$. By (1) we may suppose (at least for some subsequences of $\{a_n\}$) that either

$$\lim_{n\to\infty} \iint_{\Delta\setminus U_1^n} (f^{\#}(z))^2 g^p(z, a_n) dA(z) = +\infty \tag{3}$$

or

$$\lim_{n\to\infty} \iint_{U_1^n} (f^{\#}(z))^2 g^p(z, a_n) dA(z) = +\infty. \tag{4}$$

Suppose now that (3) is true. Since, for $z \in \Delta \setminus U_1^n$, we have $g(z, a_n) \leq 1$ and thus $g^{p'}(z, a_n) \geq g^p(z, a_n)$ for $p' < p$. Hence

$$\lim_{n\to\infty} \iint_{\Delta \setminus U_1^n} (f^{\#}(z))^2 g^{p'}(z, a_n) dA(z) = +\infty$$

and so

$$\lim_{n\to\infty} \iint_{\Delta} (f^{\#}(z))^2 g^{p'}(z, a_n) dA(z) = +\infty.$$

Suppose next that (4) is true. Then we need to consider only two cases: (i) Either there exists a sequence of points $\{b_n\}$ in U_1^n with $\sigma(a_n, b_n) \to 0$ such that

$$\lim_{n\to\infty} (1 - |b_n|^2) f^{\#}(b_n) = +\infty$$

or (ii) there exists r_0, $0 < r_0 < 1/e$, and $M > 0$ such that

$$(1 - |z|^2) f^{\#}(z) \leq M$$

for all $z \in \Delta(a_n, r_0) = \{ z : \sigma(z, a_n) < r_0 \}$. If (i) is true, then by Theorem 1, for any sequence of points $\{c_n\}$ in Δ for which $\sigma(b_n, c_n) \to 0$ and any p', we obtain that

$$\lim_{n\to\infty} \iint_{\Delta} (f^{\#}(z))^2 g^{p'}(z, c_n) dA(z) = +\infty.$$

By choosing the sequence $\{c_n\} = \{a_n\}$ we see that (2) is true in this case. On the other hand, if (ii) holds, then necessarily

$$\lim_{n\to\infty} \iint_{\Delta \setminus \Delta(a_n, \frac{r_0}{2})} (f^{\#}(z))^2 g^p(z, a_n) dA(z) = +\infty.$$

Since, for $z \in \Delta \setminus U_1^n$, $g^p(z, a_n) \leq g^{p'}(z, a_n)$ and for $z \in U_1^n \setminus \Delta(a_n, \frac{r_0}{2})$, $c g^p(z, a_n) \leq g^{p'}(z, a_n)$ for some $c > 0$,

$$\lim_{n\to\infty} \iint_{\Delta \setminus \Delta(a_n, \frac{r_0}{2})} (f^{\#}(z))^2 g^{p'}(z, a_n) dA(z) = +\infty.$$

Thus also in this case (2) is true and the proof of Theorem 3 is completed. □

Remark 3. In fact, from the proof of Theorem 3, we can see that if for a fixed r_0, $0 < r_0 < 1$,

$$\lim_{n\to\infty}\iint_{\Delta(a_n,r_0)}(f^{\#}(z))^2 g^p(z,a_n)dA(z) = +\infty,$$

then there exists a sequence of points $\{b_n\}$ in U_1^n such that

$$\lim_{n\to\infty}(1-|b_n|^2)f^{\#}(b_n) = +\infty.$$

In the opposite direction we are able to get a sharper result which is also deeper than Corollary 1.

Theorem 4. *Let f be a meromorphic function in Δ. If, for a sequence of points $\{a_n\} \subset \Delta$,*

$$\lim_{n\to\infty}(1-|a_n|^2)f^{\#}(a_n) = +\infty,$$

then for the same sequence $\{a_n\}$

$$\lim_{n\to\infty}\iint_{\Delta(a_n,r)}(f^{\#}(z))^2 g^p(z,a_n)dA(z) = +\infty$$

holds for all p, $0 < p < \infty$ and all r, $0 < r < 1$.

Proof. Suppose that

$$\lim_{n\to\infty}(1-|a_n|^2)f^{\#}(a_n) = +\infty.$$

If there exists an r_0, $0 < r_0 < 1$ and a p, $0 < p < \infty$ such that

$$\limsup_{n\to\infty}\iint_{\Delta(a_n,r_0)}(f^{\#}(z))^2 g^p(z,a_n)dA(z) = M < +\infty,$$

then there exists a subsequence $\{a_{n_k}\}$ of $\{a_n\}$ such that

$$\iint_{\Delta(a_{n_k},r_0)}(f^{\#}(z))^2 g^p(z,a_{n_k})dA(z) \le M+1$$

for k sufficiently large. Choose an r_1, $0 < r_1 < r_0$, satisfying

$$(\log 1/r_1)^{-p}(M+1) < \pi/2.$$

It follows that

$$\iint_{\Delta(a_{n_k},r_1)}(f^{\#}(z))^2 dA(z) \le (\log 1/r_1)^{-p}(M+1) < \frac{\pi}{2}.$$

By Dufresnoy's theorem (see [[7], p.83]) we have

$$(1 - |a_{n_k}|^2) f^{\#}(a_{n_k}) \leq \frac{1}{r_1},$$

which contradicts our assumption. □

Theorem 5. *Let f be a meromorphic function in Δ. Suppose that, for $0 < p < \infty$, there exists a sequence of points $\{a_n\}$ in Δ such that*

$$\lim_{n\to\infty} \iint_{\Delta} (f^{\#}(z))^2 g^p(z, a_n) dA(z) = +\infty.$$

Then, for any sequence of points $\{b_n\}$ in Δ for which $\sigma(a_n, b_n) \to 0$,

$$\lim_{n\to\infty} \iint_{\Delta} (f^{\#}(z))^2 g^p(z, b_n) dA(z) = +\infty.$$

Proof. Choose positive constants M_1 and M_2 such that $M_2 < M_1$. Let $U_1^n = \{z : g(z, a_n) > M_1\}$ and $U_2^n = \{z : g(z, a_n) > M_2\}$. Then if $w \in U_1^n$, $z \in \Delta \setminus U_2^n$,

$$cg(z, a_n) \leq g(z, w)$$

for some $c > 0$. This means for all n that

$$\iint_{\Delta\setminus U_2^n} (f^{\#}(z))^2 g^p(z, b_n) dA(z) \geq c^p \iint_{\Delta\setminus U_2^n} (f^{\#}(z))^2 g^p(z, a_n) dA(z) \tag{5}$$

for any sequence of points $\{b_n\}$ in Δ for which $\sigma(a_n, b_n) \to 0$. If

$$\limsup_{n\to\infty} \iint_{\Delta\setminus U_2^n} (f^{\#}(z))^2 g^p(z, a_n) dA(z) = +\infty,$$

then, by (6),

$$\limsup_{n\to\infty} \iint_{\Delta\setminus U_2^n} (f^{\#}(z))^2 g^p(z, b_n) dA(z) = +\infty.$$

If

$$\limsup_{n\to\infty} \iint_{U_2^n} (f^{\#}(z))^2 g^p(z, a_n) dA(z) = +\infty,$$

then continuing as in the proof of Theorem 3 we have to consider two different cases: (i) Either there exists a sequence of points $\{c_n\}$ in U_2^n for which $\sigma(a_n, c_n) \to 0$ such that

$$\lim_{n\to\infty} (1 - |c_n|^2) f^{\#}(c_n) = +\infty$$

or (ii) there exists r_0, $0 < r_0 < e^{-M_2}$, and $K > 0$ such that

$$(1 - |z|^2) f^{\#}(z) \le K$$

for all $z \in \Delta(a_n, r_0)$. If (i) is true, then, by Theorem 1, for above $\{b_n\}$, we have

$$\lim_{n\to\infty} \iint_{\Delta} (f^{\#}(z))^2 g^p(z, b_n) dA(z) = +\infty$$

since $\sigma(b_n, c_n) \to 0$. On the other hand, if (ii) holds, then using the same conclusions for Green's functions we see that necessarily for any sequence of points $\{b_n\}$ for which $\sigma(a_n, b_n) \to 0$,

$$\lim_{n\to\infty} \iint_{\Delta} (f^{\#}(z))^2 g^p(z, b_n) dA(z) = +\infty.$$

This completes the proof. □

Question. *Let $1 < p < \infty$ and suppose that*

$$\lim_{n\to\infty} \iint_{\Delta} (f^{\#}(z))^2 g^p(z, a_n) dA(z) = +\infty.$$

Is it true for p', $p < p'$,

$$\lim_{n\to\infty} \iint_{\Delta} (f^{\#}(z))^2 g^{p'}(z, a_n) dA(z) = +\infty?$$

We answer in part the question below, which is based on the fact that $N \cap M_p^{\#} = Q_p^{\#}$ for all p, $0 < p < \infty$ (see [31]).

Theorem 6. *Let $1 < p < \infty$ and suppose that*

$$\lim_{n\to\infty} \iint_{\Delta} (f^{\#}(z))^2 g^p(z, a_n) dA(z) = +\infty.$$

If the sequence $\{a_n\}$ is not a m_p-sequence, that is,

$$\lim_{n\to\infty} \iint_{\Delta} (f^{\#}(z))^2 (1 - |\varphi_a(z)|^2)^p dA(z) < +\infty,$$

then for any p', $p < p'$, we have

$$\lim_{n\to\infty} \iint_{\Delta} (f^{\#}(z))^2 g^{p'}(z, a_n) dA(z) = +\infty.$$

Acknowledgements

The first and third authors are partly supported by the Väisälä and Wihuri Funds, Finland.

References

[1] R. Aulaskari and P. Lappan, *Criteria for an analytic function to be Bloch and a harmonic or meromorphic function to be normal, Complex analysis and its applications*, Pitman Research Notes in Mathematics, 305, Longman Scientific & Technical Harlow, 1994, 136–146.

[2] R. Aulaskari, H. Wulan and R. Zhao, *Carleson measure and some classes of meromorphic functions*, Proc. Amer. Math. Soc., Vol. 128, 2000, 2329-2335.

[3] R. Aulaskari, J. Xiao and R. Zhao, *On subspaces and subsets of $BMOA$ and UBC*, Analysis, Vol. 15, 1995, 101–121.

[4] D. Campbell and G. Wickes, *Characterizations of normal meromorphic functions*, Complex Analysis, Joensuu 1978, 55-72, Lecture Notes in Mathematics, 747, Springer-Verlag, Berlin.

[5] A. Baernstein II, *Analytic functions of bounded mean oscillation, Aspects of contemporary complex analysis*, Academic Press, London, 1980, 3–36.

[6] M. Essén, *Q_p-spaces*, Lectures at the summer school on "Complex Function Spaces" at Mekrijärvi Aug.30 - Sept.1, 1999. To appear, Department of Mathematics, University of Joensuu, Finland.

[7] J. L. Schiff, *Normal families*, Springer-Verlag, New York, 1993.

[8] H. Wulan, *On some classes of meromorphic functions*, Ann. Acad. Sci. Fenn. Math. Diss., Vol. 116, 1998, 1–57.

[9] S. Yamashita, *Functions of uniformly bounded characteristic*, Ann. Acad. Sci. Fenn. Ser. A I Math., Vol. 7, 1982, 349–367.

8

RECONSTRUCTION OF AN ANALYTIC FUNCTION FROM A SEQUENCE OF VALUES: EXISTENCE AND REGULARIZATION

Dang Dinh Ang
Department of Mathematics and Informatics
HoChiMinh City, National University
225 Nguyen Van Cu, Q. 5, HoChiMinh City, Vietnam.

Vy Khoi Le
Department of Mathematics and Statistics
University of Missouri-Rolla.
Rolla, MO 65409
USA.

Dang Duc Trong
Department of Mathematics and Informatics
HoChiMinh City, National University
225 Nguyen Van Cu, Q. 5, HoChiMinh City, Vietnam.

Abstract: The authors consider the problem of recovering a holomorphic function defined in a domain Ω of the complex plane from its values at an infinite sequence of points (z_n) in Ω converging to a point in Ω. The problem, which is ill-posed, is regularized by various methods. Necessary and sufficient conditions for existence are given.

Keywords: Lagrange interpolation, holomorphic function, Hardy space, moment problem, truncated expansion, inverse problem, regularization.

AMS-classification: 35R30 30B40 35H02 41A27 41A10 41A25

Introduction

In the present paper, we consider the problem of reconstructing an analytic function u on a domain Ω of the complex plane from values (μ_n) at a given sequence of points (z_n) in Ω

$$u(z_n) = \mu_n, \quad n \in \mathbb{N}. \tag{1}$$

Now and later, without explicit mention, the sequence (z_n) is assumed to be infinite, $z_i \neq z_j$ for $i \neq j$, and to converge to a point in Ω. Note that (6) has at most one solution. However, the problem is ill-posed in the sense that solutions do not always exist and that in the case of existence there is no continuous dependence on the given data. The literature on the subject is impressive (see e.g., [1, 2, 5, 6, 7, 9, 12, 13]). However, the ill-posedness of the problem is usually not taken into account. In the classical literature, one is primarily concerned with the problem of existence of a solution. However, in practice, solutions usually do not exist, the reason being that the data are given by measurements, consisting of finitely many points that are conveniently patched up into continuous or L^2-functions. It is seen that a solution corresponding to this patched up "data" does not exist in general. Thus the problem, as it is, is intractable numerically and hence a regularization is in order. A regularized solution is a function stable with respect to variations in the given data that approximates the exact solution.

We shall consider the problem of reconstructing functions in $H(\Omega)$, the space of holomorphic functions on Ω, and in the Hardy space $H^2(U)$ of the unit disc U of the complex plane. We shall deal with the existence problem and the regularization problem.

The remainder of paper consists of two parts. Part 1 deals with the reconstruction problem in $H(\Omega)$ and Part 2 deals with that in $H^2(U)$. In both parts, we present first existence conditions and then regularization results. Convergence properties and error estimates are derived.

1. The reconstruction problem in $H(\Omega)$

1.1. Let $\Omega \subset \mathbb{C}$ be a simply connected domain whose boundary $\partial\Omega$ is the image γ^* of a closed, smooth, and nondegenerate (i.e., $|\gamma'| > 0$) Jordan curve $\gamma : [0,1] \to \mathbb{C}$. In what follows, γ^* is always oriented counter-clockwise. Let (z_n), $n = 1, 2, \ldots$, be a sequence in Ω converging to a point $z_0 \in \Omega$. We consider the problem of determining an analytic function on Ω from its values at the sequence of points (z_n), i.e., for a given sequence $(\mu_n) \subset \mathbb{C}$, find $u \in H(\Omega)$ such that

$$u(z_n) = \mu_n, \quad n \in \mathbb{N}.$$

It is known that this problem has at most one solution in $H(\Omega)$, and that solutions are unstable with respect to small changes in the μ' s. We shall transform (6) into a moment problem.

Suppose $u \in H(\Omega) \cap C(\overline{\Omega})$ satisfies (6). By the Cauchy integral formula, we have

$$\frac{1}{2\pi i} \int_0^1 \frac{u\big(\gamma(t)\big).\gamma'(t)dt}{\gamma(t) - z_n} = \frac{1}{2\pi i} \int_{\gamma^*} \frac{u(\zeta)d\zeta}{\zeta - z_n} = u(z_n) = \mu_n$$

for $n = 1, 2, \ldots$.

Putting $\varphi(t) = u\big(\gamma(t)\big)$ and $g_n = \overline{(2\pi i)^{-1}\gamma'(t)/(\gamma(t) - z_n)}$, $0 \leq t \leq 1$, $n \in \mathbb{N}$, we see that these equations can be written as

$$\int_0^1 \varphi(t).\overline{g_n(t)}dt = \mu_n, \quad n \in \mathbb{N}. \tag{2}$$

This is a moment problem on $(0, 1)$.

1.2. We consider the Hilbert space $L^2(0,1) = L^2\big((0,1);\mathbb{C}\big)$ with the usual norm $\|.\|$ and inner product $(.,.)$. Since $z_n \notin \gamma^*$, $g_n \in C([0,1]) \subset L^2(0,1)$, $\forall n$. It can be shown that $\{g_n : n \in \mathbb{N}\}$ is linearly independent (in the algebraic sense).

Since $\{g_n : n \in \mathbb{N}\}$ is linearly independent, we can define an orthonormal sequence $\{e_n : n \in \mathbb{N}\}$ associated with $\{g_n\}$ by the Gram-Schmidt process:

$$\begin{aligned} e_1 &= g_1/\|g_1\| \\ e_n &= \Big\|g_n - \sum_{j=1}^{n}(g_n, e_j)e_j\Big\|^{-1} \cdot \Big[g_n - \sum_{j=1}^{n}(g_n, e_j)e_j\Big], \quad n = 2, 3, \ldots \end{aligned} \tag{3}$$

These formulas show that there exist unique $C_{ij} \in \mathbb{C}$, $i, j \in \mathbb{N}$ such that

$$e_i = \sum_{j=1}^{i} C_{ij} g_j \tag{4}$$

With each sequence $\mu = (\mu_n) \subset \mathbb{C}$, we associate a sequence $\lambda = \lambda(\mu) = (\lambda_n)$ defined by

$$\lambda_i = \sum_{j=1}^{i} C_{ij}\mu_j, \quad i \in \mathbb{N}. \tag{5}$$

(Note that if μ is a finite sequence, then so is λ).

For $\varphi \in L^2(0,1)$, let F_φ be the Cauchy integral of $\varphi \circ \gamma^{-1}$ on γ^*:

$$F_\varphi = \frac{1}{2\pi i}\int_0^1 \frac{\varphi(t)\gamma'(t)dt}{\gamma(t)-z}.$$

By a well-known result, F_φ is analytic in $\mathbb{C}\setminus\gamma^*$. In particular, $F_\varphi \in H(\Omega)$.

For a sequence $\mu = (\mu_n) \subset \mathbb{C}$, we consider the following finite moment problem:
Find $\varphi_n = \varphi_n(\mu) \in H_n = \langle g_1, \dots, g_n \rangle$ such that

$$(\varphi_n, g_j) = \mu_j, \quad 1 \le j \le n. \tag{6}$$

It can be shown that for each n, (6) has a unique solution $\varphi_n = \varphi_n(\mu)$ depending continuously of $\mu_1, \dots, \mu_n$. The following theorem gives conditions for existence of a solution to (6)

Theorem 1. *Let $\mu = (\mu_n) \subset \mathbb{C}$.*
(i) *For each $n \in \mathbb{N}$ fixed, F_{φ_n} and its derivatives depend continuously on $\mu_1, \dots, \mu_n$ in the topology of uniform convergence on compact subsets of Ω.*
(ii) *If (6) has a solution in $H(\Omega)\cap C(\overline{\Omega})$ then*

$$\sum_{i=1}^{\infty}\Big|\sum_{j=1}^{\infty} C_{ij}\mu_j\Big|^2 < \infty. \tag{7}$$

Conversely, if (7) holds, then (6) is solvable in $H(\Omega)$.
(iii) *If (6) has a solution in $H(\Omega)\cap C(\overline{\Omega})$ then, for all $j \ge 0$, $F_{\varphi_n}^{(j)} \to u^{(j)}$ uniformly on compact subsets of Ω.*

Proof.
(i) Suppose the sequences $(\mu_j^m)_{m\in\mathbb{N}}$, $1 \le j \le n$, satisfy

$$\mu_j^m \to \mu_j \quad (m \to \infty),\ \forall j = \overline{1,n}.$$

Then $\varphi_n^m \to \varphi_n$ in $L^2(0,1)$, (φ_n^m is the solution of (6) corresponding to $\mu_1^m, \dots, \mu_n^m$).

Putting $\|\gamma'\|_\infty = \max\{|\gamma'(t)| : 0 \le t \le 1\}$, we have, for $z \in \Omega$,

$$\begin{aligned}
\big|F_{\varphi_n^m}(z) - F_{\varphi_n}(z)\big| &= \frac{1}{2\pi}\Big|\int_0^1 \frac{\gamma'(t)}{\gamma(t)-z}\big[\varphi_n^m(t)-\varphi_n(t)\big]dt\Big| \\
&\le \frac{1}{2\pi}\Big(\int_0^1\Big|\frac{\gamma'(t)}{\gamma(t)-z}\Big|^2 dt\Big)^{1/2}\big\|\varphi_n^m - \varphi_n\big\| \\
&\le \frac{1}{2\pi}\frac{\|\gamma'\|_\infty}{\operatorname{dist}(z,\partial\Omega)}\big\|\varphi_n^m - \varphi_n\big\|.
\end{aligned}$$

This implies that $F_{\varphi_n^m} \to F_{\varphi_n}$ uniformly on compact subsets of Ω.

By Weierstrass' s theorem, we have the convergence of the derivatives of $F_{\varphi_n^m}$.

(ii) Suppse $u \in H(\Omega) \cap C(\overline{\Omega})$ satisfies (6). Then $u \circ \gamma \in L^2(0,1)$ is a solution of (2). Hence (7) holds.

Now, if (7) is satisfied then (2) has a solution $\varphi \in L^2(0,1)$. We have $F_\varphi \in H(\Omega)$ and

$$F_\varphi(z_n) = \frac{1}{2\pi i}\int_0^1 \frac{\varphi(t)\gamma'(t)}{\gamma(t)-z}dt = (\varphi, g_n) = \mu_n, \quad \forall n.$$

Thus F_φ is a solution of (6)

(iii) By hypothesis, $\varphi = u \circ \gamma$ is a solution of (2) in $L^2(0,1)$ and thus $\varphi_0 = \mathrm{P}_H\varphi$ is the unique solution of (2) in H, the closed subspace of $L^2(0,1)$ generated by the g_n' s (P_H is the orthogonal projection $L^2(0,1)$ onto H). We thus have

$$\varphi_n \to \varphi_0 \text{ in } L^2(0,1).$$

On the other hand, we have, as in (i),

$$\begin{aligned}|F_{\varphi_n}(z) - F_{\varphi_0}(z)| &= \frac{1}{2\pi}\left|\int_0^1 \frac{\gamma'(t)}{\gamma(t)-z}[\varphi_n(t)-\varphi_o(t)]dt\right| \\ &\le \frac{1}{2\pi}\left(\int_0^1\left|\frac{\gamma'(t)}{\gamma(t)-z}\right|^2 dt\right)^{1/2}\|\varphi_n - \varphi_0\| \\ &\le \frac{1}{2\pi}\frac{\|\gamma'\|_\infty}{\operatorname{dist}(z,\partial\Omega)}\|\varphi_n-\varphi_0\|, \quad z \in \Omega.\end{aligned}$$

This implies that

$$F_{\varphi_n} \to F_{\varphi_0} \text{ uniformly on compact subsets of } \Omega.$$

But since φ_0 satisfies (2), (ii) shows that $F_{\varphi_0} \in H(\Omega)$ satisfies (6). By the uniqueness of the solution of (6), we have $F_{\varphi_0} = u$. Hence (iii) is proved. ∎

In the case the right hand side (μ_n) of (2) is inexact, we use the following regularization result.

Proposition A. *Let $\varphi_0 \in H$ be the solution of (2) corresponding to the sequence $\mu^0 = (\mu_n^0)$. Then we can choose $n(\delta) \in \mathbb{N}$ $(\delta > 0)$ such that for all sequences $\mu = (\mu_n)$ satisfying*

$$\|\mu - \mu^0\|_\infty \; (\equiv \sup\{|\mu_n - \mu_n^0| : n \in \mathbb{N}\}) \le \delta, \tag{8}$$

we have

$$\left\|\varphi_{n(\delta)}(\mu) - \varphi_0\right\| \leq \varepsilon.$$

Proposition A can be deduced from the following Proposition A′ that is proved in [4],

Proposition A′. *Let* $\varphi_0 \in H$ *be the solution of* (2) *corresponding to the sequence* $\mu^0 = (\mu_n^0)$. *For* $0 < \varepsilon < \left\|g_1\right\|^2$, *let*

$$n(\varepsilon) = \left[f^{-1}(\varepsilon^{-1/2})\right] \tag{9}$$

where f *is strictly increasing of* $[1, \infty)$ *onto* $[\left\|g_1\right\|^{-1}, \infty)$ *(an example of such an* f *is given in* [4]*). Then there exists a function* $\eta(\varepsilon)$ *(*$0 < \varepsilon < \left\|g_1\right\|^2$*) such that* $\lim_{\varepsilon\to 0}\eta(\varepsilon) = 0$ *and that for all sequences* μ *satisfying*

$$\left\|\mu - \mu^0\right\|_\infty < \varepsilon, \tag{10}$$

we have

$$\left\|\varphi_{n(\varepsilon)}(\mu) - \varphi_0\right\| \leq \eta(\varepsilon).$$

Proposition A implies the following regularization result for (6),

Theorem 2. *Let* $f \in H(\Omega) \cap C(\overline{\Omega})$ *be the solution of* (6) *corresponding to* $\mu^0 = (\mu_n^0)$. *Choose* $n(\delta)$ *(*$\delta > 0$*) as in Proposition A. Let* $p \in \mathbb{N}$ *and let* K *be a compact subset of* Ω. *Then for each* $\varepsilon > 0$, *there exists* $\delta > 0$ *such that for all sequences* μ *satisfying* (8), *we have, for* $0 \leq j \leq p$,

$$\sup_{z\in K}\left|F^{(j)}_{\varphi_{n(\delta)}}(z) - u^{(j)}(z)\right| \leq \varepsilon.$$

Proof. As in the proof of Theorem 1, if we put

$$\varphi = u \circ \gamma \in L^2(0,1), \quad \varphi_0 = \mathrm{P}_H\varphi,$$

then $\varphi_0 \in H$ is the unique solution of (2) in H (corresponding to μ^0) and $u = F_{\varphi_0}$ in Ω. Hence, for $j \in \mathbb{N}$, $z \in \Omega$,

$$u^{(j)}(z) = F^{(j)}_{\varphi_0}(z) = \frac{j!}{2\pi i}\int_0^1 \frac{\varphi_0(t).\gamma'(t)}{[\gamma(t) - z]^{j+1}}dt.$$

Since

$$F^{(j)}_{\varphi_n}(z) = \frac{j!}{2\pi i}\int_0^1 \frac{\varphi_n(t).\gamma'(t)}{[\gamma(t) - z]^{j+1}}dt,$$

we have, for $z \in K$,

$$\begin{aligned} \left|F^{(j)}_{\varphi_n}(z) - u^{(j)}(z)\right| &\leq \frac{j!}{2\pi}\left|\int_0^1 \frac{\gamma'(t)}{[\gamma(t)-z]^{j+1}}\left[\varphi_n(t)-\varphi_0(t)\right]dt\right| \\ &\leq \frac{j!}{2\pi}\frac{\|\gamma'\|_\infty}{\left[\operatorname{dist}(K,\partial\Omega)\right]^{j+1}}\left\|\varphi_n-\varphi_0\right\|. \end{aligned} \tag{11}$$

For $\varepsilon > 0$, choose

$$\varepsilon' = \frac{2\pi\varepsilon}{\|\gamma'\|_\infty}\min\left\{\frac{1}{j!}\left[\operatorname{dist}(K,\partial\Omega)\right]^{j+1} : 0 \leq j \leq p\right\} > 0,$$

and choose $\delta > 0$ corresponding to ε' as in Proposition A. If $\|\mu-\mu^0\|_\infty \leq \delta$ then $\left\|\varphi_{n(\delta)} - \varphi_0\right\| \leq \varepsilon'$ by Proposition A, and thus, from (11)

$$\left|F^{(j)}_{\varphi_n}(z) - u^{(j)}(z)\right| \leq \frac{j!}{2\pi}\frac{\|\gamma'\|_\infty}{\left[\operatorname{dist}(K,\partial\Omega)\right]^{j+1}}\varepsilon' \leq \varepsilon,$$

for all z in K, j in $\{0,\ldots,p\}$. ∎

2. The reconstruction problem in $H^2(U)$

2.1. In this part, we investigate the problem (6) in the case $\Omega = U$ is the unit disc. In this case $\gamma(t) = e^{it}$, $0 \leq t \leq 2\pi$.

In the sequel, we discuss the existence of solutions and the regularization of (6) in the Hardy space $H^2(U)$. The position of the problem in this space has a number of interesting points, due to the specific properties of $H^2(U)$, and to the fact that $H^2(U)$ is, in some sense, an intermediate space between $H(U)$ and $H(U) \cap C(\overline{U})$.

The definition and basic properties of $H^2(U)$ can be found in Ch. 17, [10] or Ch. 3, 5, [8]. We state here some results needed for our discussions below.

(i) $H^2(U)$ is a Hilbert space with the inner product

$$(u,v)_{H^2} = \sum_{n=0}^{\infty} a_n\overline{b_n}, \quad \forall u,v \in H^2(U),$$

$$u(z) = \sum_{n=0}^{\infty} a_n z^n,\ v(z) = \sum_{n=0}^{\infty} b_n z^n, \quad z \in U,$$

and norm $\|u\|_{H^2} = (u,u)^{1/2}_{H^2}$.

(ii) If $u \in H^2(U)$ then u has boundary values:

$$u^*(e^{i\theta}) = \lim_{r\to 1} u(re^{i\theta}),\ \text{a.e. } \theta \in [0,2\pi].$$

Moreover, $u^* \circ \gamma \in L^2(0, 2\pi)$. We consider $L^2(0, 2\pi)$ with the "normalized" inner product:

$$(\varphi, \psi) = \frac{1}{2\pi} \int_0^{2\pi} \varphi \overline{\psi} d\theta, \quad \varphi, \psi \in L^2(0, 2\pi)$$

and norm $\|\varphi\| = (\varphi, \varphi)^{1/2}$. Then the mapping $u \mapsto u^* \circ \gamma$ is norm-preserving from $H^2(U)$ to $L^2(0, 2\pi)$.

(iii) $u \in H^2(U)$ can be represented by the Cauchy integral of u^*:

$$u(z) = \frac{1}{2\pi i} \int_\gamma \frac{u^*(\zeta) d\zeta}{\zeta - z}, \quad \forall z \in U. \tag{12}$$

We now prove that

$$\int_\gamma \frac{u^*(\zeta) d\zeta}{\zeta - z} = 0, \quad \forall z \notin \overline{U}. \tag{13}$$

Indeed, put, for $0 < r < 1$, $u_r(z) = u(rz)$. Then $u_r \in H(D_r)$ ($D_r = \{z : |z| < 1/r\} \supset \overline{U}$). For $z \in \overline{U}$, we have, for r sufficiently close to 1, that $\int_\gamma [u_r(\zeta) d\zeta]/(\zeta - z) = 0$. But since $u_r \circ \gamma \to u^* \circ \gamma$ in $L^2(0, 2)$ (Theorem 17.10, [10]), letting $r \to 1$ in this equality, we get (13).

Now, we choose a sequence (ξ_n) in $\mathbb{C} \setminus \overline{U}$ such that $\xi_n \to \xi_0 \notin \overline{U}$ (for example $\xi_n = 2 + 1/n$, $n = 1, 2, \ldots$). Then, by (13), $\int_\gamma [u^*(\zeta) d\zeta]/(\zeta - \xi_n) = 0$, $\forall n$. Thus, (6) is reduced to the following moment problem

$$\begin{cases} \dfrac{1}{2\pi i} \displaystyle\int_0^{2\pi} \frac{u^*\big(\gamma(t)\big)\gamma'(t) dt}{\gamma(t) - z_n} = \mu_n \\ \dfrac{1}{2\pi i} \displaystyle\int_0^{2\pi} \frac{u^*\big(\gamma(t)\big)\gamma'(t) dt}{\gamma(t) - \xi_n} = 0, \end{cases} \quad \forall n \in \mathbb{N}. \tag{14}$$

We put

$$\zeta_n = \begin{cases} z_k & \text{if } n = 2k-1 \\ \xi_k & \text{if } n = 2k \end{cases}, \quad \tau_n = \begin{cases} \mu_k & \text{if } n = 2k-1 \\ 0 & \text{if } n = 2k \end{cases},$$

$$v_n = \overline{\gamma'(t)/\big(\gamma(t) - \zeta_n\big)}, \ n \in \mathbb{N}, \text{ and } \varphi = u^* \circ \gamma.$$

By noting that $\gamma'(t) = i\gamma(t)$, we see that (14) can be written in the form

$$(\varphi, v_n) = \tau_n, \quad \forall n \in \mathbb{N}. \tag{15}$$

As in Part 1, we approximate (15) by finite moment problems:

$$\begin{cases} (\varphi, v_j) = \tau_j, \ 1 \leq j \leq n \\ \varphi \in \langle v_1, \ldots, v_n \rangle. \end{cases} \tag{16}$$

Since the ζ_n' s are distinct, we can show that $v_1, \ldots, v_n, \ \ldots$ are linearly independent. By the linear independence of $\{v_n \mid n \in \mathbb{N}\}$, we can associate with it an orthonormal sequence (e_n) by the Gram-Schmidt method. Let C_{ij} $(i, j \in \mathbb{N})$ be defined by (4). We have the following condition for the existence of a solution of (6) in $H^2(U)$.

Theorem 3. *Let* (μ_n) *be a sequence of complex numbers. Then* (6) *is solvable in* $H^2(U)$ *if and only if* $\sum_{i=1}^{\infty} \left| \sum_{j=1}^{\infty} C_{i,2j-1} \mu_j \right|^2 < \infty$.

Proof. We show first that (6) is solvable in $H^2(U)$ if and only if (15) is solvable in $L^2(0, 2\pi)$.

Indeed, by the foregoing remarks, we know that if $u \in H^2(U)$ satisfies (6), then $u^* \circ \gamma \in L^2(0, 2\pi)$ is a solution of (15). Conversely, letting $\varphi \in L^2(0, 2\pi)$ be a solution of (15), we show that $F_\varphi \in H^2(U)$ and satisfies (6).

For $n \in \mathbb{N}$, we have

$$F_\varphi(\xi_n) = \frac{1}{2\pi i} \int_0^{2\pi} \frac{\varphi(t)\gamma'(t)}{\gamma(t) - \xi_n} dt = (\varphi, v_{2n}) = \tau_{2n} = 0.$$

Since (ξ_n) has a limit point in $\mathbb{C} \setminus \overline{U}$, and since $F_\varphi \in H(\mathbb{C} \setminus \overline{U})$, this implies that $F_\varphi(z) = 0$, $\forall z \in \mathbb{C} \setminus \overline{U}$. Hence, for $|z| > 1$,

$$\begin{aligned} 0 &= -\int_0^{2\pi} \frac{\varphi(t) e^{it} dt}{e^{it} - z} = \int_0^{2\pi} \frac{\varphi(t) e^{it} dt}{z(1 - e^{it}/z)} \\ &= \frac{1}{z} \int_0^{2\pi} \varphi(t) e^{it} \Big(\sum_{n=0}^{\infty} \frac{e^{itn}}{z^n} \Big) \quad \text{(since } |e^{it}/z| = 1/|z| < 1) \\ &= \sum_{n=0}^{\infty} \Big(\int_0^{2\pi} \varphi(t) e^{i(n+1)t} dt \Big) \frac{1}{z^{n+1}}. \end{aligned}$$

Since this holds for all $z \in \mathbb{C} \setminus \overline{U}$, the coefficients of the series must all vanish, i.e.,

$$\widehat{\varphi}(n) = \frac{1}{2\pi} \int_0^{2\pi} \varphi(t) e^{-int} dt = 0, \ n = -1, -2, \ldots$$

By Theorem 17.10 (c), [10], this implies that φ is the boundary value of some $u \in H^2(U)$, i.e.,

$$\varphi(\theta) = u^*(e^{i\theta}), \ \text{a.e. } \theta \in [0, 2\pi]. \tag{17}$$

But by the integral representation of u (see (12)), we have, for all $z \in U$

$$\begin{aligned} u(z) &= \frac{1}{2\pi i}\int_\gamma \frac{u^*(\zeta)}{\zeta - z}d\zeta = \frac{1}{2\pi i}\int_0^{2\pi} \frac{u^*(e^{it}).\gamma'(t)}{\gamma(t)-z}dt \\ &= \frac{1}{2\pi i}\int_0^{2\pi} \frac{\varphi(t)\gamma'(t)}{\gamma(t)-z}dt \quad \text{(by (17))} \\ &= F_\varphi(z). \end{aligned}$$

Hence

$$F_\varphi = u \in H^2(U). \tag{18}$$

On the other hand, we have, for $n = 1, 2, \dots$

$$F_\varphi(z_n) = (\varphi, v_{2n-1}) = \tau_{2n-1} = \mu_n.$$

Hence, F_φ is the unique the solution of (6) in $H^2(U)$.

We have proved the equivalence between problem (6) in $H^2(U)$ and problem (15) in $L^2(0, 2\pi)$. Now, from Proposition A, we know that a necessary and sufficient condition for the existence of a solution of (15) in $L^2(0, 2\pi)$ is

$$\sum_{i=1}^{\infty}\Big|\sum_{j=1}^{\infty} C_{ij}\tau_j\Big|^2 = \sum_{i=1}^{\infty}\Big|\sum_{j=1}^{\infty} C_{i,2j-1}\mu_j\Big|^2 < \infty.$$

This is also a necessary and sufficient condition for the solvability of (6) in $H^2(U)$.

According to the above proof, we see that if $\varphi_1, \varphi_2 \in L^2(0, 2\pi)$ are solutions of (15) then $F_{\varphi_1}, F_{\varphi_2}$ are solutions of (6). By uniqueness of solutions of (6), we have $F_{\varphi_1} = F_{\varphi_2}$ in U. Hence, by (17),

$$\varphi_1(\theta) = \lim_{r\to 1} F_{\varphi_1}(re^{i\theta}) = \lim_{r\to 1} F_{\varphi_2}(re^{i\theta}) = \varphi_2(\theta)$$

for a.e. $\theta \in [0, 2\pi]$, i.e., $\varphi_1 = \varphi_2$ in $L^2(0, 2\pi)$. This shows the uniqueness of solutions of (15) in $L^2(0, 2\pi)$. ■

We shall present a regularization of the reconstruction problem in $H^2(U)$. Let W be the closed subspace of $L^2(0, 2\pi)$ generated by $\{e^{int} : n = 0, 1, 2, \dots\}$. W is the orthogonal complement of the closed subspace generated by $\{e^{int} : n = -1, -2, \dots\}$. Let P_W be the orthogonal projection of $L^2(0, 2\pi)$ onto W:

$$\mathrm{P}_W\varphi = \sum_{n=0}^{\infty}(\varphi, e^{int})e^{int}, \quad \varphi \in L^2(0, 2\pi).$$

We put $\psi_n(\mu) = \mathrm{P}_W(\varphi_n(\mu))$. Then the following holds (cf. Proposition A),

Theorem 4. (i) *$F_{\psi_n}(\mu)$ is in $H^2(U)$ and depends continuously on $\mu_1, \ldots, \mu$ in the norm topology of $H^2(U)$ when n is fixed.*
(ii) *Suppose u_0 is the solution in $H^2(U)$ of (6) corresponding to μ_0. Let $n(\delta)$, $\delta > 0$, be defined as in Proposition A. Then for $\varepsilon > 0$, there exists $\delta > 0$ such that for all sequences μ with $\|\mu - \mu^0\|_\infty \leq \delta$, we have*

$$\left\|F_{\psi_{n(\delta)}(\mu)} - u_0\right\|_{H^2} \leq \varepsilon \text{ with } H^2 = H^2(U).$$

2.2. In the preceding sections, we have used the Cauchy integral representation to convert the problem into a moment problem. The regularization is given by truncated expansion. The key step is the choice of the function f giving the dimension of the approximation, i.e., the integer $n(\varepsilon) = [f^{-1}(\varepsilon^{-1/2})]$. The sequence (z_n) of points with values (μ_n) of the unknown function $u(z)$ given, can be arbitrary (provided that (z_n) is an infinite sequence accumulating at a point in Ω). In the present section, we deal with the regularization problem in $H^2(U)$. The given sequence (z_n) should lie in a disc U_α of radius α $(0 < \alpha < 1)$. On the other hand, we here use Lagrange interpolation and the dimension of the truncation is readily obtained.

Let $\mu = (\mu_n) \in \ell^\infty$ be a known "measured" data satisfying

$$\|\mu - \mu^0\|_\infty = \sup_n |\mu_n - \mu_n^0| < \varepsilon. \tag{19}$$

From μ, we shall construct a polynomial that, in a sense to be specified, approximates u_0. For $m \in \mathbb{N}$, we consider the following system of linear equations

$$\sum_{k=0}^{m-1} a_{mk} z_n^k = \mu_n, \quad n = 1, 2, \ldots, m. \tag{20}$$

Put

$$P_m(z) = \sum_{0 \leq k \leq [\ln m]} a_{mk} z^k.$$

An estimate of the error between P_m and u_0 will be given. We have

Theorem 5. *Let $\varepsilon > 0$ and assume that (19) holds. Let α be a positive number satisfying*

$$0 < \alpha < 1, \quad \frac{\sqrt{\alpha}}{(1 - \sqrt{\alpha})^4} < 1.$$

Let $(z_n) \subset U$ be such that

$$z_k \neq z_j \ \forall k \neq j; \quad |z_n| \leq \alpha \ \forall n = 1, 2, \ldots$$

and let $u_0 \in H^2(U)$ satisfy (6) corresponding to $\mu^0 \in \ell^\infty$. Let $f : [1, \infty) \to \mathbb{R}$ be a function $f = f(\theta)$ increasing to infinity as $\theta \to \infty$, and satisfying

$$f(m) \geq D_m^2 \alpha^m \ln m$$

where

$$D_m = \sup_{1 \leq n \leq m} \prod_{j \in \overline{1,m} \setminus \{n\}} |z_n - z_j|^{-1}.$$

Put

$$n(\varepsilon) = \left[f^{-1}(\varepsilon^{-1})\right].$$

Then $n(\varepsilon) \to \infty$ as $\varepsilon \downarrow 0$ and there are an $\varepsilon_0 > 0$ and a function $\eta(\varepsilon)$, $0 < \varepsilon < \varepsilon_0$ such that $\lim_{\varepsilon \downarrow 0} \eta(\varepsilon) = 0$ and that

$$\left\|P_{n(\varepsilon)} - u_0\right\|_{H^2}^2 \leq \eta(\varepsilon).$$

(where, we recall, $P_m(z) = \sum_{0 \leq k \leq [\ln m]} a_{mk} z^k$ and the a_{mk}'s are defined in (20)).

If, in addition, $u_0' \in H^2(U)$, then one has

$$\left\|P_{n(\varepsilon)} - u_0\right\|_{H^2}^2 \leq C\varepsilon + \frac{2 \ln n(\varepsilon)}{(1 - \sqrt{\alpha})^{4n(\varepsilon)}} \left(\sqrt{\alpha}\right)^{n(\varepsilon)} \|u_0\|_{H^2}^2 + \frac{\|u_0'\|_{H^2}^2}{\left(\ln n(\varepsilon)\right)^2}$$

for $0 < \varepsilon < \varepsilon_0$.

Proof. Since $u_0 \in H^2(U)$, u_0 can be represented by a series

$$u_0(z) = \sum_{k=0}^{\infty} a_k z^k \tag{21}$$

where $\sum_{k=0}^{\infty} |a_k|^2 < \infty$. One has

$$P_m(z) - u_0(z) = \sum_{0 \leq k \leq [\ln m]} c_{mk} z^k - \sum_{k \geq [\ln m] + 1} a_k z^k$$

where $c_{mk} = a_{mk} - a_k$ (a_{mk} defined in (20)). It follows that

$$\left\|P_m - u_0\right\|_{H^2}^2 = \sum_{0 \leq k \leq [\ln m]} |c_{mk}|^2 + \sum_{k \geq [\ln m] + 1} |a_k|^2. \tag{22}$$

We shall estimate $|c_{mk}|$. Using Lagrange' s polynomial

$$Q_m(z) = Q_{1m}(z) + Q_{2m}(z)$$

where

$$Q_{1m}(z) = \sum_{n=1}^{\infty}(\mu_n - \mu_n^0) \prod_{j\in\overline{1,m}\setminus\{n\}} \left(\frac{z - z_j}{z_n - z_j}\right)$$

$$Q_{2m}(z) = \sum_{l=0}^{\infty} a_{m+l} \sum_{n=1}^{m} z_n^{m+l} \prod_{j\in\overline{1,m}\setminus\{n\}} \left(\frac{z - z_j}{z_n - z_j}\right),$$

we get in view of (20), (21) that

$$\left.\begin{aligned} Q_m(z_n) &= \mu_n - \mu_n^0 + \sum_{k=m}^{\infty} a_k z_n^k \\ Q_m(z) &= \sum_{k=0}^{\infty} c_{mk} z^k \end{aligned}\right. \qquad n = 1, 2, \dots, m.$$

Using expansions of Q_{1m}, Q_{2m}, we get an explicit formula for c_{mk} (not written out here). After some lengthly computations, we get the following estimate

$$|c_{mk}| \leq \varepsilon D_m \sum_{n=1}^{m} \left|\sigma_{m-k-1}(\widehat{z}_n)\right| + \frac{\sup|a_n|}{1-\sqrt{\alpha}} \cdot \frac{(\sqrt{\alpha})^{m-k}}{(1-\sqrt{\alpha})^{2m-1}} \tag{23}$$

where

$$\begin{aligned} \widehat{z}_1 &= (z_2, \dots, z_m) \\ \widehat{z}_j &= (z_1, \dots, z_{j-1}, z_{j+1}, \dots, z_m) \text{ for } 2 \leq j \leq m-1 \\ \widehat{z}_m &= (z_1, \dots, z_{m-1}), \end{aligned}$$

and, for $t = (t_1, \dots, t_{m-1})$

$$\sigma_k(t) = \sum_{1\leq j_1<\dots<j_k\leq m-1} (t_{j_1} \dots t_{j_k}), \quad k = 1, 2, \dots, m-1.$$

Noting that $|z_j| \leq \alpha$ $(j = 1, 2, \dots)$, one has

$$\left|\sigma_{m-k-1}(\widehat{z}_n)\right| \leq \alpha^{m-k-1} C_{m-1}^{m-k-1}.$$

Hence,

$$\sum_{n=1}^{m} \left|\sigma_{m-k-1}(\widehat{z}_n)\right| \leq m\alpha^{m-k-1} C_{m-1}^{m-k-1}. \tag{24}$$

Using Stirling' s formula, one gets

$$
\begin{aligned}
C_{m-1}^{m-k-1} &= \frac{(m-1)!}{k!(m-k-1)!} \\
&\le C\Big(\frac{m-1}{k}\Big)^k\Big(1+\frac{k}{m-k-1}\Big)^{m-k-1}\sqrt{\frac{m-1}{2\pi k(m-k-1)}} \qquad (25)\\
&\le C\Big(\frac{m-1}{k}\Big)^{k+1/2}e^k.
\end{aligned}
$$

From (24), (25) we have, for $0 \le k \le [\ln m]$

$$
\begin{aligned}
\sum_{n=1}^{m}\big|\sigma_{m-k-1}(\widehat{z}_n)\big| &\le Cm\alpha^{m-k-1}\Big(\frac{m-1}{k}\Big)^{k+1/2}e^k \\
&\le Cm\alpha^{m-\ln m-1}m^{\ln m+1} \qquad (26)\\
&\le C\alpha^{m/2} \quad \text{for } m \text{ large.}
\end{aligned}
$$

Substituting (26) into (23) gives (for m large)

$$
|c_{mk}| \le \varepsilon C D_m \alpha^{m/2} + \frac{\sup|a_n|}{1-\sqrt{\alpha}}\cdot\frac{(\sqrt[4]{\alpha})^m}{(1-\sqrt{\alpha})^{2m-1}}.
$$

It follows that

$$
\sum_{0\le k\le[\ln m]}|c_{mk}|^2 \le 2\varepsilon^2 C(\ln m)D_m^2\alpha^m + \frac{2\ln m}{(1-\sqrt{\alpha})^{4m}}(\sqrt{\alpha})^m\sup_n|a_n|^2. \qquad (27)
$$

From the definition of $n(\varepsilon)$, one has $n(\varepsilon)\to\infty$ as $\varepsilon\downarrow 0$ and

$$
\varepsilon^{-1} \ge f\big(n(\varepsilon)\big) \ge D_{n(\varepsilon)}^2\alpha^{n(\varepsilon)}\ln n(\varepsilon). \qquad (28)
$$

Combining (27) with (28), we get

$$
\sum_{0\le k\le[\ln m]}|c_{mk}|^2 \le C\varepsilon + \frac{2\ln n(\varepsilon)}{(1-\sqrt{\alpha})^{4n(\varepsilon)}}(\sqrt{\alpha})^{n(\varepsilon)}\|u_0\|_{H^2}^2. \qquad (29)
$$

From (22) and (29), we get

$$
\big\|P_{n(\varepsilon)}-u_0\big\|_{H^2}^2 \le \eta(\varepsilon) \qquad (30)
$$

for

$$
\eta(\varepsilon) = C\varepsilon + \frac{2\ln n(\varepsilon)}{(1-\sqrt{\alpha})^{4n(\varepsilon)}}\big(\sqrt{\alpha}\big)^{n(\varepsilon)}\|u_0\|_{H^2}^2 + \sum_{k\ge[\ln n(\varepsilon)]+1}|a_k|^2.
$$

Now, if $u_0' \in H^2(U)$ then

$$\sum_{k=0}^{\infty} k^2 |a_k|^2 = \left\| u_0' \right\|_{H^2}^2 < \infty.$$

Hence

$$\sum_{k \geq [\ln n(\varepsilon)]+1} |a_k|^2 \leq \frac{1}{\left(\ln n(\varepsilon)\right)^2} \sum_{k \geq [\ln n(\varepsilon)]+1} k^2 |a_k|^2 \leq \frac{\left\| u_0' \right\|_{H^2}^2}{\left(\ln n(\varepsilon)\right)^2}.$$

In view of (30), this completes the proof of the theorem. ∎

References

[1] G. Alessandrini, *An Extrapolation Problem for Harmonic Functions*, Boll. UMI **17B**, 1980.

[2] S. Y. Al' per, G. I. Kalinogorskaja, *The Convergence of Lagrange Interpolation Polynomials in the Complex Domain*, Izv. Vyss. Ucebn. Zaved. Mathematika, 1969, No. 11, 13-23 MR41, 700, **II.2**.

[3] D. D. Ang and D. D. Trong, *Asymptotic Behavior of Solutions of Vandermonde Equations*, Preprint, 1993.

[4] D. D. Ang, L. K. Vy and R. Gorenflo, *A regularization Method for the Moment Problem*, in Math. Research, Vol. 74: *Inverse problems: Principles & Applications in Geophysics, Technology, and Medicine*, Akademie Verlag, 1993.

[5] J. R. Cannon and K. Miller, *Some Problems in Numerical Analytic Continuation*, SIAM J. Num. Anal. **2**, 1965.

[6] D. Gaier, *Vorlesungen über Approximation in Komplexen*, Birkhauser Verlag, Besel-Boston-Stutgart (1980).

[7] P. M. Gauthier and W. Hengartner, *Complex Approximation and Simultaneous Interpolation on Closed Set*, Canad. J. Math. **29** (1977) 701-706, MR 58, 4214.IV.3.

[8] K. Hoffman, *Banach Spaces of Analytic Functions*, Prenctice-Hall Inc., Englewood Cliffs, N. J., 1962.

[9] N. V. Huy, N. V. Nhan and D. D. Trong, *Reconstruction of Analytic Functions on the Unit Disc from a Sequence of Moments: Regularization and Error Estimates*, in *Complex Analysis and Applications*, Proceedings, International Conference in Hanoi (1997) (to appear).

[10] W. Rudin, *Real and Complex Analysis*, Mc Graw-Hill Co. (1987).

[11] A. Taylor, *Advanced Calculus*, Blaisdel Publishing Co., N. Y., Toronto, London, 1965.

[12] D. D. Trong and D. D. Ang, *Reconstruction of Analytic Functions: Regularization and Optimal Recovery*, Preprint (1997).

[13] J. L. Walsh, *Interpolation and Approximation by Rational Functions in the Complex Domain*, American Mathematical Society (1960).

9

P-ADIC INTERPOLATION AND APPLICATIONS

Ha Huy Khoai
Hanoi Institute of Mathematics
P. O. Box 631 Bo Ho, 10000 Hanoi, Vietnam
E-mail: hhkhoai@thevinh.ac.vn

Le Thi Hoai Thu
Traning teacher college of Quang Binh province, Vietnam

Abstract: This paper consists of 3 parts:

1 Interpolation of p-adic holomorphic functions.

2 Applications to find the inverse formula of the p-adic Mellin transform and Integral representations of p-adic functions.

3 Study of p-adic L-functions associated to modular forms by using Shnirelman integral.

1. P-adic interpolation of holomorphic functions in the unit disc.

Let p be a prime number, $\mathbb{Q}_p$ the field of p-adic numbers, and $\mathbb{C}_p$ the p adic completion of the algebraic closure of $\mathbb{Q}_p$. The absolute value in $\mathbb{Q}_p$ is normalized so that $|p| = p^{-1}$. We further use the notion $v(z)$ for the additive valuation on $\mathbb{C}_p$ which extends ord_p.

Let $f(z)$ be a holomophic function on $D \subset \mathbb{C}_p$ (the unit disc), it is represented by a convergent series

$$f(z) = \sum_{n=0}^{\infty} a_n z^n, \quad |z| < 1.$$

Definition 1.1. A sequence $u = \{u_i\} \subset D$ is called *an interpolating sequence* of f if the sequence of Lagrangian interpolating polynomials for f on u converges to f.

Notice that we consider *infinite discrete sequences* in D.

It is naturally to hope that a sequence u is an interpolating sequence of a function f if the number of points of the sequence u is "sufficiently large" with respect to the growth of f. To show this we introduce the notion of *height* of a sequence and a function, and then use the height function to give a necessary and sufficient condition for a sequence u to be an interpolating of a given function f.

For each n, we draw the graph Γ_n which depicts $v(a_n z^n) = v(a_n) + nv(z)$ as a function of $v(z) = t$

$$v(a_n z^n) = v(a_n) + nt.$$

This is a straight line with slope n. Since we have

$$\lim_{n\to\infty} \{v(a_n) + nt\} = \infty$$

for $t > 0$, it follows that for every $t > 0$ there exists an n for which $v(a_n) + nt$ is minimal.

Let $h(f,t)$ denote the boundary of the intersection of all of the half-planes lying under the lines Γ_n. Then in any segment $[r,s]$, $(0 < r < s)$ there are only finitely many Γ_n which appear in $h(f,t)$. Thus $h(f,t)$ is a polygonal line.

This line is what we called *the height of the function f*.

The points t at which $h(f,t)$ has vertices are called *the critical points* of $f(z)$. A finite segment contains only finitely many critical points. If t is a critical point, then $v(a_n) + nt$ contains its minimum at least at two values of n. If $v(z) = t$ is not a critical point, then

$$|f(z)| = p^{-h(f,t)}.$$

The height of a function $f(z)$ gives complete information about the number of zeros of $f(z)$. Namely, f has zeros when $v(z) = t_i$ (a critical point) and the number of zeros of f such that $v(z) = t_i$ is equal to the difference $n_{i+1} - n_i$ between slopes of $h(f,t)$ at $t_i - 0$ and $t_i + 0$.

For a descrete sequence $\{u_i\} \subset D$ we can define *the height of u* as follows. We set

$$N_u(t) = \#\{u_i \mid v(u_i) \geq t\}.$$

We can write the sequence u in the form

$$u = \{u_1, ..., u_{n_1}, u_{n_1+1}, ..., u_{n_2}, ...\},$$

where

$$v(u_i) = t_k \text{ for } n_{k-1} < i < n_k,$$

(we take $n_0 = 0$), and $\lim_{k\to\infty} t_k = 0$.

Then we choose a sequence a_k with the property

$$v(a_0) = -n_1 t_1,\ v(a_{k+1}) = v(a_k) + (n_k - n_{k+1})t_{k+1}.$$

Set

$$\Phi_u(z) = 1 + \sum_{k=1}^{\infty} a_k z^{n_k}.$$

Then the series converges for $z \in D$ and defines the holomorphic function $\Phi_u(z)$ in D. The number of zeros of $\Phi_u(z)$ in $\{z \mid v(z) > t\}$ is equal to $N_u(t)$, and

$$h(\Phi_u, t) = \int_{\infty}^{t} N_u(t)dt.$$

Then we have ([3])

Theorem 1.2. *A sequence $u = \{u_i\}$ is interpolating sequence of a function f if and only if*

$$\lim_{t\to 0}\{h(f,t) - h(\Phi_u, t)\} = \infty.$$

(this means that $||f||_r = 0(||\Phi_u||_r), \quad r \to 1$, i.e $f \in 0(\Phi_u)$).

Example 1.3. Let $f \in 0(\log)$. Then $u = \{\mu_n, \quad n = 1, 2, ...\}$ where $\mu_n = \{\text{primitive roots of unity of degree } p^n\}$ is an interplating sequence of f.

Remark 1.4. The problem of p-adic interpolation is considered in the papers of Mahler ([4]), and is investigated systematically in Amice's Thesis ([1]). However, Theorem 1.2 is the first interpolation theorem for p-adic holomorphic functions not necessarily bounded.

2. Applications to p-adic L-functions

We consider the case $p > 2$. First we recall that a p-adic Mellin transform can be regarded as a holomorphic function in the unit disc. Denote

$$\mathbb{Z}_p = \lim_{\longleftarrow}(\mathbb{Z}_p^n\mathbb{Z})^*.$$

The group of p-adic charaters is the group of continous holomorphisms of $\mathbb{Z}_p^*$ into $\mathbb{C}_p^*$

$$\chi(\mathbb{Z}_p^*) = \mathrm{Hom}_{\mathrm{cont}}(\mathbb{Z}_p^*, \mathbb{C}_p^*).$$

We set

$$U = 1 + p\mathbb{Z}_p = \{z \in \mathbb{Z}_p \mid v(z-1) \geq 1\}.$$

For every $g \in U$ such that $v(g-1) = 1$, the map $z \mapsto g^z$ is an isomorphism of $\mathbb{Z}_p$ onto U. We call g *topological generator* of U. For each generator g of U, the map

$$\chi(U) = \mathrm{Hom}_{\mathrm{cont}}(U, \mathbb{C}_p^*) \to \mathbb{C}_p^*$$

sending a continuous character χ of U into a point $\chi(g) - 1$ in the unit disc of $\mathbb{C}_p$. Also we have

$$\mathbb{Z}_p^* \simeq (\mathbb{Z}/\mathbb{Z}_p)^* \times U.$$

Thus $\chi(\mathbb{Z}_p^*)$ is a product of a finite group and $\chi(U) \simeq \bigsqcup_{\text{finite}} D$.

Hence, $\chi(\mathbb{Z}_p^*)$ becomes an analytic Lie group, and we can consider a holomorphic function on $\chi(\mathbb{Z}_p^*)$ as a holomorphic function on the unit disc D.

Definition 2.1. Let μ be a measure on $\mathbb{Z}_p^*$, i.e μ is a continuous linear functional with values in $\mathbb{C}_p$ on the space of continuous functions on $\mathbb{Z}_p^*$. Then the restriction of μ on the analytic Lie group $\chi(\mathbb{Z}_p^*)$ gives a holomorphic function

$$L(\mu, \chi) = \int_{\mathbb{Z}_p^*} \chi d\mu.$$

The function $L(\mu, \chi)$ is called the *p-adic Mellin transform* of the measure μ. The p-adic Mellin transform of a measure μ is a holomorphic function on $\chi(\mathbb{Z}_p^*)$, and then, can be regarded as a holomorphic function in the unit disc.

Hence, if we can choose a sequence $u \in D$ such that the function Φ_u growths faster enough with respect to $L(\mu, \chi)$ we can contruct $L(\mu, \chi)$ by an interpolating process. Now let $u \in D$ is an interpolating sequence of p-adic L-function $L(\mu, \chi)$ and let $\mathcal{X} \subset \chi(\mathcal{U})$ is the subset corresponding with u by the isomorphism $\chi(U) \simeq D$. Then p-adic L-function $L(\mu, \chi)$ is *uniquely defined by the values on* $\mathcal{X}$.

Example 2.2. Amice and Vélu ([2]) and M. Vishik ([7]) proved that the p-adic L-function associated to a modular form Φ of weight $w+2$ is a holomorphic function on D of class $0(\log^{w+1})$.

Recall that every $\chi \in \mathbb{Z}_p^*$ can be written in the form

$$\chi = \alpha(x)\theta(x)$$

with $\alpha(x) \in (\mathbb{Z}/p\mathbb{Z})^*$ and $\theta(x) \in \mathbb{Z}_p^*$. The map

$$\theta : x \mapsto \theta(x)$$

is an element of $\chi(\mathbb{Z}_p^*)$ which is called *the fundamental character.*

From Theorem of Interpolation and the fact that $L(\Phi,\chi) \in 0(\log^{w+1})$ we can see that p-adic L-functions associated to modular forms of weight $w+2$ are uniquely defined by values on the set $\{\chi\Theta^k\}$, $k = 0, \ldots w$ and $\chi \bmod p^n$, $n = 1, 2, \ldots$ (because the function constructed on $u = \{\chi\Theta^k\}$ has the growth as $\log^{w+1}$). Infact,

$$L(\chi\theta^k, \Phi) = \frac{1}{\lambda^n}(-1)^k A(\chi, k), \quad \text{where}$$

$$A(\chi,k) = \frac{i^{k+1}\Lambda_\Phi(\chi, k+1)}{G(\chi)\omega(-1)^k\chi(-1)}, \quad G(\chi) = \sum_{a \mod d(\chi)} \chi(a)e^{2\pi i a/d(\chi)}$$

is the Gauss sum, $\Lambda_\Phi(\chi, s) = d(\chi)^s \int_0^\infty \tau^{s-1}\Phi_\chi(\tau)d\tau.$

3. The inverse Mellin transform

Now we consider the problem: *Given a p-adic holomorphic function $f(z)$ in D, when we can give a measure μ such that f is the p -adic L-function associated to μ*, i.e

$$f(\chi) = L(\chi, \mu) \int_{\mathbb{Z}_p} \chi d\mu \ ?.$$

We give now the answer in two cases:

1 f is bounded.

2 f is of class $0(\log^h)$, $\quad h \geq 1$.

Here we related this problem with the Shnirelman integral([?]).
Definition 3.1. Let $f(z)$ be a $\mathbb{C}_p$-valued function defined on $z \in \mathbb{C}_p$ such that $|z - a| = r$, where $a \in \mathbb{C}_p, r \in \mathbb{R}$, $r > 0$ (we shall allways assume that $r \in |\mathbb{C}_p|_p$ i.e a rational of p). Let $\gamma \in \mathbb{C}_p$ be such that $|\gamma| = r$. Then *the Shnirelman integral* is defined as the following limit, if it exists

$$\int_{a,\gamma} f(z)dz = \lim_{n\to\infty}{}' \frac{1}{n} \sum_{\zeta^n=1} f(a + \zeta\gamma)$$

(this is a kind of "integral on a curve"), where lim$'$ indicates that the limit is only over $n \not| p$.

Now let

$$f(z) = \int_{\chi(U)} \chi d\mu$$

be the p-adic Mellin transform of a measure μ. Let g be a fixed topological generator of U. We set $z = \chi(g)-1$. For $x \in U$ we have $x = g^{\log x/\log g}$, then

$$\chi(x) = \chi(g^{\log x/\log g}) = (1+z)^{\log x/\log g},$$

$$f(z) = \int_U (1+z)^{\log x/\log g}\chi d\mu(x).$$

Theorem 3.2. *Let f(z) be a bounded holomorphic function in the unit dics D. For $x \in U$ and $m = 1, 2, \dots$ we consider the following function in D*

$$G_{m,x}(z) = \frac{1}{p^m - p^{m-1}} \sum_{\zeta^{p^m}, \zeta \neq 1} \zeta^{-\log x/\log g} \frac{1}{(z+1-\zeta)}.$$

Then the function f(z) is the p-adic Mellin transform of the following measure on U

$$\mu\{x + (U_m)\} = \int_{a,\gamma} f(z) G_{m,x}(z)(z-a)dz,$$

where $U_m = 1 + p^m U$, a, γ are such that $D_a(|\gamma|) \subset D$.

Thus, we have a correspondance between the space $H_\circ(D)$ of bounded holomorphic functions in D and the space $L(U)$ of continuous linear functionals on the space $C(D)$ of continuous functions on D. Namely, for a bounded measure μ on U, the p-adic Mellin transform

$$M\mu = \int_U (1+z)^{\log x/\log g} d\mu$$

defines a bounded p-adic holomorphic function in D. Conversely, for a bounded holomorphic function $f(z)$ in D we can define a continuous functional $Nf \in L(U)$ as follows

$$C(D) \ni \Phi \mapsto \lim_{n\to\infty} \sum_{i=0}^{p^n-1} f(z)\Phi(x_i) G_{m,x_i}(z)(z-a)dz,$$

where x_i runs on the set of representations of U/U_m.

The proof uses 3 steps:

1 The above formular gives a measure.

2 The obtained measure is bounded.

3 The given function $f(z)$ coincides with $L(\mu, \chi)$ on the sequence $\{\mu_n\}$,

$n = 1, 2, \ldots$; μ_n is the set of the primitive roots of 1 of degree p^n.

For the case of holomorphic functions of class $0(\log^h)$ (for examples, p-adic functions associated to modular cusp forms), we consider so-called *h-admissible measure* (M. Vishik,[7]).

Definition 3.3. A *h-admissible measure* on U is a linear functional on the space of functions on U which are locally polynomials of degree less than h and satisfy the following condition

$$|\mu\{(x-b)^k \psi_{b,m}\}| = 0(p^{(h-k)m}), \quad k = 0, \ldots, h-1; m = 1, 2, \ldots$$

where $\psi_{b,m}$ is the characteristic function of $b + (U_m)$.

A h -admissible measure can be uniquely extended to a linear form on the $\mathbb{C}_p$-space of all $(h-1)$-differentiable functions whose derivatives of order $h-1$ satisfy the Lipschitz condition. Moreover, the Mellin transform of this measure is a p-adic holomorphic function of class $0(\log^h)$.

Theorem 3.4. *Let f(z) be a holomophic function of class* $0(\log^h)$ *in the unit disc D. Then f(z) is the p-adic Mellin transform of the following h-admissible measure on U*

$$\mu\{x^k \psi_{b,m}\} = \int_{a,\gamma} f(z) G_{m,b,k}(z)(z-a)dz,$$

$m = 1, 2, \ldots; k = 0, \ldots h-1$, $G_{m,b,k}$ *are given by*

$$G_{m,b,k}(z) = \frac{1}{p^m - p^{m-1}} \sum_{\zeta^{p^m}, \zeta \neq 1} \frac{\zeta^{-\log x / \log g}}{z + g^k \zeta - 1}.$$

As before, the proof is devided into two steps

1 μ is h-admissible measure.

2 The given function $f(z)$ coincides with the p-adic Mellin transform of μ on the set $\{\chi\Theta^k\}$, $\chi \mod p^m$, $k = 0, \ldots, h-1$.

Question 1. How we can construct a measure μ such that a p-adic holomophic function $f(z)$, increasing faster than $\log^h$, is the p-adic Mellin transform of μ?

Question 2. Let Φ be a form of weight $w + 2$. Then p-adic L-function associated to Φ is a p-adic holomophic function of class $0(\log^{w+1})$. When a p-adic holomophic function of class $0(\log^{w+1})$ is a p-adic L-function of a modular form? This is a p-adic version related to Hecke's theorem.

We relate this question also to the Shnirelman integral.

Definition 3.5. A function $f(z)$ in D is said to be in class $\mathcal{S}$ if for a, γ such that $D_a(|\gamma|) \subset D$, we have

$$\int_{a,\gamma} f(z)(z-a)dz = 0.$$

Example 3.6. Krasner analytic functions in D are in the class $\mathcal{S}$.

Let $f(\chi)$ be p-adic L-function associated to a modular form. Then f satisfies the following functional equations ([5]):

$$f(\chi\theta^k) = \epsilon(\chi_0)f(\chi^{-1}\theta^{\omega-k})$$

for all characters $\chi \in X(\mathbb{Z}_p^*)$, $\chi = \chi_0\chi_1$; $\chi_0 \mod p$, $\chi_1 \in X(U), \epsilon(\chi_0) = \chi_0^{-1}(-1)^{k+1}$.

For each $\chi_0 \mod p$ we can write functional equations in the form

$$f_{\chi_0}(\chi_1\theta^k) = \epsilon(\chi_0)f_{\chi_0^{-1}}(\chi_1^{-1}\theta^{\omega-k}).$$

Since $f_{\chi_0} \in 0(\log^{\omega+1})$, the functional equations hold if they hold for all Dirichet characters of conductor p^n, $n = 1, 2, ...$ Thus, the corresponding holomophic functions $F_{\chi_0}(z)$ in D satisfy the following equations:

$$F_{\chi_0}[(\zeta g^k - 1)] = \epsilon(\chi_0)F_{\chi_0}^{-1}(\zeta^{-1}g^{\omega-k} - 1)], \ k = 0, ...\omega,$$

where g is a topological generator of U, $\zeta^{p^n} = 1$, $n = 1, 2, ...$ Set

$$G_{F,\chi_0,\zeta}(z) = F_{\chi_0}(z)\frac{1}{z - (\zeta g^k - 1)} - \epsilon_{\chi_0}F_{\chi_0^{-1}}(z)\frac{1}{z - (\zeta^{-1}g^{m-k} - 1)}.$$

We have:

Theorem 3.7. For p-adic L-functions associated to modular forms, $G_{F,\chi_0,\zeta} \in \mathcal{S}$.

References

[1] Y. Amice. *Interpolation p-adique.* Bull. Soc.Math. Fr. 92, 117-180 (1964).

[2] Y. Amice and J. Vélu. *Distributions p-adiques associées aux séries de Hecke.* Journées arithmétiques de Bordeaux, Astérisque 24-25, 1975, 119-131.

[3] Ha Huy Khoai.*On p-adic Interpolation.* AMS translation: Math. Notes 26,1, 1980, 541-549.

[4] Mahler,K. *An interpolation series for continuous functions of a p-adic variable.* J. Reine Angew.Math. 199, 23-34 (1958).

[5] Y. Manin. *Parabolic points and zeta fonctions of modular curves.* Izv.AN SSSR. vol. 6 , 1. 1972, 19-64.

[6] L. G. Shnirelman. *On functions in normed algebraically closed division rings.* Isvestija AN SSSR 2 . (1938) 417- 498.

[7] Vishik. *Some applications of the Shnirelman integral in non-archimedean analysis*. Uspehi Math. Nuak. 34 (1979) 223-224.

10

REFINED ORDER AND THE PALEY PROBLEM IN VECTOR SPACES

Chuan-Gan Hu
Department of Mathematics, Nankai University,
Tianjin 300071, P.R.China

Chao Liu
Department of Mathematics, Nankai University,
Tianjin 300071, P.R.China

Abstract: In this paper, the concepts of a refined order and a type, and the solutions of the Paley problem are extended to the cases about vector-valued functions in finite and infinite dimensional spaces.

Keywords: refined order; vector-valued entire function; type; Paley problem.

AMS-classification: 30G35, 30G30, 46G20

1. Introduction

Assume the standard notations and the Nevanlinna theory of meromorphic functions to be known.

In 1932, Paley had formulated the following conjecture for every entire function $f(z)$ the inequality

$$\liminf_{r\to\infty} \frac{\log M(r,f)}{T(r,f)} \leq \begin{cases} \pi\rho/\sin\pi\rho \quad , \quad 0 \leq \rho \leq \frac{1}{2} & (1) \\ \pi\rho \quad , \quad \rho > \frac{1}{2} & (2) \end{cases}$$

holds. E.Valiron [19] and A.Wahlund [20] had proved the first inequality, so the open problem (which is called the Paley problem) is to prove inequality (2). In 1969, N.V.Govorov presented a complete solution of the Paley problem (see [2]). In 1997, C.G.Hu and C.C.Yang extended the Nevanlinna theory to the case of Hilbert spaces (see [8]). In 1998, C.Liu and C.G.Hu extended the Nevanlinna's first fundamental theorem to a vector-valued case with range in Hilbert spaces (see [16]). Later, in 1998, C.G.Hu extended the results of [8,16] to the cases of Banach spaces (see [5]). In this paper, the classical Paley problem is extended to vector-valued cases in finite and infinite dimensional spaces. The Paley problem is an important problem about entire functions with finite order. It gave the relations between the growth of moduli of analytic functions and characteristic functions using the method of the Riemann boundary value problem. So the research on this problem connects the boundary value problem of analytic functions and the value distribution theory, and develops the value distribution theory.

2. Refined order of vector-valued functions

Let E be a complex Banach space, and $\mathbb{C}$ a complex plane.

Definition 1. *A positive function $\rho(r)$, on the nonnegative real axis, is called a refined order if $\rho(r)$ satisfies*

$$\lim_{r\to\infty} \rho(r) = \rho, \qquad \lim_{r\to\infty} r\rho'(r)\log r = 0.$$

Definition 2. *Let f be a vector-valued entire function with range in E, and $M(r,f)$ $= \max_{|z|=r} \|f(z)\|$. If a refined order $\rho(r)$ satisfies*

$$\limsup_{r\to\infty} \frac{\log M(r,f)}{r^{\rho(r)}} = \sigma_f \quad (\in (0,\infty)),$$

then $\rho(r)$ is called the refined order of f, and σ_f is called the type under the refined order $\rho(r)$ of f (type for short).

Theorem 1. *If $M(r)$ $(r \geq 0)$ is a positive continuous function and*

$$\limsup_{r\to\infty} \frac{\log M(r)}{\log r} = \rho \quad (\in (0,\infty)),$$

then there are a $\rho(r)$ and a sequence r_n with $r_n \to \infty$ for $n \to \infty$ such that

$$M(r) \leq r^{\rho(r)}, r > 0$$

and

$$M(r_n) = r_n^{\rho(r_n)}.$$

Theorem 2. *If $f : \mathbb{C} \to E$ is a vector-valued entire function, and its Taylor's expansion is*

$$f(z) = \sum_{n=0}^{\infty} c_n z^n \qquad c_n \in E$$

with its refined order $\rho(r)$ $(r > 0)$ and type σ_f, then

$$\limsup_{n\to\infty} \varphi(n)\|c_n\|^{\frac{1}{n}} = (\sigma_f \rho e)^{\frac{1}{\rho}},$$

where $\varphi(t)$ is the inverse function which is defined by $t = r^{\rho(r)}$.

Proof Firstly, we show the existence of φ. In fact,

$$(r^{\rho(r)})' = r^{\rho(r)-1}[\rho(r) + r\rho'(r)\log(r)].$$

Because

$$\lim_{r\to\infty} [\rho(r) + r\rho'(r)\log r] = \rho \quad > 0,$$

$(r^{\rho(r)})' > 0$ for r large enough, i.e. $r^{\rho(r)}$ is a strictly monotone increasing function. It follows that its inverse function $r = \varphi(t)$ exists.

Next we prove

$$\lim_{t\to\infty} \frac{\varphi(kt)}{\varphi(t)} = k^{\frac{1}{\rho}}. \tag{3}$$

Differentiating the equality $\log t = \rho(r)\log r$, we derive

$$\frac{d\log t}{d\log r} = \rho'(r) r\log r + \rho(r).$$

It follows that

$$\lim_{r\to\infty} \frac{d\log t}{d\log r} = \rho.$$

The above formula can be rewitten as

$$\lim_{r\to\infty} \frac{d\log\varphi(t)}{d\log t} = \frac{1}{\rho}.$$

Thus

$$(\frac{1}{\rho} - \varepsilon)\frac{1}{t} < \frac{\varphi'(t)}{\varphi(t)} < (\frac{1}{\rho} + \varepsilon)\frac{1}{t}$$

for sufficiently large t.

Make the integral from t to kt on the above inequality. Then

$$(\frac{1}{\rho} - \varepsilon)\log k < \log\frac{\varphi(kt)}{\varphi(t)} < (\frac{1}{\rho} + \varepsilon)\log k.$$

It follows that

$$\lim_{t\to\infty} \frac{\varphi(kt)}{\varphi(t)} = k^{\frac{1}{\rho}}.$$

On the other hand, the Cauchy's inequality yields

$$\|c_n\| \le \frac{M(r,f)}{r^n}.$$

Hence

$$\log \|c_n\| < \sigma r^{\rho(r)} - n \log r \text{ for any } \sigma > \sigma_f.$$

Let r be the root of the equation $n = \sigma \rho r^{\rho(r)}$ $(\sigma > \sigma_f)$. It follows that

$$\log \|c_n\| < \frac{n}{\rho} - n \log \varphi(\frac{n}{\rho\sigma}),$$

i.e.

$$\log[\varphi(n)\|c_n\|^{\frac{1}{n}}] < \frac{1}{\rho} + \log \frac{\varphi(n)}{\varphi(\frac{n}{\rho\sigma})},$$

or

$$\varphi(n)\|c_n\|^{\frac{1}{n}} < e^{\frac{1}{\rho}} \cdot \frac{\varphi(n)}{\varphi(\frac{n}{\rho\sigma})}.$$

Thus

$$\limsup_{n\to\infty} \varphi(n)\|c_n\|^{\frac{1}{n}} \le (\sigma\rho e)^{\frac{1}{\rho}}$$

holds after take the limits of two sides on the above inequality. Because $\sigma(\ge \sigma_f)$ is arbitrary,

$$\limsup_{n\to\infty} \varphi(n)\|c_n\|^{\frac{1}{n}} \le (\sigma_f\rho e)^{\frac{1}{\rho}} \quad \text{for} \quad \sigma \ge \sigma_f.$$

Finally we prove the opposite inequality by contradiction. Let σ' satisfy the equality

$$\limsup_{n\to\infty} \varphi(n)\|c_n\|^{\frac{1}{n}} = (\sigma'\rho e)^{\frac{1}{\rho}}.$$

Then $\sigma' \ge \sigma_f$. In fact, if $\sigma' < \sigma_f$, we have

$$\|c_n\| < \left[\frac{(\sigma_1\rho e)^{\frac{1}{\rho}}}{\varphi(n)}\right]^n \qquad \text{for any} \quad \sigma_1 \quad \text{in} \quad (\sigma', \sigma_f).$$

Relation (3) implies

$$\|c_n\| < \left[\frac{e^{\frac{1}{\rho}}}{\varphi(\frac{n}{\sigma_1\rho})}\right]^n,$$

i.e.

$$\|c_n r^n\| < \left[\frac{e^{\frac{1}{\rho}}}{\varphi(\frac{n}{\sigma_1\rho})}\right]^n r^n \quad \text{for} \quad n > n_0.$$

Let $\mu(r) = \sup\{\|c_n\|r^n : n = 0, 1, 2, \cdots\}$. It follows that

$$\mu(r) < \sup_{n>n_0}\left[\frac{e^{\frac{1}{\rho}}}{\varphi(\frac{n}{\sigma_1\rho})}\right]^n r^n.$$

The maximum of the right-hand side in the above inequality is reached at $n = \sigma_1\rho r^{\rho(r)}$. This deduces

$$\mu(r) < e^{\sigma_1 r^{\rho(r)}}.$$

It follows that there is a constant A such that

$$M(r,f) < e^{Ar^{\rho_1}} \quad \text{for any } \rho_1 > \rho \text{ and } r \text{ large enough.}$$

Indeed, by the above inequality,

$$\|c_n\| \leq \frac{M(r,f)}{r^n} < e^{Ar^{\rho_1}} r^{-n}. \tag{4}$$

It is then easy to verify that the maximum of the right-hand side of (4) is reached at

$$r = \left(\frac{n}{A\rho_1}\right)^{\frac{1}{\rho_1}},$$

i.e.

$$\|c_n\| < \left(\frac{eA\rho_1}{n}\right)^{\frac{n}{\rho_1}}.$$

It follows that

$$\|c_n r^n\| < 2^{-n} \quad \text{for} \quad n > m_r = [2^{\rho_1} eA\rho_1 r^{\rho_1}].$$

Thereby

$$\begin{aligned} \|f(z)\| &= \|\sum_{n=0}^{\infty} c_n z^n\| \\ &\leq \sum_{n=0}^{m_r} \|c_n\| r^n + \sum_{n=m_r+1}^{\infty} 2^{-n} \\ &\leq (1+m_r)\mu(r) + 2^{-m_r}. \end{aligned}$$

So

$$\begin{aligned} M(r,f) &< (2+2^{\rho_1}eA\rho_1 r^{\rho_1})\mu(r) + 2^{-m_r} \\ &< (2+2^{\rho_1}eA\rho_1 r^{\rho_1})e^{\sigma_1 r^{\rho(r)}} + 2^{-m_r} \\ &< (2+2^{\rho_1}eA\rho_1 r^{\rho_1})e^{\sigma_1 r^{\rho(r)}}. \end{aligned}$$

One easily sees that $\sigma_f \leq \sigma_1$. This is in contradiction. Consequently $\sigma' \geq \sigma_f$, i.e.

$$\limsup_{n\to\infty} \varphi(n)\|c_n\|^{\frac{1}{n}} = (\sigma'\rho e)^{\frac{1}{\rho}} \geq (\sigma_f \rho e)^{\frac{1}{\rho}}.$$

This proves the assertion.

3. Relations between order and zeros

Let $n(r, f=0)(n(r,f))$ denote the number of zeros (poles) of a vector-valued meromorphic function on a disk $|z| \leq r$, counted with multiplicities, and

$$N(r,f) = \int_0^r \frac{n(t,f) - n(0,f)}{t} \mathrm{d}t + n(0,f)\log r,$$

and let

$$m(r,f) = \frac{1}{2\pi}\int_0^{2\pi} \log^+ \|f(re^{i\theta})\| \mathrm{d}\theta$$

and

$$T(r,f) = m(r,f) + N(r,f).$$

For an entire function we have the following conclusion.

Theorem 3. *Let $f\colon \mathbb{C} \to \mathbb{C}$ be an entire function, the order $\rho > 0$ does not be an integer, $\rho(r)$ a refined order satisfying $\lim_{r\to\infty} \rho(r) = \rho$, and let*

$$\Delta_f = \limsup_{r\to\infty} \frac{n(r, f=0)}{r^{\rho(r)}}.$$

(i) *If $\Delta_f = 0$, then $\sigma_f = 0$, i.e. f of the refined order $\rho(r)$ has the minimal type.*

(ii) *If $0 < \Delta_f < \infty$, then $0 < \sigma_f < \infty$, i.e. f of the refined order $\rho(r)$ has the normal type.*

(iii) *If $\Delta_f = \infty$, then $\sigma_f = \infty$, i.e. f of the refined order $\rho(r)$ has the maximal type.*

Note. Theorem 3 does not hold for vector-valued entire functions except for case (iii). In fact, for any complex-valued entire function f_0

of a non-integral order, $f = (f_0, f_0 + 1)\colon \mathbb{C} \to \mathbb{C}^2$ is an entire function. Obviously f has no zero. So $\Delta_f = 0$. But $\sigma_f = \sigma_{f_0}$ is arbitrary.

In order to show (iii) for the vector-valued case we need the following Lemma.

Lemma 1. *If $f\colon \mathbb{C} \to E$ is a compact projection entire function* (See [8]) *and $\|f(0)\| = 1$, then*

$$n(r, f = 0) \leq \log M(er, f).$$

Proof. Generalization of the vector-valued Jensen's formula in [16] using the method in [5] yields

$$\begin{aligned} n(r, f = 0) &\leq \int_r^{er} \frac{n(t, f = 0)}{t} dt \\ &\leq \int_0^{er} \frac{n(t, f = 0)}{t} dt \\ &\leq \frac{1}{2\pi} \int_0^{2\pi} \log \|f(ere^{i\theta}\| d\theta \\ &\leq \log M(er, f). \end{aligned}$$

This proof is finished.

Theorem 4. *If $f\colon \mathbb{C} \to E$ is an entire function of a non-integer order $\rho(r)$ and if the refined order $\rho(r)$ satisfies $\lim_{r\to\infty} \rho(r) = \rho$ and $\Delta_f = \infty$, then $\sigma_f = \infty$.*

Proof. If the above result does not hold, then $\sigma_f < \infty$. It follows that for any $\varepsilon > 0$

$$\log M(r, f) < (\sigma_f + \varepsilon) r^{\rho(r)}, r > r_0.$$

Lemma 1 implies

$$n(r, f = 0) < (\sigma_f + \varepsilon)(er)^{\rho(er)}.$$

From Lemma 5 in [2] we derive

$$\begin{aligned} (er)^{\rho(er)} &= e^\rho r^\rho L(er) \\ &< (1 + \varepsilon) e^r r^\rho L(r) \\ &= (1 + \varepsilon) e^\rho r^{\rho(r)}, \text{for } r > r_1, \end{aligned}$$

where $L(r) = r^{\rho(r) - r}$. Therefore

$$\frac{n(r, f = 0)}{r^{\rho(r)}} < (\sigma_f + \varepsilon)(1 + \varepsilon) e^\rho.$$

Take $r \to \infty$, we have

$$\Delta_f \leq (\sigma_f + \varepsilon)(1+\varepsilon)e^{\rho}.$$

Using the arbitrariness of ε, we deduce $\Delta_f \leq e^{\rho}\sigma_f$. Because $\Delta_f = \infty$, $\sigma_f = \infty$. This is in contradiction with $\sigma_f < \infty$ and the proof is complete.

4. Paley problem in $\mathbb{C}^n$

Let the norm $\|x\|$ of an element $x = (x_1, \cdots, x_n)$ in $\mathbb{C}^n$ be defined by $\max\{|x_1|, \cdots, |x_n|\}$.

Theorem 5. *If $f = (f_1, \cdots, f_n)\colon \mathbb{C} \to \mathbb{C}^n$ is an entire function of finite order $\rho > 0$, then*

$$\log M(r,f) \leq \lambda^2 r^{\lambda} \int_0^{\infty} \frac{T(t,f)t^{\lambda-1}}{(t^{\lambda}+r^{\lambda})^2}dt + c\lambda, \quad r > r_0, \tag{5}$$

for any $\lambda \geq \max\{2\rho, 1\}$, where c is a constant independent of r and λ.

Proof. Firstly, because

$$|f_j(z)| \leq \|f(z)\| \qquad \text{for } j = 1, \cdots, n,$$

the order of f_j is not greater than ρ.

Let z_0 satisfy $\|f(z_0)\| = M(r,f)$ and $\|z_0\| = r$, and let $|f_{j_0}(z_0)| = \|f(z_0)\|$ for any given $\lambda \geq \max\{2\rho, 1\}$ and r. Let $g(z) = f_{j_0}(z) + 1 - f_{j_0}(0)$. Then $g(z)$ is a general entire function satisfying $g(0) = 0$ and the order of g is not greater than ρ. It follows from formula (A.1.3) in p.232

in [3], that we derive for $r > r_0 (= \max\{r_j\}$, r_j corresponding to f_j):

$$\begin{aligned}
\log M(r, f_{j_0}) &\le \log[M(r,g) + |1 - f_{j_0}(0)|] \\
&\le \log M(r,g) + \log^+ |1 - f_{j_0}(0)| + \log 2 \\
&\le \lambda^2 r^\lambda \int_0^\infty \frac{T(t,g)t^{\lambda-1}}{(t^\lambda + r^\lambda)^2} dt + \log^+ |1 - f_{j_0}(0)| + \log 2 \\
&\le \lambda^2 r^\lambda \int_0^\infty \frac{[T(t, f_{j_0}) + \log^+ |1 - f_{j_0}(0)| + \log 2] t^{\lambda-1}}{(t^\lambda + r^\lambda)^2} dt \\
&\quad + \log^+ |1 - f_{j_0}(0)| + \log 2 \\
&\le \lambda^2 r^\lambda \int_0^\infty \frac{T(t,f)t^{\lambda-1}}{(t^\lambda + r^\lambda)^2} dt \\
&\quad + \lambda^2 r^\lambda \int_0^\infty \frac{[\log^+ |1 - f_{j_0}(0)| + \log 2] t^{\lambda-1}}{(t^\lambda + r^\lambda)^2} dt \\
&\quad + \log^+ |1 - f_{j_0}(0)| + \log 2 \\
&= \lambda^2 r^\lambda \int_0^\infty \frac{T(t,f)t^{\lambda-1}}{(t^\lambda + r^\lambda)^2} dt + \lambda(\log^+ |1 - f_{j_0}(0)| + \log 2) \\
&\quad + \log^+ |1 - f_{j_0}(0)| + \log 2 \\
&\le \lambda^2 r^\lambda \int_0^\infty \frac{T(t,f)t^{\lambda-1}}{(t^\lambda + r^\lambda)^2} dt + 2\lambda[\log(1 + \|f(0)\|) + \log 2] \\
&= \lambda^2 r^\lambda \int_0^\infty \frac{T(t,f)t^{\lambda-1}}{(t^\lambda + r^\lambda)^2} dt + c\lambda.
\end{aligned}$$

This completes the proof of Theorem 5.

Theorem 6 (Paley inequality). *Let $f : \mathbb{C} \to \mathbb{C}^n$ be a vector-valued entire function of finite order, and $\rho(r)$ a refined order of f, satisfying $\lim_{r\to\infty} \rho(r) = \rho > 0$. Let*

$$\Delta_M = \limsup_{r\to\infty} r^{-\rho(r)} \log M(r, f),$$

and

$$\Delta_T = \limsup_{r\to\infty} r^{-\rho(r)} T(r, f).$$

Then

$$\Delta_M \le \begin{cases} (\pi\rho/\sin\pi\rho)\Delta_T & , \quad 0 < \rho \le 1/2 \quad , \\ \pi\rho\Delta_T & , \quad \rho > 1/2 \quad . \end{cases}$$

Proof. For any given $\varepsilon > 0$ and $0 < \rho \le \frac{1}{2}$, we have $T(r, f) \le (\Delta_T + \varepsilon) r^{\rho(r)}$ as $r > r_0$ (large enough).

Let $\lambda = 1$ in (5). It follows that

$$\log M(r,f) \le r\int_0^\infty \frac{T(t,f)}{(r+t)^2}dt + c = \int_0^\infty \frac{T(sr,f)}{(1+s)^2}ds + c$$
$$\le T(r_0,f)\int_0^{r_0/r}\frac{ds}{(1+s)^2} + (\Delta_T+\varepsilon)\int_{r_0/r}^\infty (rs)^{\rho(rs)}\frac{ds}{(1+s)^2} + c.$$

Thus

$$\begin{aligned}\Delta_M &\le (\Delta_T+\varepsilon)\lim_{r\to\infty}\int_{r_0/r}^\infty \frac{(rs)^{\rho(rs)}}{r^{\rho(r)}(1+s)^2}ds \\ &= (\Delta_T+\varepsilon)\int_0^\infty \frac{s^\rho}{(1+s)^2}ds \\ &= (\Delta_T+\varepsilon)\frac{\pi\rho}{\sin\pi\rho},\end{aligned}$$

the properties of refined orders guarantee the commutativity of the above integral and limit.

For any given $\varepsilon > 0$ and $\rho > \frac{1}{2}$ we have

$$T(r,f) \le (\Delta_T+\varepsilon)r^{\rho(r)} \quad \text{as } r > r_0 .$$

Let $\lambda = 2\rho$ in (5). Then

$$\begin{aligned}\log M(r,f) &\le (2\rho)^2 r^{2\rho}\int_0^\infty \frac{T(t,f)t^{2\rho-1}}{(t^{2\rho}+r^{2\rho})^2}dt + 2c\rho \\ &= (2\rho)^2\int_0^\infty \frac{T(sr,f)s^{2\rho-1}}{(s^{2\rho}+1)^2}ds + 2c\rho \\ &\le (2\rho)^2 T(r_0,f)\int_0^{r_0/r}\frac{s^{2\rho-1}}{(s^{2\rho}+1)^2}ds \\ &\quad +(2\rho)^2(\Delta_T+\varepsilon)\int_{r_0/r}^\infty \frac{(sr)^{\rho(sr)}s^{2\rho-1}}{(s^{2\rho}+1)^2}ds + 2c\rho.\end{aligned}$$

It follows that

$$\begin{aligned}\Delta_M &\le (2\rho)^2(\Delta_T+\varepsilon)\lim_{r\to\infty}\int_{r_0/r}^\infty \frac{(sr)^{\rho(sr)}s^{2\rho-1}}{(s^{2\rho}+1)^2}ds \\ &= (2\rho)^2(\Delta_T+\varepsilon)\int_0^\infty \frac{s^{3\rho-1}}{(s^{2\rho}+1)^2}ds \\ &= (\Delta_T+\varepsilon)\pi\rho.\end{aligned}$$

This theorem is showed because of the arbitrariness of ε.

Note. Since an n-dimensional complex Banach space is isomorph to $\mathbb{C}^n$, the Paley inequality holds for all finite dimensional complex Banach spaces.

5. Paley problem in the space of convergent sequences

Let (c) be the space of convergent sequences, $f\colon \mathbb{C} \to (c)$ an entire function which is written as $f = (f_1, f_2, \cdots)$, where $\{f_n\}$ is of compact coordinate convergence to f_0 (see [8]).If $x = (x_1, x_2, ...) \in (c)$,then $\|x\|$ is defined by $\sup\{|x_n| : n = 1, 2, ...\}$.

Clearly, the sequence

$$\left\{\frac{\log M(r, f_n)}{r^{\rho(r)}}\right\}_{n=1}^{\infty}$$

is uniformly bounded, and satisfies

$$\frac{\log M(r, f_n)}{r^{\rho(r)}} \leq 1 \quad \text{for any } n$$

and

$$\lim_{n\to\infty} \frac{\log M(r, f_n)}{r^{\rho(r)}} = \frac{\log M(r, f_0)}{r^{\rho(r)}}. \tag{6}$$

If the above convergence in (6) is uniform on any non-negative real axis $(a, +\infty)$ $(a > 0)$, then $\{f_n\}$ is called the controlled uniform convergence.

Theorem 7. *If $f : \mathbb{C} \to (c)$ is an entire function of finite order,and if $\{f_n\}$ is the controlled uniform convergence, then the Paley inequality holds, i.e.*

$$\limsup_{r\to\infty} \frac{T(r, f)}{\log M(r, f)} \geq \begin{cases} \sin \pi\rho/\pi\rho & , \quad 0 < \rho \leq 1/2 \ , \\ 1/\pi\rho & , \quad \rho > 1/2 \ . \end{cases}$$

Proof. For any given $\varepsilon > 0$ as $0 < \rho \leq 1/2$ there is a R_T such that

$$T(r, f) \leq (\Delta_T + \varepsilon) r^{\rho(r)} \quad \text{for } r > R_T,$$

and there is a N such that

$$\left| \frac{\log M(r, f_n)}{r^{\rho(r)}} - \frac{\log M(r, f_0)}{r^{\rho(r)}} \right| < \varepsilon \quad \text{for } n > N \text{ and } r > a.$$

For $n = 0, 1, \cdots, N$ and $r > R_T$ as $\lambda \geq 1$ we have

$$\begin{aligned}
\log M(r, f_n) &\leq \log M(r, f_n + (1 - f_n(0))) + \log^+ |1 - f_n(0)| + \log 2 \\
&\leq \lambda^2 r^\lambda \int_0^\infty \frac{T(t, f_n) t^{\lambda - 1}}{(t^\lambda + r^\lambda)^2} dt + \lambda \log^+ |1 - f_n(0)| + \lambda \log 2 \\
&\quad + \log^+ |1 - f_n(0)| + \log 2 \\
&\leq \lambda^2 r^\lambda \int_0^\infty \frac{T(t, f_n) t^{\lambda - 1}}{(t^\lambda + r^\lambda)^2} dt + \lambda \log^+ (1 + \|f(0)\|) + \lambda \log 2 \\
&\quad + \lambda(\log^+(1 + \|f(0)\|) + \log 2) \\
&\leq \lambda^2 r^\lambda \int_0^\infty \frac{T(t, f_n) t^{\lambda - 1}}{(t^\lambda + r^\lambda)^2} dt + c\lambda, \qquad (7)
\end{aligned}$$

here $c = 2(\log^+(1 + \|f(0)\|) + \log 2)$.

Assume $\|f(z_r)\| = M(r, f)$. It follows that there is a non-negative integer n_r such that $\|f(z_r)\| = |f_{n_r}(z_r)|$.

Let $R = \max\{R_T, R_0, R_1, \cdots, R_N\}$ and $\lambda = 1$ in (7). We have

$$\begin{aligned}
\log M(r, f) &= \log M(r, f_{n_r}) \\
&\leq r \int_0^\infty \frac{T(t, f_{n_r})}{(r + t)^2} dt + c \\
&\leq r \int_0^\infty \frac{T(r, f)}{(r + t)^2} dt + c \quad \text{for } n_r \leq N;
\end{aligned}$$

and

$$\begin{aligned}
\log M(r, f) &= \log M(r, f_{r_n}) \\
&< \log M(r, f_0) + \varepsilon r^{\rho(r)} \\
&\leq r \int_0^\infty \frac{T(t, f_0)}{(r + t)^2} dt + c + \varepsilon r^{\rho(r)} \\
&\leq r \int_0^\infty \frac{T(t, f)}{(r + t)^2} dt + c + \varepsilon r^{\rho(r)} \quad \text{for } n_r > N.
\end{aligned}$$

Combining the above two cases, we deduce

$$\log M(r, f) \leq r \int_0^\infty \frac{T(t, f)}{(r + t)^2} dt + c + \varepsilon r^{\rho(r)}.$$

Consequently

$$\begin{aligned}\log M(r,f) &\le \int_0^\infty \frac{T(sr,f)}{(1+s)^2}ds + c + \varepsilon r^{\rho(r)}\\ &\le T(R,f)\int_0^{R/r}\frac{ds}{(1+s)^2}\\ &\quad +(\Delta_T+\varepsilon)\int_{R/r}^\infty \frac{(rs)^{\rho(rs)}}{(1+s)^2}ds + c + \varepsilon r^{\rho(r)}.\end{aligned}$$

Hence

$$\begin{aligned}\Delta_M &\le (\Delta_T+\varepsilon)\lim_{r\to\infty}\int_{R/r}^\infty \frac{(rs)^{\rho(rs)}}{r^{\rho(r)}(1+s)^2}ds+\varepsilon\\ &= (\Delta_T+\varepsilon)\int_0^\infty \lim_{r\to\infty}\frac{(rs)^{\rho(rs)}}{r^{\rho(r)}(1+s)^2}ds+\varepsilon\\ &= (\Delta_T+\varepsilon)\int_0^\infty \frac{s^\rho}{(1+s)^2}ds+\varepsilon\\ &= (\Delta_T+\varepsilon)\frac{\pi\rho}{\sin\pi\rho}+\varepsilon.\end{aligned}$$

From the arbitrariness of ε we derive

$$\Delta_M \le (\pi\rho/\sin\pi\rho)\Delta_T.$$

It is easily checked that

$$\limsup_{r\to\infty}\frac{T(r,f)}{\log M(r,f)} \ge \frac{\sin\pi\rho}{\pi\rho} \text{ for } 0<\rho\le 1/2.$$

A similar technique yields

$$\limsup_{r\to\infty}\frac{T(r,f)}{\log M(r,f)} \ge \frac{1}{\pi\rho} \text{ for } \rho>1/2.$$

Q.E.D.

6. Remark

Research on the Paley problem in vector spaces is begining for this paper. Some of its related problems remain to be resolved, for instance, the Paley problem on Banach spaces or locally convex spaces is much more difficult.

Acknowledgment

The authors would like to thank Prof. LE HUNG SON and Prof. CHUNG-CHUN YANG for their support in this work.

References

[1] C. Chuang and C.C. Yang, *The Fix-Points and Factorization Theory of Meromorphic Functions*, Peking University Press, Beijing,1988.

[2] N.V. Govorov, *Palye's hypothesis*, Functional Analysis and its Applications, 3 (1969), 41-45.

[3] N.V. Govorov, *Riemann's Boundary Problem with Infinite Index*, Birkhäuser Verlag, Basel, Boston, Berlin, 1994.

[4] C.G. Hu, *Vector-Valued Analysis and Its Aplications*, Nankai University Press, Tianjin, 1985.

[5] C.G. Hu, *Nevanlinna's theory in a Banach space*, in *"Proceedings of the Fifth International Colloquium on Finite or Infinite Dimensional Complex Analysis"*, 1998, 109-116.

[6] C.G. Hu, J.H. Jia and C. Liu, *Stability of solutions for singular integral equations for two classes in locally convex spaces*, in *"Finite or Infinite Dimensinal Complex Analysis"*, Lecture Notes in Pure and Applied Mathematics 214, Marcel Dekker, Inc.,New York,Basel,2000, 139-148.

[7] C.G. Hu and C.C. Yang, *Vector-Valued Functions and their Applications*, Kluwer Academic Publishers, Dordrecht, Boston, London, 1992.

[8] C.G. Hu and C.C. Yang, *Some remarks on Nevanlinna's theory in a Hilbert space*, Bulletin of the Hong Kong Mathematical Society, 1(1997), 267-272.

[9] C.G. Hu and C.C. Yang, *The Nevanlinna's second fundamental theorem in a Hilbert space*, in *"Recent Developments in Complex Analysis and Computer Algebra"*, Kluwer Academic Publishers, 1999, 373-384.

[10] C.G. Hu and X.F. Ye, *Some remarks on the Green-Residue theorem and differential polynomails in infinite dimensional spaces*, in *"Proceedings of the 6th International Colloquium on Complex Analysis"*, 1998, 81-86.

[11] C.G. Hu and X.F. Ye, *Nevanlinna's first main theorem for holomorphic Hermitian line bundles*, in *" Finite or Infinite Dimensional complex Analysis "*, Lecture Notes in Pure and Applied Mathematics 214, Marcel Dekker, Inc.,New York,Basel, 2000, 149-156.

[12] C.G. Hu, X.F. Ye and T.H. Yue, *Meromorphic maps and the index of a map in Banach spaces*, in *"Proceedings of the International Conference in Boundary Value Problems, Integral Equations and Related Problems"*, World Scientific Publishing Co. Pte. Ltd.,Singapore,New Jersey,London,Hong Kong, 2000, 78-85.

[13] C.G. Hu and T.H. Yue, *Normal families of holomorphic mappings*, J. Math. Anal. Appl., 171(1992), 436-447.

[14] J.M. Isidro and L.L. Stacho, *Holomorphic Automorphim Groups in Banach Spaces: An Elementary Introduction*, North-Holland Mathematics Studies 105, Amsterdam, New York, Oxford, 1985.

[15] B.Ja. Levin, *Distribution of Zeros of Entire Functions*, Amer.Math.Soc. Second Printing, 1972.

[16] C. Liu and C.G. Hu, *The Nevanlinna's first fundamental theorem in a Hilbert space*, Acta Scientiarum Neturalium Universitatis Nankaiensis, 31(1998), 1-14.

[17] J.K. Lu, *The Boundary Value Problems for Analytic Functions*, World Scientific, Singapore, New Jersey, London, Hong Kong, 1993.

[18] J. Mujica, *Complex Analysis in Banach Spaces*, in "North-Holland Math. Studies", Vol.120, North-Holland, Amsterdam, 1986.

[19] G. Valiron, *Sur un thèoremè de M.Wiman*, Opuscula Math.A.Wiman dedicate, 1930, 1-12.

[20] A. Wahlund, *Über einen Zusammenhang zwischen dem Jensensche Theoreme*, Arkiv för Mat., 21(1929), 1-34.

11

INTEGRAL REPRESENTATION OF P-ADIC FUNCTIONS

Le Thi Hoai Thu
Hanoi Institute of Mathemmatics
P.O. Box 631 Bo Ho, 10000 Hanoi, Vietnam
E-mail: hoaithu_99@yahoo.com

Abstract: Using the Shnirelman integral, we give an inverse formula for the p-adic Mellin-Mazur transform of a bounded measure and a h-addmissble measure. Then we apply this result to construct an integral representation of p-adic analytic functions, in particular we obtain the integral representation of Morita's p-adic Γ-function.

1. Introduction

One of basic problems of p-adic analysis and number theory is to construct p-adic analog of arithmetic functions. D. Barsky and M. Vishik proved that each Krasner analytic function in $\mathbb{C}_p - \mathbb{Z}_p^*$ subject to some growth conditions is the Cauchy transform of a measure on $\mathbb{Z}_p^*$, and they given an inverse formula for the Cauchy transform ([4] ,[11]). Here we consider similar problem for the Mellin-Mazur transform : Given a p-adic holomorphic function $f(z)$ in the unit disc D, we give a measure μ such that f is the Mellin-Mazur transform of μ, that means f is the p-adic L-function associated to μ. We give the answer in two cases:

1) f is a bounded function,
2) f is a function of the class $0(\log^h)$, $\quad h \geq 1$.

From that we obtain an integral representation of a p-adic analytic function in the unit disc, and we apply this result for the p-adic Γ-function.

2. Preliminaries

2.1 P-adic group of characters

Let p be a fixed prime number, $\mathbb{Q}_p$ the field of p-adic numbers , $\mathbb{C}_p$ the p-adic completion of the algebraic closure of $\mathbb{Q}_p$, and $\mathbb{C}_p^* = \mathbb{C}_p - \{0\}$. The absolute value in $\mathbb{C}_p$ is normalized so that $|p| = p^{-1}$. We use the notion $v(z)$ for the additive valuation on $\mathbb{C}_p$ which extends ord_p.

Let Δ_0 be an integer prime to p and let

$$q = \begin{cases} 4 & \text{if } p = 2 \\ p & \text{otherwise.} \end{cases}$$

We set $\Delta_0 q = \Delta$ and denote:

$$\mathbb{Z}_\Delta^* = \lim_{\longleftarrow}(\mathbb{Z}/\Delta p^m \mathbb{Z})^*.$$

The group of *$p-$adic characters* is the group of continuous holomorphisms of Z_Δ^* into $\mathbb{C}_p^*$:

$$X(\mathbb{Z}_\Delta^*) = \mathrm{Hom}_{cont}(\mathbb{Z}_\Delta^*, \mathbb{C}_p^*).$$

We set

$$U = 1 + q\mathbb{Z}_p = \{z \in \mathbb{Z}_p, \quad v(z-1) \geq v(q)\},$$

then, for every $g \in U$ such that $v(g-1) = v(q)$, the map $z \mapsto g^z$ is an isomorphism of $\mathbb{Z}_p$ onto U. We call g is a topological generator of the group U. For each generator g of the group U the map

$$X(U) = \mathrm{Hom}_{cont}(U, \mathbb{C}_p^*) \longrightarrow \mathbb{C}_p^*$$

transforming a continuous character χ of the group U into a point $\chi(g) - 1$ in the unit disc D of $\mathbb{C}_p$. Also we have the isomorphisms

$$\mathbb{Z}_\Delta^* \cong (\mathbb{Z}/\Delta_0 \mathbb{Z})^* \times \mathbb{Z}_p^* \tag{1}$$

$$\mathbb{Z}_p^* \cong (\mathbb{Z}/q\mathbb{Z})^* \times U$$

From the isomorphisms (1) it follows that $X(\mathbb{Z}_\Delta^*)$ is a product of a finite group and $X(U)$, while the last is isomorphic to D. Since D is an open disc of $\mathbb{C}_p$, this isomorphism makes $X(\mathbb{Z}_\Delta^*)$ into an analytic group. A function $f(\chi)$ is said to be holomorphic function on the analytic group $X(\mathbb{Z}_\Delta^*)$ if its restriction on each component isomorphic to D is a holomorphic function. Thus, we can regard every holomorphic function in the group $X(\mathbb{Z}_\Delta^*)$ as a holomorphic function in the unit disc D. Note that each Dirichlet character χ of conductor Δp^m is an element of the group $\mathrm{Hom}((\mathbb{Z}/\Delta p^m\mathbb{Z})^*, \mathbb{C}_p^*)$ for each $m \geq n$ and is prolonged to an unique element of the group $X(\mathbb{Z}_\Delta^*)$ which is again denoted by χ. Thus, the torsion subgroup of the group $X(\mathbb{Z}_\Delta^*)$ is identified with the group of Dirichlet characters of conductors dp^n where $d|\Delta_0$. Let χ be such a character and let $\chi = \chi_0\chi_1$ be its decomposition in $\chi_0 \in X(\mathbb{Z}/\Delta\mathbb{Z})^*$ and $\chi_1 \in X(U)$. Then χ_1 takes values in the group $\mu_{p^\infty} = \bigcup_n \mu_{p^n}$ where by μ_{p^n} denote the group of p^n-roots of unity in $\mathbb{C}_p$. On the other hand, every $x \in \mathbb{Z}_\Delta^*$ can be written in the form

$$x = \alpha(x)\theta(x)$$

with $\alpha(x) \in (\mathbb{Z}/\Delta_0\mathbb{Z})^*$ and $\theta(x) \in \mathbb{Z}_p^*$. Then map $\theta :\ x \mapsto \theta(x)$ is an element of $X(\mathbb{Z}_\Delta^*)$ which is called *the fundamental character.*

2.2 The Mellin transform

Definition. Let μ be a measure on $\mathbb{Z}_p^*$, i.e μ is a continuous linear functional with values in $\mathbb{C}_p$ on the space of continuous functions in $\mathbb{Z}_p^*$. Then the restriction of μ on the analytic group $X(\mathbb{Z}_p^*)$ gives a function

$$L(\mu, \chi) = \int_{\mathbb{Z}_p^*} \chi d\mu.$$

The function $L(\mu, \chi)$ is called the *p-adic Mellin-Mazur transform* of the measure μ.

Example. p-adic L-functions associated to modular form are p-adic Mellin transforms of measures associated to modular forms ([2], [11]).

3. The inverse Mellin-Mazur transform

3.1 The Mellin-Mazur transform of a bounded measure.

It is well-known that the Mellin-Mazur transform of a bounded measure on $\mathbb{Z}_p^*$ is a holomorphic function in the unit disc. In this section, for each holomorphic function $f(z)$ in the unit disc, we are going to give a bounded measure μ such that $f(z)$ is the Mellin transform of μ. As an application we have an integral representation of Morita's p-adic Γ-function. The main tool is the Shnirelman integral. Note that the Shnirelman integral is used by Vishik to find an inverse formula for the Stieltjes transform ([11],[7]).

Let

$$F(\chi) = \int_{\mathbb{Z}_\Delta^*} \chi d\mu.$$

be the p-adic Mellin transform of a measure μ. Then by the isomorphism (1) we can regard F as a holomorphic function on the analytic group $X(\mathbb{Z}_\Delta^*)$. This means that for every character $\chi_0 \in X(\mathbb{Z}/\Delta\mathbb{Z})^*$ we have a holomorphic function $F_{\chi_0}(\chi_1)$ on the group $X(U)$. Thus every p-adic Mellin transform on $X(\mathbb{Z}_\Delta^*)$ corresponds to a collection of holomorphic functions on the group $X(U)$. Now let $f(\chi_1)$ be such a " branch" of the function $F(\chi)$. Let g be a fixed topological generator of the group U. We set

$$z = \chi_1(g) - 1.$$

For each $x \in U$ we have $x = g^{\log x/\log g}$ and hence that

$$\chi_1(x) = (1+z)^{\log x/\log g}.$$

Then the p-adic Mellin transform of the measure μ corresponds to the function

$$f(z) = \int_U (1+z)^{\log x/\log g} d\mu(x). \tag{2}$$

Thus for every measure μ on U, the Mellin transform given by formula (2) is a holomorphic function in the unit disc D. Conversly, we have

Theorem 1. *Let $f(z)$ be a bounded holomorphic function in the unit disc D. For $x \in U$ and $m = 1, 2, \ldots$ we consider the following functions in D*

$$G_{m,x}(z) = \frac{1}{p^m - p^{m-1}} \sum_{\xi^{p^m}=1\ \xi\neq 1} \xi^{-\log x/\log g} \frac{1}{z+1-\xi}.$$

Then the function $f(z)$ is the Mellin-Mazur transform of the following measure on U

$$\mu\{x + U_m\} = \int_{a,\gamma} f(z) G_{m,x}(z)(z-a)dz \tag{3}$$

where $U_m = 1 + p^m U$, $a, \gamma \in \mathbb{C}_p$ such that $D(a, |\gamma|) \subset D$ and the points $1 - \xi$ belong to this disc.

Proof. We first show formula (3) defines a bounded measure on U. We have:

$$\sum_{k=0}^{p-1} \mu\{x + kp^m + (U_{m+1})\} = \int_{a,\gamma} f(z) \sum_{k=0}^{p-1} G_{m+1,x+kp^m}(z)(z-a)dz$$

Note that:

$$\sum_{k=0}^{p-1} G_{m+1,x+kp^m}(z) = \sum_{k=0}^{p-1} \frac{1}{p^{m+1}-p^m} \sum_{\xi^{p^{m+1}}=1, \xi\neq 1} \xi^{-\log x/\log g} \frac{1}{z+1-\xi}$$
$$= G_{m,x}(z)$$

From this it follows that

$$\mu\{x+(U_m)\} = \sum_{k=0}^{p-1} \mu\{x+kp^m+(U_{m+1})\}.$$

Obviously, the function $f(z)_{m,x}G(z)$ is the quotion of Krasner analytic functions in the disc $D(a,|\gamma|)$. Then we have:

$$\begin{aligned}\mu\{x+(U_m)\} &= \sum_{\xi^{p^m}=1,\ \xi\neq 1} \operatorname{res}_{\xi-1}(f, G_{m,x}) \\ &= \frac{1}{p^m - p^{m-1}} \sum_{\xi^{p^m}=1,\ \xi\neq 1} f(\xi-1)\xi^{-\log x/\log g}.\end{aligned}$$

From this formula, the boundedness of μ will be proved in Theorem 2. Indeed we will concern with more general situations.

It remains to prove that $f(z)$ is the p-adic Mellin transform of the measure μ.

We set

$$F(z) = \int_U (1+z)^{\log x/\log g} d\mu(x).$$

Since μ is bounded, $F(z)$ is a bounded holomorphic function, $f(z)$ and $F(z)$ are coincide on the set $\{\xi - 1\}$, where ξ are roots of unity of degree p^m, $m = 1, 2\ldots$ ([2], [5],[6]).

We have:

$$\begin{aligned}F(\xi-1) &= \int_U \xi^{\log x/\log g} d\mu(x) \\ &= \sum_{\xi^{p^m}=1,\ \xi\neq 1} (\xi^{-1})\zeta^{-\log x/\log g} f(\zeta-1) \\ &= f(\xi-1).\end{aligned}$$

Theorem is proved.

■

3.2 The Mellin-Mazur transform of non-bounded measure

Y. Amice and J. Vélu and M. Vishik defined the p-adic Mellin transform of non-bounded measures and applied it to construct p-adic L-functions associated to modular forms ([2], [11]). We recall here the definition and give an inverse formula for the p-adic Mellin transform of so called *h-addmissible measures* . These measures are defined on $\mathbb{Z}_\Delta^*$ and corresponding Mellin transforms are holomorphic functions on the group $X(\mathbb{Z}_\Delta^*)$. As in the case of bounded measures we consider the measures on U and corresponding holomorphic functions on the unit disc D.

Definition. Let f and g be two holomorphic functions in the unit disc D. We say f *is of class* $o(g)$ if

$$\sup_{|z|\leq r}|f(z)| = o(\sup_{|z|\leq r}(|g(z)|))$$

where $r \to 1-0$.

Definition. A *h-addmissible* on U is a linear functional on the space of functions on U which are locally polynomials of degrees less than h and satisfy the following condition:

$$\left|\mu\{(x-b)^k\psi_{b,m}\}\right| = 0(p^{(h-k)m}), \quad k = 0, 1, \ldots, h-1,$$

where $\psi_{b,m}$ is the characteristic function of the set $b + U_m$.

It is proved in [11] that a *h-addmissible* measure is prolonged to a continuous linear functional on the space of $(h-1)$-differentiable functions whose derivates of order $h-1$ satisfy the Lipschitz condition. The restriction of such a functional on the group $X(U)$ is a holomorphic function of class $0(\log^h)$ and called *the Mellin transform* of the measure μ. The class of such measures contains, for example, the measures associated to modular forms.

Obviously, a *h-addmissible measure* is defined by giving its values on the set $\{x^k\psi_{b,m}\}$ with $b \in U$, $m = 1, 2, \cdots$ and $\psi_{b,m}$ is the characteristic function of the set $b + U_m$.

Theorem 2. *Let $f(z)$ be a holomorphic function of class $0(\log^h)$ in the unit disc D. Then $f(z)$ is p-adic Mellin transform of the following h-addmissible measure on U:*

$$\mu\{x^k\psi_{b,m}\} = \int_{a,m} f(z)G_{m,b,k}(z)(z-a)dz \tag{4}$$

where $m = 1, 2, \cdots$; $k = 0, 1, \cdots h-1$ and $G_{m,b,k}$ are given by the following formula:

$$G_{m,b,k}(z) = \frac{1}{p^m - p^{m-1}} \sum_{\xi^{p^m}=1, \xi \neq 1} \frac{\xi^{-\log b/\log g}}{z + g^k\xi - 1}$$

We recall that g is a fixed topological generator of the group $X(U)$, $v(g-1) = v(q)$.

Proof. It is easy to see that the formula (4) correctly defines a linear functional μ on the space of functions on $\mathbb{Z}_p$ which are locally polynomials of degrees less than h. To show that μ satisfies the conditions of *h-addmissiblility,* we note that:

$$\mu\{x^k \psi_{b,m}\} = \frac{1}{p^m - p^{m-1}} \sum_{\xi^{p^m}=1, \xi \neq 1} \xi^{-\log b/\log g} + f(g^k\xi - 1)$$

We first recall some notations. For each holomorphic function $f(z)$ on D and each $t_0 > 0$ we set

$$||f||_{t_0} = \sup_{v(z)=t_0} |f(z)|.$$

Then we obtain:

$$||\log^h(1+z)||_{t_m} = p^{mh},$$

where $t_m = \dfrac{1}{p^m - p^{m-1}}$, $m = 1, 2, \ldots,$ ($||\log^h(1+z)||$ is calculated by the Newton polygon ([5]). From the hypothesis we have:

$$||f||_{t_m} = 0(p^{mh}) \quad (m \to \infty).$$

Let u be the sequence $\{g^i\xi - 1\}$, $i = 0, 1, \cdots h-1$, where$\{\xi\}$ is the sequence of primitive roots of unity of degree p^m $(m = 1, 2, \cdots .)$ Since the function $f(z)$ is of class $0(\log^h)$, one infers u is an interpolating sequence of $f(z)$ ([2], [5]). We denote $\{S_m\}$ the sequence of Lagrange's interpolation polynomials for the function $f(z)$ on the sequence u. Then $\{S_m(z)\}$ is defined by the following conditions:

$$deg S_m(z) \leq hp^m - 1$$

$$S_m(g^i\xi - 1) = f(g^i\xi - 1), \quad i = 0, 1, \cdots, h-1.$$

By Lazard's lemma ([8]) we represent $f(z)$ in the form:

$$f(z) = \phi(z) \prod_{\xi \in \mu_m, i=0,1,\cdots h-1} \frac{1-z}{g^i\xi - 1} + Q_m(z). \tag{5}$$

Where μ_m is the set of primitive roots of unity of degree p^m, $Q_m(z)$ are polynomials of order hp^m satisfying the condition:

$$||Q_m||_{t_m} \leq ||f||_{t_m}$$

Since the representation (5) is unique, we have $S_m(z) \equiv Q_m(z)$, and hence that

$$||S_m||_{t_m} \leq ||f||_{t_m}.$$

From this it follows

$$||S_m||_{t_m} = 0(p^{mh}).$$

Supposing $S_m(z)$ is written in the form

$$S_m(z) = \sum_{l=0}^{hp^m-1} b_l(m) z^l$$

we have then

$$\begin{aligned} ||S_m||_{t_m} &= \max_{0 \leq l \leq hp^m} \{|b_l(m) z^l|t_m\} \\ &= \max_l \{|b_l(m)| p^{-1/(p^m - p^{m-1})}\} \\ &> p^{-hp/(p-1)} \max_l \{|b_l(m)\}. \end{aligned}$$

Thus we have $\max|b_l(m)| = 0(p^{mh}) \quad (m \to \infty)$. Note that if we write

$$S_m(z-1) = \sum_{l=0}^{hp^m-1} a_l(m) z^l,$$

then we obtain also $\max|a_l(m)| = 0(p^{mh})$. By definition of the measure μ we have:

$$\begin{aligned}
&\mu\{(x-b)^k\psi_{b,m}\} = \\
&= \sum_{j=0}^{k}(-b)^{k-j}\binom{k}{j}\frac{1}{p^m - p^{m-1}}\sum_{\xi^{p^m}=1,\ \xi\neq 1}\xi^{-\log b/\log g} f(g^j\xi - 1) \\
&= \sum_{j=0}^{k}(-b)^{k-j}\binom{k}{j}\frac{1}{p^m - p^{m-1}}\sum_{\xi^{p^m}=1,\ \xi\neq 1}\xi^{-\log b/\log g} S_m(g^j\xi - 1) \\
&= \sum_{j=0}^{k}(-b)^{k-j}\binom{k}{j}\frac{1}{p^m - p^{m-1}}\sum_{\xi^{p^m}=1,\ \xi\neq 1}\xi^{-\log b/\log g}\sum_{l=0}^{p^m-1} a_l(m)g^{jl}\xi^l \\
&= \sum_{j=0}^{k}(-b)^{k-j}\binom{k}{j}\sum_{l=0}^{p^m-1} a_l(m)g^{jl}\frac{1}{p^m - p^{m-1}}\sum_{\xi^{p^m}=1,\ \xi\neq 1}\xi^{l-\log b/\log g} \\
&= \sum_{j=0}^{k}(-b)^{k-j}\binom{k}{j}\sum_{l=0}^{p^m-1} a_l(m)g^{jl} \\
&= \sum_{g^l\equiv b} a_l(m)(g^l - a)^j.
\end{aligned}$$

Since $\max|a_l(m)| = 0(p^{mh})$. From this it follows the *h-addmissibility* of μ. ■

4. Integral representation of Morita's p-adic Γ-function

We now apply Theorem 1 to Morita's p-adic Γ- function. In [3] it is proved that we may consider the function $\Gamma_p(x)$ as restriction on $\mathbb{Z}_p$ of a locally analytic function $\Gamma_p(z)$ of local analyticity ratio $\rho = p^{-\frac{1}{p}-\frac{1}{p-1}}$. This means for each point $x \in \mathbb{Z}_p$ there exists ρ_x such that on $D(x,\rho_x)\cap\mathbb{Z}_p$ the function $\Gamma_p(x)$ is the restriction of $F(x) = \sum_{n>0} a_n(z-x)^n$ which is holomorphic on $D(x,\rho_x)$. The local analyticity ratio, by definition, is the number

$$\rho = \inf_{x\in\mathbb{Z}_p}\rho_x > 0.$$

Thus, in the disc $D(0, p^{-\frac{1}{p}-\frac{1}{p-1}})$ the function $\Gamma_p(x)$ is represented by a convergent power series. We set

$$f(z) = \Gamma_p(p^{-\frac{1}{p}-\frac{1}{p-1}}z),$$

then $f(z)$ is a bounded holomorphic function on the unit disc D. We have an integral representation of the function $f(z)$:

$$f(z) = \int_U (1+z)^{\log x/\log g} d\mu,$$

where the measure μ is defined by the formula (3). Hence, for Morita's p-adic Γ-function we have the following integral representation:

$$\Gamma_p(z) = \int_U (1+\alpha z)^{\log x/\log g} d\mu, \tag{6}$$

where $\alpha = p^{\frac{1}{p}+\frac{1}{p-1}}$ and the measure μ is defined by the formula:

$$\mu\{x+(p^m U)\} = \frac{1}{p^m - p^{m-1}} \int_{a,\alpha} \Gamma_p(\alpha z) G_{m,x}(z)(z-a)dz.$$

References

[1] Y. Amice. *Interpolation p-adique.* Bull. Soc.Math. Fr. 92, 117-180 (1964).

[2] Y. Amice and J. Vélu. *Distributions p-adiques associées aux séries de Hecke.* Journées arithmétiques de Bordeaux, Astérisque 24-25, 1975, 119-131.

[3] D. Barsky. *On Morita's p-adic Γ-function.* Groupe d' Etude d'Analyse ultrme'trique, Paris-1977.

[4] D. Barsky. *Transformation de Cauchy p-adique et algèbre d'Iwasawa.* Math. Ann. 232. 1978. 255-266.

[5] Ha Huy Khoai. *On p-adic Interpolation.* AMS translation: Math. Notes 26,1, 1980, 541-549.

[6] Ha Huy Khoai. *On p-adic meromorphic functions.* Duke Math. J, 50, 1983, 695-711.

[7] N. Koblitz. *p-adic Analysis.*London Mathematical Society Lecture Note Series.

[8] M. Lazard. *Les zéros d'une fonction analytique d'une variable sur un corps value complet.* Publ. Math. IHES, N^0 14.

[9] Y. Morita. *A p-adic analogue of the Γ-function,* J. Fac. Sci. Univ. Tokyo, 22, 1975, 255-266.

[10] L.G. Shnirelman. *On functions in normed algebraicaly closed division rings.* Izv. AN SSSR 2, 1938, 487-498.

[11] M.M. Vishik. *Non-archimedean measures connected with Dirichlet series.* Math. Sb. 99, 1976, 248-260.

[12] M. M. Vishik. *Some applications of the Schnirelman intergral in non-archimedean analysis.* Uspehi Math. Nuak. 34 (1979) 223-224.

12

MALMQUIST-YOSIDA TYPE THEOREMS FOR ALGEBRAIC DIFFERENTIAL EQUATIONS

Liangwen Liao
Department of Mathematics
Nanjing University
Nanjing, China

Chung-Chun Yang
Department of Mathematics
The Hong Kong University of Science & Technology
Clear Water Bay, Kowloon, Hong Kong

Abstract: In this paper, we discussed the Malmquist-Yosida type theorems for algebraic differential equations.

Keywords: Malmquist-Yosida type theorem, algebraic differential equations.

AMS-classification: Primary 33E30, 30D35, 30D99.

1. Malmquist-Yosida type theorems for the first-order algebraic differential equations

The Malmquist-Yosida type theorem is to investigate when an algebraic differential equation can be reduced into some standard forms if the algebraic differential equation admits a transcendental meromorphic solution. For first order algebraic differential equations, Malmquist [12] proved the following theorem in 1913,

Theorem 1. *Let $R(z, f)$ be birational. If a differential equation of the form*

$$f'(z) = R(z, f(z)),$$

admits a transcendental meromorphic solution, then the equation can be reduced into a Riccati differential equation

$$f'(z) = a_0(z) + a_1(z)f(z) + a_2(z)f^2(z),$$

where $a_i(z)(i = 0, 1, 2)$ are rational functions.

Malmquist's proof was independent of the Nevanlinna theory. In 1933, Yosida [16] proved the Malmquist's theorem by using the Nevanlinna theory. In seventies, Laine [7], Yang [15] and Hille [6] generalized the Malmquist's theorem into the case of $R(z, f)$ rational in f with meromorphic coefficients. Steinmetz [14] extended Theorem A into the following result in his thesis:

Theorem 2. *Let $R(z, f)$ be birational. If the following differential equation*

$$f'(z)^n = R(z, f(z)),$$

admits a transcendental meromorphic solution, then the differential equation can be reduced into

$$f'(z)^n = \sum_{i=0}^{2n} a_i(z) f(z)^i,$$

where $a_i(z)(i = 0, 1, ..., 2n)$ are rational functions and at least one of them does not vanish.

Theorem 2 was finally extended to the case of $R(z, f)$ rational with meromorphic coefficients by Rieth [13] and He-Laine [5].

2. Malmquist-Yosida type theorems for the second-order algebraic differential equations

Malmquist-Yosida type theorems for some second order algebraic differential equations of special forms, for instance $f'' = R(z, f)$, has been obtained, see [8]. However, Malmquist-Yosida type theorem for an arbitrary second order algebraic differential equation remains open. For the second-order differential equations

$$f'' = R(z, f, f'), \tag{1}$$

where R is rational in z, f and f', a classical and yet unsolved conjecture is the following:

Conjecture. *(see [8], p251) If the equation (1) has a transcendental meromorphic solution, then the equation can be transformed into the form:*

$$f'' = L(z,f)(f')^2 + M(z,f)f' + N(z,f), \tag{2}$$

where L, M, N are rational functions in their arguments.

Liao-Yang [10] considered the growth of the meromorphic solutions of the equations (2) and obtained the following results:

Theorem A. *[10] Let f be a solution of (2). Further assume that $L(z,f) \not\equiv 0$ in the equation (2) has the form:*

$$L(z,f) = \frac{P(z,f)}{Q(z,f)} = \frac{a_n(z)f^n + a_{n-1}(z)f^{n-1} + \dots + a_s(z)f^s}{b_m(z)f^m + b_{m-1}(z)f^{m-1} + \dots + b_r(z)f^r},$$

where $a_i(z), b_j(z) (s \le i \le n, r \le j \le m)$ are rational functions. If $m - n < 1$ or $r - s > 1$, then $\rho(f) < \infty$.

Theorem B. *[10] Let f be a transcendental entire solution of the following equation*

$$f'' = M(z,f)f' + N(z,f),$$

where $N(z,f)$, $M(z,f)$ are polynomials in f with rational functions as the coefficients. Then the order of f is a positive integral multiple of $\frac{1}{2}$.

Bergweiler [4] studied the growth of the meromorphic solution of the first-order algebraic differential equations, by using Zalcman's Lemma in Normal families. In the same paper, he also discussed the growth of the meromorphic solutions of the differential equations (2). Also, the growth estimates of meromorphic solutions of certain classes of higher order differential equations were discussed and obtained by Barsegian ([1], [2], [3]), by using arguments, which are quite different from Bergweiler's. In this paper, we shall partially prove the conjecture as follows.

Theorem 3. *If the second order differential equation (1) admits a meromorphic solution f of infinite order, then, either f satisfies a second order algebraic differential equation of the form*

$$f'' = L(z,f)(f')^2,$$

or the equation (1) reduces into

$$f'' = L_2(z,f)(f')^2 + L_1(z,f)f' + L_0(z,f), \tag{3}$$

where $L_i(z,f) (i = 0,1,2)$ are rational functions in z and f.

Remark 3. *The meromorphic solutions of the equation of form (3) may be of finite order or infinite order. For example:*

$$f'' = (f')^2 + f^2 - f - 1,$$

admits the solution $f(z) = \sin z$, *and* $f(z) = e^{e^z}$ *satisfies the following algebraic differential equation*

$$f'' = \frac{(f')^2}{f} - f'.$$

Open question: *Under what forms or additional conditions on* $L_i(z, f)$ *in the equation (3) can assure that all the possible non-rational meromorphic solutions of the equation are of finite order or infinite order, respectively ?*

In this paper, we denote the spherical derivative of meromorphic function f by $f^{\#}(z)$,

$$f^{\#}(z) = \frac{|f'(z)|}{1 + |f(z)|^2},$$

and

$$A(r, f) = \frac{1}{\pi} \iint_{|z|\leq r} [f^{\#}(z)]^2 dxdy,$$

The following lemmas will be needed in the proof of our result.

Lemma 2. *(Zalcman [17]) Let* $\mathcal{F}$ *be a family of meromorphic functions on D:* $|z| < 1$ *and* α *be a real number satisfying* $-1 < \alpha < 1$. *Then* $\mathcal{F}$ *is not normal in D iff there exist*

(i) a number r, $0 < r < 1$;

(ii) a sequence points z_k, $|z_k| < r$;

(iii) a positive sequence ρ_k, $\rho_k \to 0$ *and*

(iv) a sequence $\{f\}_{n\in N} \subset \mathcal{F}$, *such that* $\rho_k^{\alpha} f_k(z_k + \rho_k\zeta) \to g(\zeta)$ *spherically uniformly on compact subsets of* $\mathbb{C}$, *where* g *is a non-constant meromorphic function. Specially, One can choose* w_k *and* ρ_k *properly such that,*

$$\rho_k \leq \frac{2}{f_k^{\#}(w_k)^{\frac{1}{1+|\alpha|}}}, \quad f_k^{\#}(w_k) \geq f_k^{\#}(0).$$

Now following Makhmutov [11] , W_p will denote the classes of meromorphic functions

$$W_p = \{f(z) \mid \overline{\lim_{|z|\to+\infty}} |z|^{2-p} f^{\#}(z) < \infty\}, \quad (p \geq 1).$$

Lemma 3. *(Makhmutov [11]) A meromorphic function $f(z)$ on $\mathbb{C}$ is not in W_p, $p \geq 1$, if and only if there exist a sequence $\{z_n\}$ with $\lim_{n\to\infty} z_n = \infty$ and a sequence of positive numbers $\{\varepsilon_n\}$ with $\lim_{n\to\infty} \varepsilon_n = 0$ such that $g_n(z) = f(z_n + \varepsilon_n |z_n|^{2-p} z)$ converges uniformly to a non-constant meromorphic function $g(z)$ on compact subsets of $\mathbb{C}$.*

The following Lemma was proved in [9], we repeat its proof here for the convenience and interest of the reader.

Lemma 4. *If $f \in W_p$, then $\rho(f) \leq 2p - 2$.*

Proof: In fact, if $f \in W_p$, then we can find a constant M such that

$$f^{\#}(z) \leq M|z|^{p-2}.$$

If $p = 1$, then $A(r, f) = \frac{1}{\pi} \int\int_{|z|\leq r} [f^{\#}(z)]^2 dxdy \leq 2M \log r$. Thus $T(r, f) \leq M (\log r)^2$, and $\rho(f) = 0$.

If $p > 1$, then

$$A(r, f) = \frac{1}{\pi} \int\int_{|z|\leq r} [f^{\#}(z)]^2 dxdy \leq \frac{2M}{2p-2} r^{2p-2},$$

$$T(r, f) \leq \frac{2M}{(2p-2)^2} r^{2p-2}.$$

Lemma 5. *(Makhmutov [11]) If for any five values $a_1, a_2, a_3, a_4, a_5 \in \hat{\mathbb{C}}$,*

$$sup\{|z|^{2-p} f^{\#}(z); z \in \mathbb{C}, f(z) = a_k, k = 1, 2, 3, 4, 5\} < \infty,$$

then $f(z) \in W_p, p > 1$.

The following lemma seems to provide a useful and new tool in studying the growth of solutions of differential equations in complex domain.

Lemma 6. *(Barsegian [1], [2] and [3]) Suppose w is a meromorphic function in $\mathbb{C}$, $a_v (1 \leq v \leq q, q > 4)$ are different complex constants, and $\psi(r)$ is a monotonic decreasing function on $[0, \infty)$ with $\psi(r) \to 0$ as $r \to \infty$. Then there exists a set E of finite logarithmic measure in $[0, \infty)$, such that for every $r \notin E$ there exists a subset*

$$\{z_j(a_n) | n = 1, 2, ..., q, j = 1, 2, ..., n_0(r, a_n)\}$$

of the a_n-points of w in $|z| \leq r$ such that

$$\sum_{n=1}^{q} n_0(r, a_n) \geq (q-4)A(r) - o|A(r)|, r \to \infty, r \notin E, \tag{4}$$

$$|w'(z_j(a_n))| \geq \psi(r) A^{\frac{1}{2}}(r) r^{-1},$$
$$n = 1, 2, ..., q, j = 0, 1, 2, ..., n_0(r, a_n), r \notin E. \tag{5}$$

Moreover, for any integer $k \geq 2$ there exists a constant

$$C = C(k, a_1, ..., a_q) < \infty$$

such that

$$|w^{(k)}(z_j(a_n))| \leq C |w'(z_j(a_n))|^k,$$
$$n = 1, 2, ..., q, j = 0, 1, 2, ..., n_0(r, a_n), r \notin E. \tag{6}$$

The inequalities (4) and (5) were proved in [1], the inequality (6) was proved in [3] and the present lemma is stated as in [2].

3. Proof of Theorem 3

Proof. Let

$$R(z, f, f') = \frac{L_n(z, f)(f')^n + ... + L_1(z, f) f' + L_0(z, f)}{M_m(z, f)(f')^m + ... + M_1(z, f) f' + M_0(z, f)},$$

where $L_i(z, f)(i = 0, 1, ..., n), M_j(z, , f)(j = 0, 1, ..., m)$ are rational functions in z and f. Assume that f does not satisfy the second order algebraic differential equation of the form

$$f'' = L(z, f)(f')^2.$$

We first prove $m = 0$. In fact, to prove this, we assume, on the contrary that $m \geq 1$ and based on this, we discuss three cases separately.

Case 1: $n > m + 2$. Since $\rho(f) = +\infty$, then there exists a sequence $\{z_k\}_{k=1}^{\infty}$ such that

$$\frac{\log f^{\#}(z_k)}{\log |z_k|} \to +\infty (k \to +\infty). \tag{7}$$

This implies that the family $\{f(z_k + z)\}_{k \in N}$ is not normal at zero. Then by Lemma 2, there exist a sequence $\{\beta_k\}$ and a positive sequence $\{\rho_k\}$ such that

$$|z_k - \beta_k| < 1, \rho_k \to 0,$$

and $g_k(\zeta) = f(\beta_k + \rho_k \zeta)$ converges locally spherically uniformly to a nonconstant meromorphic function $g(\zeta)$. Moreover ρ_k and β_k can be chosen so that

$$\rho_k \leq \frac{2}{f^{\#}(\beta_k)}, \qquad f^{\#}(\beta_k) \geq f^{\#}(z_k).$$

It follows from this and (7) that for any positive constants ϵ and M

$$\lim_{k\to\infty} \beta_k^M \rho_k^\epsilon = 0 \tag{8}$$

By replacing z by $\beta_k + \rho_k\zeta$ in (1), we have

$$\frac{g''_k(\zeta)}{\rho_k^2} = \frac{\sum_{j=0}^n L_j(\beta_k + \rho_k\zeta, g_k(\zeta))\frac{g_k'(\zeta)^j}{\rho_k^j}}{\sum_{j=0}^m M_j(\beta_k + \rho_k\zeta, g_k(\zeta))\frac{g_k'(\zeta)^j}{\rho_k^j}}.$$

We rewrite this

$$g_k'(\zeta)^n = \frac{M_m(\beta_k + \rho_k\zeta)}{L_n(\beta_k + \rho_k\zeta, g_k(\zeta))}{\rho_k}^{n-m-2} g_k''(\zeta) g_k'(\zeta)^m + o({\rho_k}^{n-m-2}).$$

Noting $n > m+2$, we can conclude from this and (8), by letting $k \to \infty$, $g'(\zeta) \equiv 0$. This is a contradiction.

Case 2: $n = m + 2$. Now we assume that

$$M_m(z, f)f'' - L_n(z, f)(f')^2 \not\equiv 0.$$

$$M_{m-1}(z, f)f'' - L_{n-1}(z, f)(f')^2 \not\equiv 0.$$

Otherwise, nothing need to prove. Since $\rho(f) = \infty$, by Lemma 4, $f \notin W_p$, for any $p > 1$. By Lemmas 5 and 6, for any positive constant λ, we can choose $c \in \mathbb{C}$ and a sequence $\{z_k\}_{k=1}^\infty$, which tends to infinity, such that

$$f(z_k) = c, \tag{9}$$

$$\frac{|z_k|^\lambda}{|f'(z_k)|} \to 0, \tag{10}$$

$$|f''(z_k)| \leq B|f'(z_k)|, \tag{11}$$

$$M_m(z_k, c)f''(z_k) - L_n(z_k, c)f'(z_k)^2 \neq 0, \tag{12}$$

$$M_{m-1}(z_k, c)f''(z_k) - L_{n-1}(z_k, c)f'(z_k)^2 \neq 0, \tag{13}$$

Thus, we have

$$f''(z_k) = \frac{L_n(z_k, c)f'(z_k)^n + ... + L_1(z_k, c)f'(z_k) + L_0(z_k, f)}{M_m(z_k, c)f'(z_k)^m + ... + M_1(z_k, c)f'(z_k) + M_0(z, c)}. \tag{14}$$

Specially, we can choose λ such that

$$\lim_{k\to\infty} \frac{L_i(z_k, c)}{z_k^\lambda} = 0, \qquad i = 0, 1, ..., n,$$

$$\lim_{k\to\infty} \frac{M_j(z_k, c)}{z_k^\lambda} = 0, \qquad j = 0, 1, ..., m.$$

Thus

$$\lim_{k\to\infty} \frac{L_i(z_k, c)}{f'(z_k)} = 0, \qquad i = 0, 1, ..., n,$$

$$\lim_{k\to\infty} \frac{M_j(z_k, c)}{f'(z_k)} = 0, \qquad j = 0, 1, ..., m.$$

It follows from this and (14) that

$$\lim_{k\to\infty} \frac{f''(z_k)}{f'(z_k)^2} = \lim_{k\to\infty} \frac{L_n(z_k, c)}{M_m(z_k, c)} = \alpha.$$

By (11), we have $\alpha \neq 0$. Now we rewrite the equation (14) into

$$\begin{aligned}&(M_m(z_k, c)f''(z_k) - L_n(z_k, c)f'(z_k)^2)f'(z_k)^m + (M_{m-1}(z_k, c)f''(z_k) -\\ &L_{n-1}(z_k, c)f'(z_k)^2)f'(z_k)^{m-1} + ... + (M_0(z_k, c)f''(z_k) -\\ &L_2(z_k, c)f'(z_k)^2) + L_1(z_k, c)f'(z_k) + L_0(z_k, c) = 0\end{aligned} \tag{15}$$

Noting for some $q > 0$ that

$$\left| M_p(z_k, c)\frac{f''(z_k)}{f'(z_k)^2} - L_{p+2}(z_k, c)\right| \leq r_k^q, \qquad (p = 0, 1, ..., m),$$

and

$$L_i(z_k, c) \leq r_k^q.$$

It follows from above estimates, (10) and (15) that

$$\lim_{k\to\infty} (M_m(z_k, c)\frac{f''(z_k)}{f'(z_k)^2} - L_n(z_k, c)) = 0.$$

Hence, noting that $M_m(z_k, c), L_n(z_k, c)$ are rational functions, we have, for some $s > 0$,

$$M_m(z_k, c)\frac{f''(z_k)}{f'(z_k)^2} - L_n(z_k, c) = O(\frac{1}{r_k^s}).$$

Similarly, we can obtain

$$\lim_{k\to\infty} I_k = 0$$

where I_k is defined by

$$I_k = (M_m(z_k, c)\frac{f''(z_k)}{f'(z_k)^2} - L_n(z_k, c))f'(z_k) + M_{m-1}(z_k, c)\frac{f''(z_k)}{f'(z_k)^2} - L_{n-1}(z_k, c$$

and

$$M_{m-1}(z_k, c)\frac{f''(z_k)}{f'(z_k)^2} - L_{n-1}(z_k, c) = O(r_k^t), \qquad (t \in \mathbb{Z}).$$

Thus we have

$$f'(z_k) = O(r_k^{s+t}). \tag{16}$$

Noting that s and t are only dependent of $L_n, L_{n-1}, M_m, M_{m-1}$, not on c, we can choose $\lambda > s + t$. This contradicts with (10).

Case 3: $n < m + 2$. Let $g = \frac{1}{f}$. Since $\rho(f) = \infty$, we have $\rho(g) = \infty$ and

$$f' = -\frac{g'}{g^2}, \qquad f'' = -\frac{g''}{g^2} + 2\frac{(g')^2}{g^3}.$$

Thus the equation (1) is changed into the following form

$$g'' = \frac{M_m^*(z, g)g'^{m+2} + L_n^*(z, g)g'^n + ... + L_0^*(z, g)}{gM_m{}^*(z, g)g'^m + ... + M_0^*(z, g)}.$$

This is case 2.

By combining the above three cases, we see that m must be zero. Hence the equation (1) is reduced into

$$f'' = L_n(z, f)(f')^n + L_{n-1}(z, f)(f')^{n-1} + ... + L_0(z, f).$$

If $n > 2$, then by similar argument with case 1, we obtain a contradiction. Hence $n \leq 2$, i.e., the equation (1) must reduce to

$$f'' = L_2(z, f)(f')^2 + L_1(z, f)f' + L_0(z, f).$$

This also completes the proof of the Theorem. ■

4. The first-order algebraic differential equations with polynomials as coefficients

In this section, we will discuss the first-order algebraic differential equations of the following form:

$$f'(z) = \sum_{k=0}^{n} P_k(z)f(z)^k, \tag{17}$$

where $n \geq 3$, $P_k(z)$ are polynomials. In fact, we proved the following theorem,

Theorem 4. *If f is a meromorphic solution of the equation (17), then f satisfies the following Ricatti differential equation,*

$$f' = Q_0 + Q_1 f + Q_2 f^2,$$

where $Q_0 = P_0$, $Q_1 = P_1$ and Q_2 are polynomials with $degQ_2 \leq \max_{0 \leq k \leq n}\{degP_k\} = d$.

Proof. Let f be a meromorphic solution of the equation (17). By Malmquist-Yosida theorem, f is a rational function. From (17), we see that any pole of f of multiplicity μ has to be a zero of P_n of multiplicity μ. Therefore $P_n f$ must be a polynomial, denoted by Q. Let $f = \frac{R}{S}$, where R and S are co-prime polynomials. Then $degQ \leq degP_n + degR - degS$. If $degR \leq degS$, then $degQ \leq degP_n \leq d$. If $degR > degS$, which implies that $|f(z)| \to \infty$ as $|z| \to \infty$, we now write (17) in the form,

$$f' - \sum_{k=0}^{n-1} P_k f^k = P_n f^n,$$

then, $\frac{|P_n(z)f(z)^n|}{|z|^{d_n + n(d_R - d_S)}}$ tends to a finite non-zero number as $|z| \to \infty$, where $d_n = degP_n$, $d_R = degR$ and $d_S = degS$. Thus, at least one of the numbers

$$d_R - d_S - 1, d_0, d_1 + d_R - d_S, ..., d_{n-1} + (n-1)(d_R - d_S)$$

must be $\geq d_n + n(d_R - d_S)$. This mean that there exists $k (0 \leq k \leq n-1)$ such that

$$0 < d_R - d_S \leq \frac{d_k - d_n}{n - k} < d - d_n.$$

Thus $degQ = d_n + d_R - d_S < d$. Hence, the equation (17) can be rewrite as the following form

$$f' = P_0 + P_1 f + ... + (P_{n-1} + Q) f^{n-1},$$

where $P_{n-1} + Q$ is a polynomial with degree $\leq d$. Continuing the above argument, we can obtain the conclusion. ■

References

[1] G. Barsegian, *Estimates of derivatives of meromorphic functions on sets of a-points*, J. London Math Soc. (2) **34**(1986), 534-540.

[2] G. Barsegian, *On a method of study of algebraic differential equations*, Bull. Hong Kong Math. Soc. **2**, 159-164.

[3] G. Barsegian, *Estimates of derivatives of higher orders of meromorphic functions on sets of a-points and the second fundamental theorem of R. Nevanlinna*, Bulletin of Hong Kong Math. Soc., Vol. 2(1999), No. 2, 341-346.

[4] W. Bergweiler, *On a theorem of Gol'dberg concerning meromorphic solutions of algebraic differential equations, Complex Variables*, **37**(1998), 93-96.

[5] Y. Z. He & I. Laine, *The Hayman-Miles theorem and the differential equation* $(y')^n = R(z, y)$, Analysis **10**(1990), 387-396.

[6] E. Hille, *On some generalizations of the Malmquist theorem*, Math. Scand. **39**(1976), 59-79.

[7] I. Laine, *On the behaviour of the solutions of some first order differential equations*, Ann. Acad. Sci. Fenn. Ser. A I **497**(1971).

[8] I. Laine, *Nevanlinna Theory and Complex Differential Equations*, Walter de Gruyter, Berlin-NewYork, 1993.

[9] L. W. Liao, *Applications of Value Distribution Theory of Meromorphic Functions*, Ph.D Thesis of HKUST, 2000.

[10] L. W. Liao & C. C. Yang, *On the growth of meromorphic and entire solutions of algebraic differential equations*, to appear in Annali Di Matematica Pura ed Applicata.

[11] Sh. A. Makhmutov, *Classes of meromorphic functions characterized by the spherical derivative*, Soviet Math. Dokl., **33**(1986), 450-455.

[12] J. Malmquist, *Sur les fonctions á un nombre fini des branches définies par les équations différentielles du premier ordre*, Acta Math. **36**(1913),297-343.

[13] J. V. Rieth, *Untersuchungen gewisser Klassen gewöhnlicher Differentialgleichungen erster und zweiter Ordnung im Komplexen*, Aachen, 1986.

[14] N. Steinmetz, *Eigenschaften eindeutiger Lösungen gewöhnlicher Differentialgleichungen im Komplexen*, Karlsruhe, 1978.

[15] C. C. Yang, *A note on Malmquist's theorem on first order differential equations*, Yokohama Math. J. **20**(1972), 115-123.

[16] K. Yosida, *A generalisation of Malmquist's theorem*, J. Math **9**(1933), 253-256.

[17] L. Zalcman, *Normal families: New perspectives*, Bull. Amer. Math. Soc., **35**(1998), 215-230.

13

TRANSFORMATION TECHNIQUES FOR PARTIAL DIFFERENTIAL EQUATIONS

Jürgen Püngel
Technical University of Graz
Department of Mathematics
Steyrergasse 30/III, A-8010 Graz, Austria
E-mail: puengel@matd.tu-graz.ac.at

Abstract: The great importance of transformations for solutions of linear and nonlinear partial differential equations (pde's) is demonstrated by various topics. Families of linear differential transformations for linear complex formally hyperbolic pde's and the associated transformations of the complex Riemann functions by integral-differential operators are constructed. A special transformation technique for initial value problems is applicable to the construction of certain Riemann functions as well as to an addition formula for Riemann functions. Further we construct families of nonlinear differential transformations for nonlinear pde's and an extension principle onto systems.

Keywords: Complex partial differential equations, differential transformations, Riemann functions, nonlinear transformations

AMS-classification: 35A22,35C05

1. Differential transformations for linear equations

The aim of this section is to describe families of linear transformations T which map all holomorphic solutions u of linear second-order partial differential equations $Lu = o$ with holomorphic coefficients into solutions $v = Tu$ of associated equations $Mv = o$ of the same type. This transformations are given by linear differential operators which depend on an arbitrary number of solutions $\alpha_1, \dots, \alpha_n$ of the first equation $L\alpha_j = o$.

For a domain $D \subset \mathbb{C}^n$ let $H_n(D) := \{f \mid f : D \to \mathbb{C} \text{ holomorphic}\}$ denote the ring of all functions of n complex variables holomorphic in a common domain $D \subset \mathbb{C}^n$. The group of units is given by $E_n(D) := \{f \in H_n(D) \mid 0 \notin f[D]\}$. Let o denote the zero element.

Let $\partial_j : H_n(D) \to H_n(D)$ denote the operators of partial derivations $(\partial_j f)(z) := (\partial f/\partial z_j)(z)$ for $z = (z_1, \dots, z_n) \subset D$. Further we use the notation $f_{,j} := \partial_j f$ for $f \in H_n(D)$.

Now we consider "formally hyperbolic" equations

$$w_{,12} + Aw_{,2} + Bw_{,1} + Cw = o$$

with $A, B, C \in H_2(D)$ for $w \in H_2(D), D \subset \mathbb{C}^2$. We mention that the restrictions $z_1 = x + iy, z_2 = x - iy$ lead to real elliptic equations in domains $G \subset \mathbb{R}^2$ if $D = G \times G^*, G^* = \{(x, -y) \mid (x, y) \in G\}$. If $\eta \in E_2(D)$ is a solution of $\eta_{,12} + A\eta_{,2} + B\eta_{,1} + C\eta = o$ then $w/\eta \in H_2(D)$ is a solution of the "abbreviated" equation

$$u_{,12} + au_{,2} + bu_{,1} = o.$$

Hence we consider for simplification the set

$$\Lambda := \{\partial_1\partial_2 + a\partial_2 + b\partial_1 \mid a, b \in H_2(D)\}$$

of abbreviated formally hyperbolic equations.

We consider now linear differential operators T with holomorphic coefficients such that

$$Lu = o \quad \wedge \quad v = Tu \quad \Rightarrow \quad Mv = o$$

for all $u \in H_2(D)$, where $L, M \in \Lambda$. Hence we have

$$T[\ker L] \subset \ker M \tag{1}$$

where $\ker L := \{u \in H_2(D) \mid Lu = o\}$ for $L \in \Lambda$.

We start with transformations of first order.

Lemma 1. *Let $L, M \in \Lambda$ be given by*

$$L := \partial_1\partial_2 - a_{,1}\,\partial_2 - b_{,2}\,\partial_1 \qquad \text{and} \qquad M := \partial_1\partial_2 - p_{,1}\,\partial_2 - q_{,2}\,\partial_1$$

and let $\alpha, \beta \in \ker L$. Then (1) holds in the following cases:

(i) $Tu = u/\alpha$, $p = a - \ln\alpha$, $q = b - \ln\alpha$, $\alpha \in E_2(D)$
(ii) $Tu = u_{,1}/\alpha_{,1}$, $p = a - \ln(\alpha_{,1}/a_{,1})$, $q = b - \ln\alpha_{,1}$, $a_{,1}, \alpha_{,1} \in E_2(D)$
(iii) $Tu = u_{,2}/\alpha_{,2}$, $p = a - \ln\alpha_{,2}$, $q = b - \ln(\alpha_{,2}/b_{,2})$, $b_{,2}, \alpha_{,2} \in E_2(D)$,
(iv) $Tu = (\alpha_{,1}u_{,2} - \alpha_{,2}u_{,1})/d$, $p = a - \ln(\alpha_{,1}/d)$, $q = b - \ln(\alpha_{,2}/d)$ with $d = \alpha_{,1}\beta_{,2} - \alpha_{,2}\beta_{,1}$, $\alpha_{,1}, \alpha_{,2}, d \in E_2(D)$

Proof. (i),(ii),(iii) can be verified by direct calculation and (iv) by composition of (ii) and (iii), see [10]. ∎

All first-order differential transformations between $L, M \in \Lambda$ can be represented by suitable compositions $S \circ T$ of operators S and T of type (i),(ii),(iii) and two further transformations (see [10]).

Of course, transformations of any order can be constructed by compositions of first-order transformations. Let

$$[T_1, \dots, T_k / f_1, \dots, f_k] := \det(T_i f_j \mid 0 \le i, j \le k)$$

for lineare operators $T_i : H_2(D) \to H_2(D)$ and functions $f_j \in H_2(D)$ for $0 \le i, j \le k$. Further we use the determinants

$$D^{n,m}(f_1, \dots, f_{n+m}) := [\partial_1^1, \dots, \partial_1^n, \partial_2^1, \dots, \partial_2^m / f_1, \dots, f_{n+m}],$$
$$D^{n,0}(f_1, \dots, f_n) := [\partial_1^1, \dots, \partial_1^n / f_1, \dots, f_n],$$
$$D^{0,m}(f_1, \dots, f_m) := [\partial_2^1, \dots, \partial_2^m / f_1, \dots, f_m]$$

and $D^{0,0} := 1$ for $f_1, \dots, f_{n+m} \in H_2(D)$.

Theorem 1. *Suppose $L, M \in \Lambda$ as in Lemma 1, $\alpha_1, \dots, \alpha_{n+m} \in \ker L$ and $d_1 d_2 d \in E_2(D)$ with $d_1 := D^{n-1,m}(\alpha_1, \dots, \alpha_{n+m-1})$, $d_2 := D^{n,m-1}(\alpha_1, \dots, \alpha_{n+m-1})$, $d := D^{n,m}(\alpha_1, \dots, \alpha_{n+m})$.*

Then (1) holds with

$$Tu = D^{n,m}(\alpha_1, \dots, \alpha_{n+m-1}, u)/d$$

and

$$p = a + \ln(d_2/d) \qquad \text{and} \qquad q = b + \ln(d_1/d).$$

For the proof see [10].

The question if all linear differential transformations (between operators in Λ) are decomposable into linear differential transformations of first order is in the author's opinion an unsolved problem.

2. Transformations of Riemann functions

A transformation T with $T[\ker L] = \ker M$ instead of (1) is the exceptional case. The fact that the complex Riemann function of operators $L \in \Lambda$ can be transformed into the Riemann function of $M \in \Lambda$ by integral-differential operators associated to those T, which are decomposible into products of first-order transformations, makes it possible to transform all solutions of $Lu = o$ into *all* solutions of $Mv = o$ also in the case of (1) with $T[\ker L] \neq \ker M$.

Let $D := G \times G^*$ where $G \subset \mathbb{C}$ is a simply connected domain and $G^* := \{\bar{z}; \mid z \in G\}$.

Lemma 2. *(see [11]) Let $a, b, c \in H_2(D)$ and*

$$L := \partial_1\partial_2 - a_{,1}\,\partial_2 - b_{,2}\,\partial_1 + c_{,12}\,I,$$
$$L^+ := \partial_1\partial_2 + a_{,1}\,\partial_2 + b_{,2}\,\partial_1 + (a + b + c)_{,12}\,I$$

(I is the identity operator). Then there exists a unique solution $r \in H_4(D \times D)$ of the initial value problem

$$L^+ r = o$$

in $D \times D \subset \mathbb{C}^4$ with

$$r(z, \zeta; z, \tau) = \exp[b(z, \tau) - b(z, \zeta)], \qquad r(z, \zeta; t, \zeta) = \exp[a(t, \zeta) - a(z, \zeta)]$$

for all $z, t \in G$ and $\zeta, \tau \in G^*$.

This function r associated to L is called *(complex) Riemann function* of L (or of the equation $Lu = o$). Using the Riemann function of L, all solutions $u \in H_2(D)$ can be represented by integral operators, see [11, 8, 9].

Associated to any first-order differential transformation T with (1) there exists an integral-differential operator $J : H_4(D \times D) \to H_4(D \times D)$ such that $s = Jr$ is the Riemann function of M if r is the Riemann function of L . In case of the transformations of Lemma 1 we have the

Theorem 2. *Let L, M be as in Lemma 1. Let r be the Riemann function of L and $\alpha \in \ker L$. Then s is the Riemann function of M with coefficients p, q and α according to Lemma 1 (i), (ii), (iii) resp. and*

(i) $s(z, \zeta; t, \tau) = \omega(z, \zeta; t, \tau) r(z, \zeta; t, \tau)$ *with* $\omega(z, \zeta; t, \tau) = \alpha(z, \zeta)/\alpha(t, \tau)$

(ii) $s(z, \zeta; z, \tau) = \omega_1(z, \zeta; z, \tau) \frac{\partial}{\partial t} \int_t^z \eta_1(s; z, \zeta) r(s, \zeta; t, \tau) ds$
with $\omega_1(z, \zeta; t, \tau) = \alpha_z(z, \zeta)/\alpha_t(t, \tau)$ *and* $\eta_1(s; z, \zeta) = a_s(s, \zeta)/a_z(t, \tau)$

(iii) $s(z, \zeta; z, \tau) = \omega_2(z, \zeta; z, \tau) \frac{\partial}{\partial \tau} \int_t^z \eta_2(\sigma; z, \zeta) r(z, \sigma; t, \tau) d\sigma$
with $\omega_2(z, \zeta; t, \tau) = \alpha_\zeta(z, \zeta)/\alpha_\tau(t, \tau)$ *and* $\eta_2(\sigma; z, \zeta) = b_\sigma(z, \sigma)/b_\zeta(z, \zeta)$.

The Riemann function of M in the case of (iv) and the transformations of Theorem 1 can be obtained by suitable compositions of (ii) and (iii), c. f. [8, 9].

3. Representations of Riemann functions by associated initial value problems

In this section we consider self-adjoint operators of the class

$$\Lambda_0 := \{\partial_1\partial_2 - cI \mid c \in H_2(D)\}$$

with $D := G \times G^*$ as in Section 2. The Riemann function $r \in H_4(D \times D)$ of $L = \partial_1\partial_2 - cI$ with $L \in \Lambda_0$ is defined as the (unique) solution of the initial value problem

$$r_{z\zeta}(z,\zeta;t,\tau) = c(z,\zeta)r(z,\zeta;t,\tau)$$
$$r(t,\zeta;t,\tau) = r(z,\tau;t,\tau) = 1.$$

Now we suppose the coefficient c to be of the type

$$c(z,\zeta) = f(z+\zeta).$$

Lemma 3. *Let* $A := \{z+\zeta \mid (z,\zeta) \in D\}$, $\Omega := \{(z-t)(\zeta-\tau) \mid (z,\zeta),(t,\tau) \in D\}$ *and*

$$(Lu)(z,\zeta) := u_{z\zeta}(z,\zeta) - f(z+\zeta)u(z,\zeta).$$

Then the Riemann function of L *can be represented by*

$$r(z,\zeta;t,\tau) = R(\alpha,\alpha^*;\omega)$$

with

$$\alpha = z+\zeta, \quad \alpha^* = t+\tau, \quad \omega = (z-t)(\zeta-\tau)$$

where $R \in H_3(A \times A \times \Omega)$ *is a solution of*

$$R_{\alpha\alpha} + (\alpha-\alpha^*)R_{\alpha\omega} + (\omega R_\omega)_\omega = f(\alpha)R, \quad R(\alpha,\alpha^*;0) = 1. \tag{2}$$

Proof. We have $r_{,12} = f(z+\zeta)r$ which leads to $r_{,112} = fr_{,1} + f'r$ and $r_{,122} = fr_{,2} + f'r$. Let $\rho := r_{,1} - r_{,2}$. Hence $\rho_{,12} = f\rho$ with $\rho(t,\zeta;t\tau) = \rho(z,\tau;t,\tau) = 0$. The unique solution is $\rho = o$. Then (2) is a consequence of $Lr = o$. ∎

Because R is holomorphic in a neighbourhood of $\omega = 0$ we have a power series expansion

$$R(\alpha,\alpha^*;\omega) = \sum_{\mu=0}^{\infty} p_\mu(\alpha,\alpha^*)\frac{\omega^\mu}{(\mu!)^2} \tag{3}$$

with $p_\mu \in H_2(A \times A)$. The initial value problem (2) leads to the recurrence formula

$$(\alpha-\alpha^*)\partial_1 p_{\mu+1} + (\mu+1)p_{\mu+1} = (\mu+1)[f(\alpha)p_\mu - \partial_1^2 p_\mu] \tag{4}$$

with $p_0 = 1$. Let us now assume that the power series

$$S(\alpha,\alpha^*;\omega) = \sum_{\mu=0}^{\infty} p_\mu(\alpha,\alpha^*)\frac{\omega^\mu}{\mu!} \tag{5}$$

converges for an $\epsilon > 0$ in $|\omega| < \epsilon$ (then R must be an entire function with respect to ω).

From (4) and (5) we deduce the initial value problem

$$S_{\alpha\alpha} + \omega^{-1}(\alpha - \alpha^*)S_\alpha + S_\omega = f(\alpha)S, \quad S(\alpha, \alpha^*; 0) = 1 \tag{6}$$

Using Cauchy's integral formula in order to represent the coefficients p_μ, we are finally led to

Theorem 3. *Let $S(\alpha, \alpha^*; \omega)$ be a holomorphic solution of (6) for $\alpha, \alpha^* \in A$ and $|\omega| < \epsilon, \epsilon > 0$. Then r defined by the integral*

$$r(z, \zeta; t, \tau) = \frac{1}{2\pi i} \oint_\gamma \frac{1}{\eta} e^{(z-\zeta)(t-\tau)/\eta} S(z + \zeta, t + \tau; \eta) d\eta$$

along $\gamma(\varphi) = \epsilon_0 e^{i\varphi}, 0 \leq \varphi \leq 2\pi, \epsilon_0 < \epsilon$ is the Riemann function of

$$u_{z\zeta}(z, \zeta) - f(z + \zeta)u(t, \zeta) = 0.$$

Explicit representations of Riemann functions (e. g. in case of $f(\alpha) = k\alpha + l, f(\alpha) = k\alpha^2 + l, f(\alpha) = k\alpha^{-2}$ can be obtained by this method. In many cases it is convenient to interpret (5) as a formal power series and (6) as a formal initial value problem.

4. Generalized seperability

By means of the transformation of the initial value problem (2) for R onto the formal initial value problem (6) for S a further method to construct explicite representations of Riemann functions can be obtained.

Let $r^+, r^-, r \in H_4(D \times D)$ be the Riemann functions of

$$u^+_{z\zeta} = f(z + \zeta)u^+, \; u^-_{z\zeta} = -g(z - \zeta)u^-, \; u_{z\zeta} = [f(z + \zeta) - g(z - \zeta)]u \tag{7}$$

respectively. Then the functions

$$r^+(z, \zeta; t, \tau) = R^+(\alpha, \alpha^*; \omega), \quad r^-(z, \zeta; t, \tau) = R^-(\beta, \beta^*; \omega),$$

$$r(z, \zeta; t, \tau) = R(\alpha, \alpha^*; \beta, \beta^*; \omega)$$

with $\alpha = z + \zeta, \; \alpha^* = t + \tau, \; \beta = z - \zeta, \; \beta^* = t - \tau, \; \omega = (z - t)(\zeta - \tau)$ are solutions of

$$R^+_{\alpha\alpha} + (\alpha - \alpha^*)R^+_{\alpha\omega} + (\omega R^+_\omega)_\omega = f(\alpha)R^+,$$

$$R^-_{\beta\beta} - (\beta - \beta^*)R^-_{\beta\omega} + (\omega R^-_\omega)_\omega = g(\beta)R^-$$

$$R_{\alpha\alpha} - R_{\beta\beta} + (\alpha - \alpha^*)R_{\alpha\omega} + (\beta - \beta^*)R_{\beta\omega} + (\omega R_\omega)_\omega = [f(\alpha) - g(\beta)]R \tag{8}$$

with $R^+(\alpha, \alpha^*; 0) = R^-(\beta, \beta^*; 0) = R(\alpha, \alpha^*; \beta, \beta^*; 0) = 1$. Equation (8) cannot be seperated into nontrivial factors

$$R(\alpha, \alpha^*; \beta, \beta^*; \omega) = S(\alpha, \alpha^*; \omega)T(\beta, \beta^*; \omega).$$

However by use of the (formal) transformations of the power series

$$R(\alpha, \alpha^*; \beta, \beta^*; \omega) = \sum_{\mu=0}^{\infty} p_\mu(\alpha, \alpha^*; \beta, \beta^*)\frac{\omega^\mu}{(\mu!)^2}$$

$$R^+(\alpha, \alpha^*; \omega) = \sum_{\mu=0}^{\infty} p_\mu^+(\alpha, \alpha^*)\frac{\omega^\mu}{(\mu!)^2}, \quad R^-(\beta, \beta^*; \omega) = \sum_{\mu=0}^{\infty} p_\mu^-(\beta, \beta^*)\frac{\omega^\mu}{(\mu!)^2}$$

into the (formal) power series

$$S(\alpha, \alpha^*; \beta, \beta^*; \omega) = \sum_{\mu=0}^{\infty} p_\mu(\alpha, \alpha^*; \beta, \beta^*)\frac{\omega^\mu}{\mu!}$$

$$S^+(\alpha, \alpha^*; \omega) = \sum_{\mu=0}^{\infty} p_\mu^+(\alpha, \alpha^*)\frac{\omega^\mu}{\mu!}, \quad S^-(\beta, \beta^*; \omega) = \sum_{\mu=0}^{\infty} p_\mu^-(\beta, \beta^*)\frac{\omega^\mu}{\mu!}$$

we obtain the (formal) initial value problems

$$\begin{gathered} S_{\alpha\alpha} - S_{\beta\beta} + \omega^{-1}(\alpha - \alpha^*)S_\alpha + \omega^{-1}(\beta - \beta^*)S_\beta + S_\omega = [f(\alpha) - g(\beta)]S, \\ S(\alpha, \alpha^*; \beta, \beta^*; 0) = 1 \end{gathered} \tag{9}$$

$$S^+_{\alpha\alpha} + \omega^{-1}(\alpha - \alpha^*)S^+_\alpha + S^+_\omega = f(\alpha)S^+, \ S(\alpha, \alpha^*; 0) = 1 \tag{10}$$

$$S^-_{\beta\beta} - \omega^{-1}(\beta - \beta^*)S^-_\beta + S^-_\omega = g(\beta)S^-, \ S(\beta, \beta^*; 0) = 1 \tag{11}$$

Unlike (8) equation (9) is seperable:

$$S(\alpha, \alpha^*; \beta, \beta^*; \omega) = S^+(\alpha, \alpha^*; \omega)S^-(\beta, \beta^*; \omega)$$

is a solution of (9) if S^+, S^- are solutions of (10),(11). Hence we have the result

Theorem 4. *Let $r^+(z,\zeta;t,\tau) = R^+(z+\zeta,t+\tau;(z-t)(\zeta-\tau))$ be the Riemann function of $u^+_{z\zeta} = f(z+\zeta)u^+$ and let $r^-(z,\zeta;t,\tau) = R^-(z-\zeta,t-\tau;(z-t)(\zeta-\tau))$ be the Riemann function of $u^-_{z\zeta} = -g(z-\zeta)u^-$. Then*

$$r(z,\zeta;t,\tau) = \sum_{\mu=0}^{\infty} p_\mu(z+\zeta,t+\tau;z-\zeta,t-\tau)\frac{(z-t)^\mu(\zeta-\tau)^\mu}{\mu!}$$

with the coefficients

$$p_\mu(\alpha,\alpha^*;\beta,\beta^*) = \sum_{\mu=0}^{\infty} \frac{\partial^\lambda R^+}{\partial\omega^\lambda}(\alpha,\alpha^*;0)\frac{\partial^{\mu-\lambda}R^-}{\partial\omega^{\mu-\lambda}}(\beta,\beta^*;0)$$

is the Riemann function of $u_{z\zeta} = [f(z+\zeta) - g(z-\zeta)]u$.

One can obtain similar results for Bergman operators (see [6]). In [7] generalizations of this method are developed.

5. Nonlinear differential transformations

In this section we turn to nonlinear differential transformations for (in general) nonlinear partial differential equations. For reference we list a number of nonlinear first-order transformations for second-order equations (see also [3, 4, 5]). Let us assume $u(z,\zeta)$ and $v(z,\zeta)$ to be holomorphic functions in (variable suitable) domains $D \subset \mathbb{C}^2$.

$$u_{z\zeta} = au_\zeta + bu_z \quad (a,b \in H_2(D)) \quad \wedge \quad v = \ln(u_\zeta/u_z)$$

$$\Longrightarrow v_{z\zeta} = (a+be^{-v})_\zeta - (b+ae^v)_z \tag{12}$$

$$u_{z\zeta} = cu \quad (c \in H_2(D)) \quad \wedge \quad v = \ln(u_\zeta/u)$$

$$\Longrightarrow \quad v_{z\zeta} = (ce^{-v})_\zeta - (e^v)_z \tag{13}$$

$$u_{z\zeta} = o \quad \wedge \quad v = \ln(2u_z u_\zeta/u^2) \quad \Longrightarrow \quad v_{z\zeta} = e^v \tag{14}$$

$$u_{z\zeta} = uu_z \quad \wedge \quad v = \ln u_\zeta \quad \Longrightarrow \quad v_{z\zeta} = e^v \tag{15}$$

$$u_\zeta = u_{zz} \quad \wedge \quad v = u_z/u \quad \Longrightarrow \quad v_\zeta = v_{zz} + 2vv_z \tag{16}$$

Some of these examples can be generalized in different ways. (16) turns out to be a special case of

$$u_\zeta = uh(u_{zz}/u, u_z/u) \quad \wedge \quad v = u_z/u \quad \Longrightarrow \quad v_\zeta = \partial_z h(v_z + v^2, v) \tag{17}$$

with arbitrary functions h, holomorphic in some suitable domains.

More general, we now ask for holomorphic functions f, γ, g such that

$$u_\zeta = f(u, u_z, u_{zz}) \quad \wedge \quad v = \gamma(u, u_z) \quad \Longrightarrow \quad v_\zeta = g(v, v_z, v_{zz}) \tag{18}$$

holds for all $u \in H_2(D)$ that ensure $v \in H_2(D^*)$ in suitable domains $D, D^* \subset \mathbb{C}^2$.

Let us assume that $\partial_3 f \neq o, \partial_2\gamma \neq o$ and $\partial_3 g \neq o$, then (18) is locally equivalent to a Bäcklund system

$$u_z = \varphi(u, v) \tag{19}$$

$$u_\zeta = F(u, v, v_z) \tag{20}$$

Substituting v from equation (19) and v_z into equation (B2), we have the integrability condition $u_\zeta = f(u, u_z, u_{zz})$ for u again.

Now we have to determine an integrability condition of the type $v_\zeta = g(v, v_z, v_{zz})$ for v too. In order to do this, we eliminate u from (19), (20) using $u_{z\zeta} = u_{\zeta z}$ and interpret u, v and $w := v_z$ as three independent variables. That way we get

$$v_\zeta = P_w(v, v_z)v_{zz} + Q(v, v_z) \tag{21}$$

where $P(v, w), Q(v, w)$ must be independent of u and fulfill the equations

$$F_w = \varphi_v P_w, \quad P_w \neq o \tag{22}$$

$$wF_v + \varphi F_u - \varphi_u F = \varphi_v Q. \tag{23}$$

From (22) we get

$$F(u, v, w) = \varphi_v(u, v)P(v, w) + \psi(u, v).$$

Sustituting this result into (23) we obtain

$$[a(u, v)w + b(u, v)]P(v, w) = c(u, v)w + d(u, v) + S(v, w) \tag{24}$$

with $S(v, w) := Q(v, w) - wP_v(v, w)$ and

$$\begin{gathered} a := \varphi_{vv}/\varphi_v, \quad b := \varphi\varphi_{uv}/\varphi_v - \varphi_u, \quad c := -\psi_v/\varphi_v, \\ d := (\varphi_u\psi - \varphi\psi_u)/\varphi_v. \end{gathered} \tag{25}$$

Derivating (24) with respect to u we get

$$[a_u(u,v)w + b_u(u,v)]P(v,w) = c_u(u,v)w + d_u(u,v). \tag{26}$$

Here we always have to guarantee $P_u = o$ and to distinguish between

$$case(I) : a_u w + b_u = o \qquad \text{and} \qquad case(II) : a_u w + b_u \neq o$$

In case (II) $P(v,w)$ turns out to be a fractional linear map with respect to w.

In case (I) we obtain the Bäcklund system

$$u_z = v - bu, \quad u_\zeta = P(v, v_z) + ku \tag{27}$$

with arbitrary $b, k \in \mathbb{C}$ and arbitrary holomorphic $P(v,w)$ with $P_w \neq 0$ corresponding to the transformation

$$u_\zeta = P(u_z + bu, u_{zz} + bu_z) + ku \quad \wedge \quad v = u_z + bu$$

$$\Longrightarrow \; v_\zeta = \partial_z P(v, v_z) + bP(v, v_z) + kv \tag{28}$$

with $\partial_z = \partial/\partial z$. Up to point transformations $u \to \sigma(u)$,$v \to \rho(v)$ with $\sigma', \rho' \neq o$, case (I) is completely characterized by this result.

For case (II), assuming $a_u = o$, we obtain (up to point transformations) a complete characterisation by the Bäcklund system

$$u_z = v + e^u, \quad u_\zeta = (v + e^u)G(v) + \partial_z G(v) \tag{29}$$

and the transformation

$$u_\zeta = e^{-u}\partial_z[e^u G(u_z - e^u)] \quad \wedge \quad v = u_z - e^u$$

$$\Longrightarrow \quad v_\zeta = \partial_z[vG(v) + \partial_z G(v)] \tag{30}$$

respectively.

In case (II) with $a_u \neq o$ we can obtain an extensive set of transformations. We present only the following result.

Choosing $u_z = \alpha(v) + vu$, α an arbitrary holomorphic function with $a'' \neq o$, equation (20) turns out to be of the form $u_\zeta = A(v, v_z) + uB(v, v_z)$. Assuming an integrability condition for v according to (18), we obtain

$$u_z = \alpha(v) + vu, \quad u_\zeta = [\alpha'(v) + u]B(v, v_z) - \beta(v) \tag{31}$$

with arbitrary holomorphic β and

$$B(v,w) = [\beta'(v)w - v\beta(v)]/[\alpha''(v)w + \alpha(v) - v\alpha'(v)].$$

This Bäcklund system corresponds to

$$u_\zeta = f(u, u_z, u_{zz}) \quad \wedge \quad v = \gamma(u, u_z) \quad \Longrightarrow \quad v_\zeta = \partial_z B(v, v_z) \qquad (32)$$

where f and γ are defined by computing v form the first equation in (31) and substituting into the second equation in (31).

6. An extension principle onto systems

In [1, 2] we developed a general method to get various products $\mathbb{C}^n \times \mathbb{C}^n \to \mathbb{C}^n$ such that $\mathbb{C}^n$ is a commutative ring. That way we obtain a useful principle for numerous extensions of scalar ($\mathbb{C}$-valued) holomorphic functions into $\mathbb{C}^n$-valued holomorphic functions of several complex variables which form a suitable differential ring in order to transpose scalar differential transformations into differential transformations for systems immediately.

For the sake of clarity we set $n = 2$. The mentioned extension principle for $\mathbb{C}$-valued holomorphic functions of one or several complex variables can be based on the family of products $\mathbb{C}^2 \times \mathbb{C}^2 \to \mathbb{C}^2$ defined by

$$\begin{pmatrix} \alpha \\ \beta \end{pmatrix} \begin{pmatrix} \lambda \\ \mu \end{pmatrix} := \begin{pmatrix} \alpha\lambda - pq\beta\mu \\ \alpha\mu + \beta\lambda + (p+q)\beta\mu \end{pmatrix}$$

where $p, q \in \mathbb{C}$ are arbitrary fixed constants. Using

$$\begin{pmatrix} \alpha \\ \beta \end{pmatrix}^n = \frac{1}{p-q} \begin{pmatrix} p(\alpha + q\beta)^n - q(\alpha + p\beta)^n \\ (\alpha + p\beta)^n - (\alpha + q\beta)^n \end{pmatrix}$$

and

$$\begin{pmatrix} \alpha \\ \beta \end{pmatrix}^n = \begin{pmatrix} (\alpha + p\beta)^n - np(\alpha + p\beta)^{n-1} \\ n\beta(\alpha + p\beta)^{n-1} \end{pmatrix}$$

in case $p \neq q$ and $p = q$ respectively, we get the following extension principle for functions of two variables.

Associated to a domain $D \subset \mathbb{C}^2$ we define the domain $D_{p,q} := D_p \cap D_q \subset \mathbb{C}^4$ where

$$D_s := \left\{ \left(\begin{pmatrix} \alpha \\ \beta \end{pmatrix}, \begin{pmatrix} \gamma \\ \delta \end{pmatrix} \right) \in \mathbb{C}^2 \times \mathbb{C}^2 \,\middle|\, \begin{pmatrix} \alpha + s\beta \\ \gamma + s\delta \end{pmatrix} \in D \right\}.$$

Now let the map

$$\begin{aligned} H(D, \mathbb{C}) &\longrightarrow H(D_{p,q}, \mathbb{C}^2) \\ f &\longmapsto f_{p,q} \end{aligned}$$

be defined by

$$f_{p,q}\left(\begin{pmatrix}\alpha\\ \beta\end{pmatrix},\begin{pmatrix}\gamma\\ \delta\end{pmatrix}\right):=$$

$$=\frac{1}{p-q}\begin{pmatrix} pf(\alpha+q\beta,\gamma+q\delta)-qf(\alpha+p\beta,\gamma+p\delta)\\ f(\alpha+p\beta,\gamma+p\delta)-f(\alpha+q\beta,\gamma+q\delta)\end{pmatrix}$$

if $p \neq q$ and

$$f_{p,q}\left(\begin{pmatrix}\alpha\\ \beta\end{pmatrix},\begin{pmatrix}\gamma\\ \delta\end{pmatrix}\right):=$$

$$=\begin{pmatrix} pf(\alpha+p\beta,\gamma+p\delta)-p\delta f_{,2}(\alpha+p\beta,\gamma+p\delta)-p\beta f_{,1}(\alpha+p\beta,\gamma+p\delta)\\ \delta f_{,2}(\alpha+p\beta,\gamma+p\delta)+\beta f_{,1}(\alpha+q\beta,\gamma+q\delta)\end{pmatrix}$$

if $p = q$.

Replacing all functions in the above mentioned transformations (e.g. (12)-(16),(17), (27)-(32) and the linear transformations of Section 1) by $\mathbb{C}^2$-valued functions in the described way we immediately get differential transformations for numerous systems of (linear and nonlinear) partial differential equations.

References

[1] Püngel, J. *Transmutations in differential rings. Florian, H. (ed.) et al., Generalized analytic functions: theory and applications to mechanics.* Proceedings of the conference, Graz, Austria, January 6–10, 1997. Dordrecht: Kluwer Academic Publishers. Int. Soc. Anal. Appl. Comput. 1, 55-63 (1998).

[2] Florian, H.; Püngel, J.; Tutschke, W. *Matrix differential transformations.* Appl. Anal. 65, No.1-2, 103-117 (1997).

[3] Florian, H.; Püngel, J.; Tutschke, W. *Solution of boundary value problems for nonlinear partial differential equations using differential transformations.* Anz., Abt. II, Oesterr. Akad. Wiss., Math.-Naturwiss. Kl. 131, 3-16 (1994).

[4] Florian, H.; Püngel, J. *Differentialtransformationen bei quasilinearen Gleichungen zweiter Ordnung. (Differential transformations for quasilinear second order equations)* (German). Begehr, H. (ed.) et al., Partial differential equations with complex analysis. Dedicated to Robert P. Gilbert on the occasion of his 60th birthday. Harlow: Longman Scientific & Technical. Pitman Res. Notes Math. Ser. 262, 14-22 (1992).

[5] Florian, H.; Püngel, J. *Differentialtransformationen von Lösungen partieller Differential-gleichungen. (Differential transformations of*

solutions of partial differential equations) (German). Complex Variables, Theory Appl. 18, No.1-2, 125-134 (1992).

[6] Püngel, J. *Additionsformeln fuer Riemannfunktionen und Bergmanerzeugende.* (German) Complex methods on partial differential equations, Proc. Int. Symp. Complex Anal., Graz/Austria 1988, Math. Res. 53, 126-136 (1989).

[7] Püngel, J. *Verallgemeinerte Separierbarkeit formaler Anfangswertprobleme. (Generalized separability of formal initial value problems).* (German). Rostocker Math. Kolloq. 45, 35-42 (1991).

[8] Püngel, J. *Riemann functions for generalized Euler equations I: General results.* Bull. Appl. Math. 100-109, 77-93 (1982).

[9] Püngel, J. *Riemann functions for generalized Euler equations II: Applications.* Bull. Appl. Math. 137-144, 23-36 (1982).

[10] Püngel, J. *Zur Darstellung von Lsungen einer Klasse linearer partieller Differentialgleichungen.* (German). Rend. Ist. Mat. Univ. Trieste 14, 71-84 (1982).

[11] Vekua, I. N. *New methods for solving elliptic equations.* North-Holland Series in Applied Mathematics and Mechanics. 1. Amsterdam: North-Holland Publishing Company; New York: John Wiley&Sons, Inc. XII, 358 p. (1967).

14

BESOV-TYPE CHARACTERIZATIONS FOR QUATERNIONIC BLOCH FUNCTIONS

L. F. Reséndis O.*

Universidad Autónoma Metropolitana, Unidad Azcapotzalco,
C.B.I. Apartado Postal 16-306 C.P. 02200 México 16, D.F.
Area de Análisis Matemático y sus Aplicaciones.
E-mail: lfro@correo.azc.uam.mx

L. M. Tovar S.†

Escuela Superior de Física y Matemáticas del IPN
Edif. 9, Unidad ALM, Zacatenco del IPN.,
C.P. 07300, D.F., México.
E-mail: tovar@esfm.ipn.mx

Abstract: In this paper we obtain a Besov-type characterization for quaternionic Bloch functions that generalizes a Stroethoff theorem.

AMS-classification: 30C45.

1. Introduction.

Consider the open unit disk

$$\Delta := \{z \in \mathbf{C} : |z| < 1\},$$

and its boundary by $T = \partial\Delta$.

*Partially supported by CONACYT

†Partially supported by CONACYT, Becario de COFAA-IPN

For $z,\ w \in \Delta$, we denote Green's function of Δ, with logarithmic singularity at w, by

$$g(z,w) = \log \frac{|1-\overline{w}z|}{|w-z|}.$$

Let $\mathcal{A}$ denote the set of analytic functions in Δ. In 1994, Aulaskari and Lappan [2] (cf. also [1], [7]), introduced the $\mathcal{Q}_p$ spaces for $p > 1$ and in 1995, Aulaskari, Xiao and Zhao for $0 < p < 1$, as the sets of functions $f \in \mathcal{A}$ that satisfy the condition:

$$\{f\}^2_{\mathcal{Q}_p} := \sup_{a\in\Delta} \int_\Delta |f'(z)|^2 g(z,a)^p \, dx\, dy < \infty\ . \tag{1}$$

Already earlier Xiao considered the especial case $p = 2$ in [10].
It is well-known that $\mathcal{Q}_p$ is a Banach space under the norm

$$\begin{aligned} \| f \|_{\mathcal{Q}_p} &= |f(0)| + \left(\sup_{a\in\Delta} \int_\Delta |f'(z)|^2 g(z,a)^p \, dx\, dy \right)^{1/2} \\ &= |f(0)| + \{f\}_{\mathcal{Q}_p}. \end{aligned}$$

Let $\phi_a : \mathbf{C} \to \mathbf{C}$ be the Möbius transformation,

$$\phi_a(z) = \frac{a-z}{1-\overline{a}z}, \qquad |a| < 1,$$

with the pole at $z = 1/\overline{a}$. When ϕ_a is restricted to Δ, then ϕ_a is an inner function on Δ such that $\phi_a(\phi_a(z)) = z$ and with the Jacobian determinant

$$|\phi_a'(z)|^2 = \frac{(1-|\phi_a(z)|^2)^2}{(1-|z|^2)^2} = \frac{1-|a|^2}{|1-\overline{a}z|^4}.$$

Aulaskari, Xiao and Zhao have shown [3] that $f \in \mathcal{Q}_p$, if and only if,

$$[f]^2_{\mathcal{Q}_p} := \sup_{a\in\Delta} \int_\Delta |f'(z)|^2 \left(1-|\phi_a(z)|^2\right)^p \, dx\, dy < \infty. \tag{2}$$

It follows immediately that if $p < q$ then $\mathcal{Q}_p \subset \mathcal{Q}_q$.
Consider now the well-known *Bloch space*,

$$B = \{f \in \mathcal{A} : B(f) := \sup_{z\in\Delta} (1-|z|^2)|f'(z)| < \infty\}, \tag{3}$$

the *Dirichlet space*,

$$D = \{f \in \mathcal{A} : \int_\Delta |f'(z)|^2 \, dx \, dy < \infty\}, \tag{4}$$

and the $BMOA$ space,

$$BMOA = \{f \in \mathcal{A} : \sup_{a \in \Delta} \int_\Delta |f'(z)|^2 \left(1 - |\phi_a(z)|^2\right) \, dx \, dy < \infty\}. \tag{5}$$

Thus, for $0 \le p \le 1$, the $\mathcal{Q}_p$ spaces form a scale of spaces between D and $BMOA$; Aulaskari and Lappan [2] also have shown that $\mathcal{Q}_p = B$ for $p > 1$. It is clear from their definition, that the $\mathcal{Q}_p$ spaces are an interesting subject in the study of analytic functions in the unit disk, from the point of view of their behavior near the boundary T.
For a point $a \in \Delta$ and $0 < r < 1$ the pseudo-hyperbolic disc $D(a, r)$ with pseudo-hyperbolic centre a and pseudohyperbolic radius r is defined by $D(a, r) = \phi_a(r\Delta)$. The pseudo-hyperbolic disc $D(a, r)$ is also a euclidean disc: its euclidean centre and euclidean radius are $(1 - r^2)a/(1 - r^2|a|^2)$ and $(1 - |a|^2)r/(1 - r^2|a|^2)$, respectively.
Let A denote the normalised Lebesgue area measure on Δ, and for a Lebesgue measurable set $K \subset D$, let $|K|$ denote the measure of K with respect to A. It follows immediately that

$$|D(a, r)| = \frac{(1 - |a|^2)^2}{(1 - r^2|a|^2)^2} r^2 \; . \tag{6}$$

Two quantities A_f and B_f, both depending on an analytic function f on Δ, are said to be equivalent, written as $A_f \approx B_f$, if there exists a finite positive constant C not depending on f such that for every analytic function f on Δ we have:

$$\frac{1}{C} B_f \le A_f \le C B_f \; .$$

If the quantities A_f and B_f are equivalent, then in particular we have $A_f < \infty$ if and only if $B_f < \infty$.
Contained in the Bloch space is the *little Bloch space* B_0, which is by definition the set of all analytic functions f on Δ for wich

$$\lim_{|z| \to 1^-} (1 - |z|^2)|f'(z)| = 0 \; .$$

In 1989, K. Stroethoff [9] proved two important theorems about several characterizations for the Bloch space B and little Bloch space B_0.

Theorem 1. *Let $0 < p < \infty$, $0 < r < 1$, and $n \in \mathbf{N}$. Then for an analytic function $f : \Delta \to \mathbf{C}$ the following quantities are equivalent:*

(a) $\| f \|_B$;

(b)

$$\sup_{a\in\Delta}\left(\frac{1}{|D(a,r)|^{1-\frac{np}{2}}}\int_{D(a,r)}|f^{(n)}(z)|^p\,dA(z)\right)^{\frac{1}{p}}+\sum_{k=1}^{n-1}|f^{(k)}(0)|<+\infty;$$

(c)

$$\sup_{a\in\Delta}\left(\int_{D(a,r)}|f^{(n)}(z)|^p(1-|z|^2)^{np-2}\,dA(z)\right)^{\frac{1}{p}}+\sum_{k=1}^{n-1}|f^{(k)}(0)|<+\infty;$$

(d)

$$\sup_{a\in\Delta}\left(\int_{\Delta}|f^{(n)}(z)|^p(1-|z|^2)^{np-2}(1-|\phi_a(z)|^2)^2\,dA(z)\right)^{\frac{1}{p}}+\sum_{k=1}^{n-1}|f^{(k)}(0)|$$

$$<+\infty.$$

Note that for $n=1$, $p=2$ quantity (D) of the theorem means

$$\| f \|_B \approx \sup_{a\in D}\left(\int_{\Delta}|f'(z)|^2(1-|\phi_a(z)|^2)^2\right)^{\frac{1}{2}}.$$

In the case $n=1$ quantities (C) and (D) are of interest because the quantity

$$\int_{\Delta}|f'(z)|^p(1-|z|^2)^{p-2}\,dA(z)$$

is invariant under Möbius transformations of f.
The equivalence of Theorem 1 carry over the little Bloch space, as is shown the following theorem.

Theorem 2. *Let $0<p<\infty$, $0<r<1$, and $n\in\mathbf{N}$. Then for an analytic function $f:\Delta\to\mathbf{C}$ the following quantities are equivalent:*

(a) $f\in B_0$;

(b)

$$\lim_{|a|\to 1^-}\frac{1}{|D(a,r)|^{1-\frac{np}{2}}}\int_{D(a,r)}|f^{(n)}(z)|^p\,dA(z)=0\ ;$$

(c)

$$\lim_{|a|\to 1^-} \int_{D(a,r)} |f^{(n)}(z)|^p (1-|z|^2)^{np-2}\, dA(z) = 0 \ ;$$

(d)

$$\lim_{|a|\to 1^-} \int_{\Delta} |f^{(n)}(z)|^p (1-|z|^2)^{np-2} (1-|\phi_a(z)|^2)^2\, dA(z) = 0 \ .$$

2. The Quaternionic extension of $\mathcal{Q}_p$

There are several attempts to generalize the previous ideas and the corresponding approach to higher dimensions [9],[4], [11]. Independently on the method these approaches are connected with the unit ball $\boldsymbol{D}$ in $\mathbf{C}^n$ and not in $\mathbf{R}^n$.

Using an invariant Green's function some results similar to the complex case can be proved. The most important results are that

$$\mathcal{Q}_p = B \qquad \text{for } 1 < p < \frac{n}{n-1}$$

and $\mathcal{Q}_1 = BMOA(\partial \boldsymbol{D})$. But, for $p \notin (\frac{n-1}{n}, \frac{n}{n-1})$ all $\mathcal{Q}_p$-spaces are trivial.

In [5] the authors present the hypercomplex generalization of $\mathcal{Q}_p$-spaces. Instead of holomorphic functions in the unit disc they studied monogenic functions (that is, solutions of generalized Cauchy-Riemann systems), which are a higher-dimensional generalization of holomorphic functions also in the case of odd real dimension of the Euclidean space. Such important function classes like the solutions of the $div - rot$ system are included in the theory of monogenic functions. These functions can be considered in all real space dimensions.

In what follows we will work in $\mathbf{H}$, the skew field of quaternions. Also, we will work in $B_1(0) \subset \mathbf{R}^3$ the real ball in the real three-dimensional space. $B_1(0)$ is a bounded, simply connected domain with a C^∞-boundary $S_1(0)$. Moreover, we will consider functions f defined on $B_1(0)$ with values in $\mathbf{H}$. We now define the generalized Cauchy-Riemann operator by,

$$Df = \frac{\partial f}{\partial x_0} + i\frac{\partial f}{\partial x_1} + j\frac{\partial f}{\partial x_2}$$

and its conjugate operator

$$\overline{D}f = \frac{\partial f}{\partial x_0} - i\frac{\partial f}{\partial x_1} - j\frac{\partial f}{\partial x_2} \ .$$

For these operators we have that,

$$D\overline{D} = \overline{D}D = \Delta_3.$$

Functions belonging to ker D are called hyperholomorphic or monogenic regular functions. Observe that we generalize the notation of previous section. For $a \in B_1(0)$ we will denote by,

$$\phi_a(x) := (a - x)(1 - \overline{a}x)^{-1}$$

the Möbious transform, which maps the unit ball onto itself. Furthermore, let

$$g(x, a) = \frac{1}{4\pi}\left(\frac{1}{|\phi_a(x)| - 1}\right)$$

be the modified fundamental solution of the Laplacian in $\mathbf{R}^3$ composed with our Möbious transform $\phi_a(x)$. Especially, we denote for all $p \geq 0$

$$g^p(x, a) = \frac{1}{4^p\pi^p}\left(\frac{1}{|\phi_a(x)| - 1}\right)^p .$$

Let $f : B_1(0) \to \mathbf{H}$ be a hyperholomorphic function. Then we can introduce the seminorms

$$B(f) = \sup_{x \in B_1(0)} (1 - |x|^2)^{\frac{3}{2}} |\overline{D}f(x)| \,,$$

and

$$\mathcal{Q}_p(f) = \sup_{a \in B_1(0)} \int_{B_1(0)} |\overline{D}f(x)|^2 g^p(x, a)\, dB_x \,,$$

which lead to the following definitions

Definition 1. *The spatial (or three-dimensional) Bloch space B is the right $\mathbf{H}$-module of all hyperholomorphic functions $f : B_1(0) \to \mathbf{H}$ with $B(f) < \infty$.*

Definition 2. *The right module of all quaternion-valued functions f defined on the unit ball, wich are hyperholomorphic and satisfy $\mathcal{Q}_p(f) < \infty$, is called the $\mathcal{Q}_p$-space.*

Remark 1. *Because the special structure of $g(x, a)$ the seminorms $\mathcal{Q}_p(f)$ make sense for $p < 3$ only. Consequently, we consider in this paper $\mathcal{Q}_p$-spaces only for $p < 3$.*
Obviously, these spaces are not Banach spaces. Nevertheless, if we consider a small neighbourhood of the origen U_ϵ, with an arbitrary but fixed

$\epsilon > 0$, then we can add the L_1-norm of f over U_ϵ to our seminorms and B as well as $\mathcal{Q}_p$ will become Banach spaces.

Definition 3. *The right* **H***-module of all quaternion-valued functions f defined on the unit ball, which are hyperholomorphic and satisfy the condition*

$$\int_{B_1(0)} |\overline{D}f(x)|^2 \, dB_x < \infty,$$

is called spatial (or three-dimensional) Dirichlet space D.

Remark 2. *Since $g(x,a)$ is non-negative in $B_1(0)$ we have, obviously,*

$$D \subset \mathcal{Q}_p, \qquad for\ 0 \leq p < 3.$$

The next set of results are due to K. Gürlebeck, U. Kähler, M. V. Shapiro and L. M. Tovar, [5].

Lemma 1. *Let $f : B_1(0) \to \mathbf{H}$ be a hyperholomorphic funtion. Then the following conditions are equivalent:*
i) $f \in B$;
ii) $\mathcal{Q}_p(f) < \infty$ for all $2 < p < 3$;
iii) $\mathcal{Q}_p(f) < \infty$ for some $p \in (2,3)$.

Theorem 3. *Let $f : B_1(0) \to \mathbf{H}$ be a hyperholomorphic funtion. Then for $1 \leq p < 2.99$, f belongs to the space $\mathcal{Q}_p$ if and only if*

$$\sup_{a \in B_1(0)} \int_{B_1(0)} |\overline{D}f(x)|^2 (1 - |\phi_a(x)|^2)^p \, dB_x < \infty \ .$$

Theorem 4. *Let $f : B_1(0) \to \mathbf{H}$ be a hyperholomorphic funtion. Then for $0 < p \leq 1$, f belongs to the space $\mathcal{Q}_p$ if and only if*

$$\sup_{a \in B_1(0)} \int_{B_1(0)} |\overline{D}f(x)|^2 (1 - |\phi_a(x)|^2)^p \, dB_x < \infty \ .$$

Proposition 1. *For $0 < p < q$ holds the inclusion $\mathcal{Q}_p \subset \mathcal{Q}_p$.*

Recently K. Gürlebeck [6] have proved that even, as in the usual complex case the inclusions are strict.
Let $q > 0$ be and $f : B_1(0) \to \mathbf{H}$ be a hyperholomorphic funtion. We say that $f \in B^q$ if

$$\sup_{a \in B_1(0)} \int_{B_1(0)} |\overline{D}f(x)|^q (1 - |x|^2)^{\frac{3q}{2} - 2} (1 - |\phi_a(x)|^2)^3 \, dB_x < \infty \ .$$

Note that for $q = 2$, $B^2 = \mathcal{Q}_3$, that is the limit case of quaternionic $\mathcal{Q}_p$-spaces. In this paper we present several characterizations of the quaternionic Bloch and little Bloch spaces for monogenic functions on the unit ball $B_1(0) \subset \mathbf{R}^3$, which extend -for this kind of functions- the Theorems 1, 2 of Stroethoff's paper for any $p > 0$ and $n = 1$. In particular we prove that for any $q > 0$, B^q is equivalent to be Bloch. The main problem to consider the case $n > 1$, is that in general the composition of hyperholomorphic functions is not hyperholomorphic. The authors and K. Gürlebeck are preparing an article about this matter.

Lemma 2. *Let $f : B_1(0) \to \mathbf{H}$ be a hyperholomorphic funtion. Let $0 < r < 1$, $1 < p$ and be $\Gamma_r := \{x \in \mathbf{R}^3 : \| x \| = r\}$. Then*

$$|f(0)|^p \leq \frac{1}{4\pi r^2} \int_{\Gamma_r} |f(y)|^p d\Gamma_{r,y}$$

Proof. If we apply Cauchy's integral formula to $f(0)$, through Γ_r, we have

$$\begin{aligned} |f(0)| &= \left| \int_{\Gamma_r} e(-y)\alpha(y)f(y)\, d\Gamma_{r,y} \right| \\ &= \frac{1}{4\pi} \left| \int_{\Gamma_r} \frac{\overline{-y}}{|y|^3} \alpha(y) f(y)\, d\Gamma_{r,y} \right| \\ &\leq \frac{1}{4\pi r^2} \int_{\Gamma_r} |f(y)|\, d\Gamma_{r,y} , \end{aligned}$$

where

$$e(y) = \frac{1}{4\pi} \frac{\overline{y}}{|y|^3}$$

is the usual Cauchy kernel and $\alpha(y)$ is the outward pointing normal vector at y.
If we apply Hölder inequality to the last expression for p and $1 - \frac{1}{p}$, we obtain

$$\begin{aligned} \frac{1}{4\pi r^2} \int_{\Gamma_r} |f(y)|\, d\Gamma_{r,y} &\leq \frac{1}{4\pi r^2} \left(\int_{\Gamma_r} |f(y)|^p\, d\Gamma_{r,y} \right)^{\frac{1}{p}} \left(\int_{\Gamma_r} d\Gamma_{r,y} \right)^{1-\frac{1}{p}} \\ &= \frac{(4\pi r^2)^{1-\frac{1}{p}}}{4\pi r^2} \left(\int_{\Gamma_r} |f(y)|^p\, d\Gamma_{r,y} \right)^{\frac{1}{p}} \\ &= \frac{K}{(4\pi r^2)^{\frac{1}{p}}} \left(\int_{\Gamma_r} |f(y)|^p\, d\Gamma_{r,y} \right)^{\frac{1}{p}} . \end{aligned}$$

So we get our desired result. ■

Now we are going to apply this result to obtain an estimation envolving the $\overline{D}$-derivative of f.
Let $R < 1$ and $D(a, R) := \{x \in \mathbf{R}^3 : |\phi_a(x)| < R\}$ be the pseudohyperbolic ball with radius R. Let A denote the normalized Lebesgue area measure on $B_1(0)$. For a Lebesgue measurable set $K \subset B_1(0)$, let $|K|$ denote the measure of K with respect to A. It follows immediately that

$$|D(a, R)| = \frac{(1 - |a|^2)^3}{(1 - R^2|a|^2)^3} R^3 \ .$$

Lemma 3. *Let $f : B_1(0) \to \mathbf{H}$ be a hyperholomorphic funtion. Let $0 < r < 1$, $1 < p$. Then for every $a \in D_1(0)$*

$$|\overline{D}f(a)|^p \leq \frac{5 \cdot 4^{p+1}}{\pi R^5 (1 - R)^{2p} (1 - |a|^2)^3} \int_{D(a,R)} |\overline{D}f(x)|^p \, dV_x \ .$$

Proof. The function $\overline{D}f(x)$ is hyperholomorphic, but after the change of variables $\overline{D}f(\phi_a(w))$ in general is not hyperholomorphic. However we know from Ryan [8] that

$$F(w) = \frac{1 - \overline{w}a}{|1 - \overline{a}w|^3} f(\phi_a(w))$$

is again hyperholomorphic in $B_1(0)$. Then if we apply the previous lemma to $F(w)$ in $\Gamma_r(0)$ we obtain

$$|F(0)|^p \leq \frac{1}{4\pi r^2} \int_{\Gamma_r} |F(y)|^p \, d\Gamma_{r,y},$$

so

$$|f(a)|^p \leq \frac{1}{4\pi r^2} \int_{\Gamma_r} \frac{|f(\phi_a(y))|^p}{|1 - \overline{a}y|^{2p}} d\Gamma_{r,y} \ .$$

Now, multiply both sides of the inequality by r^4 and integrate from 0 to R and consider spherical coordinates

$$|f(a)|^p \int_0^R r^4 \, dr \leq \frac{1}{4\pi} \int_0^R \int_0^\pi \int_0^{2\pi} \frac{|f(\phi_a(y))|^p}{|1 - \overline{a}y|^{2p}} r^2 \sin^2\theta \, d\theta \, d\phi \, dr \ .$$

We have then

$$|f(a)|^p \leq \frac{5}{4\pi R^5} \int_{B_R(0)} \frac{|f(\phi_a(y))|^p}{|1 - \overline{a}y|^p} dB_r \ .$$

Let us apply this inequalyty to $\overline{D}f$.

$$|\overline{D}f(a)|^p \leq \frac{5}{4\pi R^5} \int_{B_R(0)} \frac{|\overline{D}f(\phi_a(y))|^p}{|1-\overline{a}y|^p} dB_r \ .$$

By applying the next formula of change of variable for a Lebesgue measurable function h on $B_r(0)$

$$\int_{B_r(0)} h(\phi_a(w))\, dB_w = \int_{D(a,R)} h(x) \left(\frac{1-|\phi_a(x)|^2}{1-|x|^2}\right)^3 dB_x \ .$$

We have

$$\begin{aligned} |\overline{D}f(a)|^p &\leq \frac{5}{4\pi R^5} \int_{D(a,R)} \frac{|\overline{D}f(x)|^p}{|1-\overline{a}\phi_a(x)|^{2p}} \left(\frac{1-|\phi_a(x)|^2}{1-|x|^2}\right)^3 dB_x \\ &\leq \frac{5}{4\pi R^5(1-R)^{2p}} \int_{D(a,R)} |\overline{D}f(x)|^p \left(\frac{1-|\phi_a(x)|^2}{1-|x|^2}\right)^3 dB_x \end{aligned}$$

But we can use the next estimation

$$\frac{1-|\phi_a(x)|^2}{1-|x|^2} \leq \frac{4}{1-|a|^2}$$

so we obtain finally the estimation

$$|\overline{D}f(a)|^p \leq \frac{5 \cdot 4^2}{\pi R^5(1-R^2)(1-|a|^2)^3} \int_{D(a,R)} |\overline{D}f(x)|^p \, dV_x \ . \blacksquare$$

Now we are going to apply the previous Lemmas to give a set of equivalences for quaternionic Bloch functions. Besides this theorem generalize the Theorem 1 of [9] for the case of hyperholomorphic functions on $B_1(0)$, wih $p > 0$ and $n = 1$.

Theorem 5. *Let $0 < R < 1$. Then for a hyperholomorphic function f on $B_1(0)$ the following conditions are equivalents*

a) $f \in B$,

b) For each $p > 0$

$$\sup_{a\in B_1(0)} \int_{B_1(0)} |\overline{D}f(x)|^p(1-|x|^2)^{\frac{3p}{2}-3}(1-|\phi_a(x)|^2)^3\, dV_x < +\infty$$

c) For each $p > 0$

$$\sup_{a\in B_1(0)} \int_{D(a,R)} |\overline{D}f(x)|^p(1-|x|^2)^{\frac{3p}{2}-3}\, dV_x < +\infty$$

d) For each $p > 0$

$$\sup_{a\in B_1(0)} \frac{1}{|D(a,R)|^{1-p/2}} \int_{D_{(a,R)}} |\overline{D}f(x)|^p \, dV_x < +\infty$$

e) For some $p > 1$

$$\sup_{a\in B_1(0)} \frac{1}{|D(a,R)|^{1-p/2}} \int_{D_{(a,R)}} |\overline{D}f(x)|^p \, dV_x < +\infty \, .$$

Proof a) implies b)

$$\sup_{a\in B_1(0)} \int_{B_1(0)} |\overline{D}f(x)|^p (1-|x|^2)^{\frac{3p}{2}-3} (1-|\phi_a(x)|^2)^3 \, dV_x$$

$$\leq \sup_{x\in B_1(0)} [(1-|x|^2)^{3/2} |\overline{D}f(x)|]^p \int_{B_1(0)} \frac{(1-|\phi_a(x)|^2)^3}{(1-|x|^2)^3} dV_x$$

$$= (\| f \|_B)^p.$$

b) implies c). For $x \in D(a,R)$ we have $(1-R^2)^3 < (1-|\phi_a(x)|^2)^3$, so

$$(1-R^2)^3 \int_{D_{(a,R)}} |\overline{D}f(x)|^p (1-|x|^2)^{\frac{3p}{2}-3} \, dV_x$$

$$\leq \int_{B_1(0)} |\overline{D}f(x)|^p (1-|x|^2)^{\frac{3p}{2}-3} (1-|\phi_a(x)|^2)^3 \, dV_x$$

c) if and only if d) it follows from the fact that $(1-|x|^2)^3 \approx |D(a,R)|$
d) implies e) Is trivial
e) implies a) By Lemma 2, we have

$$[(1-|a|^2)^{3/2} |\overline{D}f(a)|]^p \leq$$

$$\frac{5\cdot 4^2}{\pi R^5 (1-R)^{2p} (1-|a|^2)^{3-3p/2}} \int_{D(a,r)} |\overline{D}f(x)|^p \, d\Gamma_x$$

$$= \frac{5\cdot 4^2}{\pi R^5 (1-R)^{2p} (1-|a|^2)^{3-3p/2}} \frac{(1-|a|^2R^2)^{3-3p/2}}{(1-|a|^2R^2)^{3-3p/2}} \frac{R^{3-3p/2}}{R^{3-3p/2}} \times$$

$$\times \int_{D(a,R)} |\overline{D}f(x)|^p d\Gamma_x$$

$$\leq \frac{5\cdot 4^2}{\pi R^5 (1-R)^{2p} |D(a,R)|^{1-p/2} R^{-3+3p/2} (1-R^2)^{3-3p/2}} \times$$

$$\times \int_{D(a,R)} |\overline{D}f(x)|^p \, d\Gamma_x$$

$$= \frac{5\cdot 4^2}{\pi R^{2+3p/2} (1-R)^{3+p/2} |D(a,R)|^{1-p/2}} \int_{D(a,R)} |\overline{D}f(x)|^p \, d\Gamma_x$$

so the result follows.
In the same way as in the complex case, we can define the *little quaternion Bloch space* B_0, as the set of hyperholomorphic functions on $B_1(0)$ for which

$$\lim_{|x|\to 1^-}(1-|x|^2)|\overline{D}f(x)|=0.$$

So $B_0 \subset B$ and B_0 contains for instance all the hyperholomorphic functions on $B_1(0)$ and continuous on $\overline{B_1(0)}$. Then with the same arguments used to prove the previous theorem, we can prove

Theorem 6. *Let* $0<R<1$. *Then for an hyperholomorphic function* f *on* $B_1(0)$ *the following conditions are equivalents*

a) $f \in B_0$,

b) For each $p>0$

$$\lim_{|a|\to 1^-}\int_{B_1(0)}|\overline{D}f(x)|^p(1-|x|^2)^{\frac{3p}{2}-3}(1-|\phi_a(x)|^2)^3\,dV_x<+\infty$$

c) For each $p>0$

$$\lim_{|a|\to 1^-}\int_{D_{(a,R)}}|\overline{D}f(x)|^p(1-|x|^2)^{\frac{3p}{2}-3}\,dV_x<+\infty$$

d) For each $p>0$

$$\lim_{|a|\to 1^-}\frac{1}{|D(a,R)|^{1-p/2}}\int_{D_{(a,R)}}|\overline{D}f(x)|^p\,dV_x<+\infty$$

e) For some $p>1$

$$\lim_{|a|\to 1^-}\frac{1}{|D(a,R)|^{1-p/2}}\int_{D_{(a,R)}}|\overline{D}f(x)|^p\,dV_x<+\infty\ .$$

References

[1] R. Aulaskari, *On Q_p functions*, Complex Analysis and Related Topics, Operator Theory, Advances and Applications, Vol 114, Birkhauser, 2000.

[2] R. Aulaskari and P. Lappan, *Criteria for an analytic function to be Bloch and harmonic or meromorphic function to be normal*, Complex analysis and its applications, Pitman Research Notes in Mathematics, 305, Longman Scientific & Technical, Harlow (1994), 136-146.

[3] R. Aulaskari, J. Xiao and R. Zhao *On subspaces and subsets of BMOA and UBC*, Analysis 15 (1995), 101-121.

[4] J. Choa, H.O. Kim, Y.Y. Park, *A Bergmann-Carleson measure characterization of Bloch funtions in the unit ball of* $\mathbf{C}^n$, Bull. Korean Math. Soc. 29 (1992).

[5] K. Gürlebeck, U. Kähler, M. V. Shapiro, L.M. Tovar, *On* $\mathcal{Q}_p$ *spaces of Quaternionic-Valued Functions*, Journal of Complex Variables, Vol. 39, (1999) pp. 115-135.

[6] K. Gürlebeck, H.R. Malonek, *On strict inclusions of* $\mathcal{Q}_p$*-spaces of quaternionic valued functions*, Preprint.

[7] P. Lappan, *A survey of* Q_p *spaces*, Complex Analysis and Related Topics, Operator Theory, Advances and Applications, Vol 114, Birkhauser, 2000.

[8] J. Ryan, *Conformally covariant operators in Clifford analysis,* ZAA, 14, (1995), pp. 677-704.

[9] K. Stroethoff, K., *Besov-type characterizations for the Bloch space,* Bull. Austral. Math. Soc. 39 (1989), pp. 405-420.

[10] J. Xiao, *Carleson measures, atomic decomposition and free interpolation from Bloch space*, Ann. Acad. Sci. Fenn. Ser. AI Math. 19 (1994), pp. 35-46.

[11] C. Ouyang, W. Yang, R. Zhao *Möbius invariant* $\mathcal{Q}_p$*-spaces associated with the Green's function on the unit ball of* $\mathbf{C}^n$, Pacific J. Math. (1998)

15

BERGMAN AND BAUER OPERATORS FOR ELLIPTIC EQUATIONS IN TWO INDEPENDENT VARIABLES

K.W. Tomantschger

University Graz of Technology, Department of Mathematics,

A 8010 Graz, Steyrergasse 30, Austria, Europe

E-mail: toman@weyl.math.tu-graz.ac.at

Abstract: Object of this talk is to discuss some new results in solving the linear second order elliptic partial differential equation given in the canonical form

$$\Delta u + \alpha(x,y)u_x + \beta(x,y)u_y + \gamma(x,y)u = 0$$

with real analytic coefficients on a simply connected domain. First the equation is transformed in a suitable complex form, the formally hyperbolic differential equation

$$v_{z\zeta} + a_1(z,\zeta)v_z + a_2(z,\zeta)v_\zeta + a_3(z,\zeta)v = 0,$$

in which it can be solved by different integral operators and the differential operators of K. W. Bauer. In this paper the integral operators of St. Bergman are discussed. M. Kracht and E. Kreyszig could prove that Bauer's operators, which possess quite another structure than these of Bergman, are a special case of Bergman's operator with polynomial kernels (Lemma 3). Such differential and integral operators for representing solutions of our equation provide a way of applying function theoretic methods and results to the study of various general properties of solutions. The rate of success of this approach depends on the construction of suitable kernels.

In this paper the kernels of the first kind of the Bergman operators are calculated for two differential equations which are adjoint to each other. By using Lemma 3 the differential operator solutions are obtained.

Keywords: elliptic differential equation; formally hyperbolic differential equation in the complex analysis; representations of solutions by the integral operator of St. Bergman and the differential operator of K. W. Bauer.

AMS-classification: 35C15

1. Introduction

In this paper we shall study one of the most popular integral operators and the differential operator which were introduced in the last 70 years of the 20th century. They associate in a one-to-one manner the solutions of linear, elliptic partial differential equations,

$$\Delta u + \alpha(x,y)u_x + \beta(x,y)u_y + \gamma(x,y)u = 0, \qquad u = u(x,y)\,, \tag{1}$$

with holomorphic functions of one complex variable. The coefficients α, β, γ are real-analytic functions on a simple connecting domain. For continuing these coefficients analytically to the complex we extend x and y to complex variables and introduce the new variables

$$z = x + iy \qquad \text{and} \qquad \zeta = x - iy$$

which are independent. (Note that $\zeta = \bar{z}$ if and only if x and y are real). By introducing the differential operators of Wirtinger

$$\frac{\partial}{\partial z} = \frac{1}{2}\left(\frac{\partial}{\partial x} - i\frac{\partial}{\partial y}\right), \qquad \frac{\partial}{\partial \zeta} = \frac{1}{2}\left(\frac{\partial}{\partial x} + i\frac{\partial}{\partial y}\right)$$

we obtain

$$\Delta = \frac{\partial^2}{\partial x^2} + \frac{\partial^2}{\partial y^2} = 4\,\frac{\partial^2}{\partial z\,\partial \zeta}\,.$$

Under the above transformation the elliptic equation (1) becomes the formally hyperbolic equation

$$L_1 v = v_{z\zeta} + a_1(z,\zeta)v_z + a_2(z,\zeta)v_\zeta + a_3(z,\zeta)v = 0, \qquad v = v(z,\zeta). \tag{2}$$

Here

$$a_1(z,\zeta) = \frac{1}{4}\left[\alpha\left(\frac{z+\zeta}{2}, \frac{z-\zeta}{2}\right) + i\beta\left(\frac{z+\zeta}{2}, \frac{z-\zeta}{2}\right)\right],$$

$$a_2(z,\zeta) = \frac{1}{4}\left[\alpha\left(\frac{z+\zeta}{2}, \frac{z-\zeta}{2}\right) - i\beta\left(\frac{z+\zeta}{2}, \frac{z-\zeta}{2}\right)\right],$$

$$a_3(z,\zeta) = \frac{1}{4}\gamma\left(\frac{z+\zeta}{2}, \frac{z-\zeta}{2}\right) \qquad \text{and} \qquad v(z,\zeta) = u\left(\frac{z+\zeta}{2}, \frac{z-\zeta}{2}\right).$$

Note:

$a_2(z,\bar{z}) = \overline{a_1(z,\bar{z})} = \overline{a_1}(\bar{z},z)$ and $a_3(z,\bar{z}) = \overline{a_3(z,\bar{z})}$, that means a_3 is real, iff x, y are real variables and the coefficients of (1) are real functions.

$(z,\zeta) \in G = G_1 \times \bar{G_1} \subset \mathbb{C}^2$, where G_1 denotes a simple connected domain of the z plane and $\bar{G_1}$ is the image of G_1 in the real axis. G_1 we call a *fundamental domain.* a_k are analytic regular (holomorphic) functions in our bicylinder G. We can eliminate one of the two derivatives of first order of (2), e.g., the derivative with respect to z. In this case we set $v = w \cdot exp(-\int a_1(z,\zeta)\, d\zeta)$ and obtain

$$Lw = 0, \quad L := \frac{\partial^2}{\partial z \partial \zeta} + B(z,\zeta)\frac{\partial}{\partial \zeta} + C(z,\zeta), \quad w = w(z,\zeta) \quad (3)$$

$$B = a_2 - \int \frac{\partial a_1(z,\zeta)}{\partial z}\, d\zeta, \qquad C = a_3 - a_1 a_2 - \frac{\partial a_1}{\partial z}.$$

2. Bergman's Integral Operators

Stefan Bergman in a series of papers has investigated the differential equation (1) and obtained methods for constructing solutions. This happened between 1930 and 1943.

Bergman's methods is concerned with obtaining an integral representation of solutions for equation (3) in the form

$$(T_1 f)(z,\zeta) := \int_{\substack{-1 \\ \Gamma}}^{1} E(z,\zeta,t)\, f\left(\frac{z-z_0}{2}(1-t^2)\right) \frac{dt}{\sqrt{1-t^2}}. \qquad (4)$$

E is a suitable choosen *Bergman kernel* (*generating function*), and $\Gamma \subset B_0 \cup \partial B_0$ is an oriented rectifiable arc from $t = -1$ to $t = 1$, which usually is taken not to pass through the origin. B_0 is the open unit disk in the complex t-plane. f is the *associated function* of $w(z,\zeta)$. E is independent of the choice of f and only depends on the coefficients of $Lw = 0$, whereas the holomorphic function $f(z)$ is independent of the coefficients of (3). z_0, called the *point of reference* of T_1, is an arbitrarily fixed point of G_1. Very often we can choose $z_0 = 0$, mainly in order to simplify formulas. The *Bergman operator* T_1 transforms the class of f into particular solutions of (3). If we substitute (4) into (3) and integrate by parts, we obtain the following sufficient conditions:

Lemma 1.
Let the coefficients $B(z,\zeta)$ and $C(z,\zeta)$ of the equation (3) be holomorphic functions in $[\mid z - z_0 \mid \le r,\ \mid \zeta - \zeta_0 \mid \le r,\ r > 0,\ fixed]$. Suppose that E is a holomorphic solution of

$$(1-t^2)E_{\zeta t} - \frac{1}{t}E_\zeta + 2(z-z_0)tLE = 0, \qquad E \not\equiv 0, \qquad (5)$$

on $G \times \Gamma$ satisfying the conditions

$$(1-t^2)^{-1/2} E_\zeta \to 0 \quad as \quad t \to \pm 1$$

(uniformly on G) and

$$(z-z_0)^{-1}t^{-1}E_\zeta \quad \textit{is a continuous function on } G\times\Gamma.$$

Then $w = T_1 f$ satisfies $Lw = 0$ locally.

At first glance it might appear that the problem of representing solutions of (2) resp. (3) as integrals of the form (4) has been made more difficult because one must solve a more complicated partial differential equation (5) in three variables than $Lw = 0$ with two. However, this is misleading since only one single particular solution of (5), fulfilling the two conditions is needed to obtain a continuous operator T_1, which constructs infinitely much solutions $w = T_1 f$, because $f(z)$ can be any arbitrary holomorphic function. If G_1 is a simple connected fundamental domain, then since $Re(T_1 f)$ is continuous we may generate a complete system of real solutions for (1) by mapping the powers of z onto solutions, i.e., $\Psi_n(z,\zeta) \equiv (T_1 z^n)$, $n = 1,2,3,\cdots$. This follows from the well-known fact that the powers of z form a complete system of holomorphic functions for any simple connected region. If the $\{\Psi_n\}$ were not complete then the $\{z_n\}$ would not be complete.

Demanding the restriction $\zeta = \bar{z}$, $Re\{w\}$ and $Im\{w\}$ generally imply two independent real solutions of the corresponding real elliptic equation of (3), defined in the fundamental domain G_1.

Bergman has given a constructive method for obtaining solutions for the kernel equation (5). For instance, if we seek a solution $E(z,\zeta,t)$, called *Bergman kernel of the first kind*, defined as

$$E(z,\zeta,t) = 1 + \sum_{k\geq 1}(z-z_0)^k P_k(z,\zeta)\, t^{2k} \quad \text{with} \quad E(z,\zeta_0,t) = 1. \quad (6)$$

ζ_0 represents an arbitrary fixed point of $\bar{G}_1$.

Remark:

1) The series (6) does not contain odd powers of t. This is no loss of generality, as can be seen from (4).
2) Bergman has shown by using the method of dominants that the series (6) converge to a holomorphic function of three complex variables z,ζ,t in

$$[\,|\,z-z_0\,| < r/3, \quad |\,\zeta-\zeta_0\,| < r/3, \quad |\,t\,| \leq 1\,], \qquad r > 0,$$

if the coefficients of (2) resp. (3) are holomorphic functions in the bicylinder

$$[\,|\,z-z_0\,| \leq r, \quad |\,\zeta-\zeta_0\,| \leq r\,].$$

Inserting (6) into (5) we find that the P_k satisfy the following equations:

$$\frac{(2k+1)}{2} P_{k+1,\zeta} + LP_k = 0, \qquad P_0 = 1, \qquad k = 0, 1, 2, \cdots. \tag{7}$$

It can be proved, that a kernel of the first kind is *uniquely* and that we obtain *all* solutions by two kernels of that type.

The second solution we get in the same way from the following equation with the z derivative

$$\tilde{L}W = W_{z\zeta} + \tilde{A}(z,\zeta)W_z + \tilde{C}(z,\zeta)W = 0, \qquad W = W(z,\zeta), \tag{8}$$

using the integral operator

$$(T_1^* g)(z,\zeta) := \int_{\substack{-1 \\ \Lambda}}^{1} E^*(z,\zeta,t)\, g\left(\frac{\zeta-\zeta_0}{2}(1-t^2)\right) \frac{dt}{\sqrt{1-t^2}}. \tag{9}$$

$$E^*(z,\zeta,t) = 1 + \sum_{k\geq 1} (\zeta-\zeta_0)^k q_k(z,\zeta)\, t^{2k} \qquad \text{with} \qquad E^*(z_0,\zeta,t) = 1 \tag{10}$$

is the *Bergman kernel of the first kind of* T_1^*. Λ represents an oriented rectifiable path in $B_0 \cup \partial B_0$. $\zeta_0 \in \bar{G}_1$ is the reference point of T_1^*. $g(\zeta)$ is an arbitrary holomorphic function.

Lemma 2.

a) For each equation $L_1 v = 0$ with holomorphic coefficients in $G \ni (z_0, \zeta_0)$ there exist Bergman kernels E, E^* of the first kind, which are uniquely determined, so that

$$v = exp\left(-\int_{\tilde{\zeta}_0}^{\zeta} a_1(z,\rho)d\rho\right) T_1 f + exp\left(-\int_{\tilde{z}_0}^{z} a_2(\sigma,\zeta)d\sigma\right) T_1^* g\,.$$

$f(z)$ and $g(\zeta)$ arearbitrary holomorphic functions in G_1 resp. $\bar{G}_1$; $(\tilde{z}_0, \tilde{\zeta}_0) \in G$ is an arbitrarily fixed point.

b) Conversely, each holomorphic solution v of $L_1 v = 0$ can be represented in this form with suitable holomorphic functions $f(z)$ and $g(\zeta)$ and Bergman kernels $E(z,\zeta,t)$, $E^*(z,\zeta,t)$ of the first kind.

Example 1. We consider first the following equation given in form (2):

$$L_1 v = v_{z\zeta} - \frac{(\mu+1)K}{\eta H} v_z - \frac{(\lambda+1)H}{\eta K} v_\zeta = 0, \tag{11}$$

$$H(z,\zeta) = (\mu+1)\zeta-\lambda z, \quad K(z,\zeta) = (\lambda+1)z-\mu\zeta, \quad \eta = z+\zeta, \quad \lambda,\mu \in \mathbb{C}.$$

In our first step we put the equation into the reduced form

$$Lw = w_{z\zeta} + B(z,\zeta)w_\zeta + C(z,\zeta)w = 0.$$

Here $w = w(z,\zeta;\lambda,\mu) = T_1 f$ is the Bergman integral with the kernel $E(z,\zeta,t;\lambda,\mu)$; see (4). The coefficients of $Lw = 0$ are given by

$$\begin{aligned} B(z,\zeta;\lambda,\mu) &= \frac{\lambda-\mu}{\eta} - \frac{\lambda+1}{K} - \frac{\lambda}{H}, \\ C(z,\zeta;\lambda,\mu) &= -\frac{\lambda(\mu+1)}{\eta^2} + \frac{\lambda(\mu+1)}{H^2}. \end{aligned} \tag{12}$$

For obtaining a second solution we calculate the kernel $E^*(z,\zeta,t)$ of the Bergman integral (9). This integral $T_1^* g$ satisfies equation

$$\tilde{L}W = W_{z\zeta} + \tilde{A}(z,\zeta)W_z + \tilde{C}(z,\zeta)W = 0.$$

After determining the coefficients of this equation we get

$$\tilde{A}(z,\zeta;\lambda,\mu) = \frac{\mu-\lambda}{\eta} - \frac{\mu+1}{H} - \frac{\mu}{K},$$

$$\tilde{C}(z,\zeta;\lambda,\mu) = -\frac{\mu(\lambda+1)}{\eta^2} + \frac{\mu(\lambda+1)}{K^2}.$$

Note: H becomes K resp. K turnes to H if we exchange λ with μ and z with ζ.

Comparing the functions $\tilde{A}$ and $\tilde{C}$ with that of (12) and applying the notation, yields the relations

$$\tilde{A}(z,\zeta;\lambda,\mu) = B(\zeta,z;\mu,\lambda), \qquad \tilde{C}(z,\zeta;\lambda,\mu) = C(\zeta,z;\mu,\lambda).$$

This implies

$$W(z,\zeta;\lambda,\mu) = w(\zeta,z;\mu,\lambda).$$

Since E resp. E^* are the Bergman kernels of the solving operators $T_1 f$ resp. $T_1^* g$ of equation $Lw = 0$ resp. $\tilde{L}W = 0$, we may deduce

$$E^*(z,\zeta,t;\lambda,\mu) = E(\zeta,z,t;\mu,\lambda). \tag{13}$$

That means, we only have to calculate *one* Bergman kernel for obtaining a general solution of (11). And we even get *all* solutions of $L_1 v = 0$, if this kernel is a Bergman kernel of the first kind.

Now we compute the Bergman kernel of the first kind (6). Starting with $P_0 = 1$ the recursive scheme (7) implies

$$\begin{aligned} P_1 &= \left[-2\,\frac{(\mu+1)}{\eta} + \frac{2}{H}\right]\lambda\,, \\ P_2 &= \left[\frac{2}{3}\,\frac{(\mu+2)}{\eta^2} - \frac{4}{3}\,\frac{1}{\eta H}\right](\lambda-1)\lambda(\mu+1)\,, \\ P_3 &= \left[-\frac{4}{45}\,\frac{(\mu+3)}{\eta^3} + \frac{4}{15}\,\frac{1}{\eta^2 H}\right](\lambda-2)(\lambda-1)\lambda(\mu+1)(\mu+2)\,. \end{aligned}$$

Concerning these three functions we make the following assumption:

$$P_k(z,\zeta;\lambda,\mu) = \frac{\alpha_k}{\eta^k} + \frac{\beta_k}{\eta^{k-1}H}\,, \qquad \alpha_k,\beta_k \in \mathbb{C}\,, \quad k \in \mathbb{N}\,. \tag{14}$$

Substituting (14) into (7) and using the relation $H + K = \eta$ we obtain

$$\eta^{-k-2}D_1 + \frac{\eta^{-k-1}}{K}\,D_2 + \frac{\eta^{-k-1}}{H}\,D_3 + \frac{\eta^{-k}}{H^2}\,D_3 = 0$$

with

$$\begin{aligned} D_1 &= -\frac{1}{2}(2k+1)(k+1)\alpha_{k+1} + (k+\mu+1)(k-\lambda)\alpha_k, \\ D_2 &= k\alpha_k + (k+\mu)\beta_k, \\ D_3 &= -\frac{1}{2}(2k+1)\beta_{k+1} + (k+\mu)\beta_k + \lambda\alpha_k. \end{aligned}$$

Setting $D_1 = 0$ we get a recursive relation for α_k. Solving this and defining $\alpha_0 = 1$ yields [1]

$$\alpha_k = \frac{(-\lambda)_k\,(\mu+1)_k}{k!\,\left(\frac{1}{2}\right)_k}\,, \qquad \lambda,\mu \in \mathbb{C}, \quad k \in \mathbb{N}. \tag{15}$$

Taking $D_2 = 0$ and using (15) we get the relation

$$\beta_k = 2\lambda\,\frac{(1-\lambda)_{k-1}\,(\mu+1)_{k-1}}{(k-1)!\,\left(\frac{3}{2}\right)_{k-1}}\,, \qquad \lambda,\mu \in \mathbb{C}, \quad k \in \mathbb{N}. \tag{16}$$

Demanding $D_3 = 0$, this equation is identically satisfied by (15) and (16). So the functions P_k are uniquely determined. Inserting (14) with (15) and (16) into (6), taking $z_0 = 0$, then introducing the new variable

$$s = \frac{zt^2}{\eta}\,, \tag{17}$$

[1] $(c)_k = c(c+1)\cdots(c+k-1)$, $(c)_0 = 1$, $c \in \mathbb{C}$, $k \in \mathbb{N}$.

we obtain the Bergman kernel of the first kind [2]

$$E(z,\zeta,t;\lambda,\mu) = {}_2F_1(-\lambda,\mu+1;\frac{1}{2};s) + \frac{2\lambda\eta s}{H}\,{}_2F_1(1-\lambda,\mu+1;\frac{3}{2};s)\,. \tag{18}$$

Using relation (13) and setting $\zeta_0 = 0$ the second Bergman kernel of the first kind is given by

$$E^*(z,\zeta,t;\lambda,\mu) = {}_2F_1(-\mu,\lambda+1;\frac{1}{2};S) + \frac{2\mu\eta s}{K}\,{}_2F_1(1-\mu,\lambda+1;\frac{3}{2};S) \tag{19}$$

with

$$S = \frac{\zeta t^2}{\eta}.$$

Remark:

A second way of determining the kernel is the following. If one seeks a solution in the form of a Bergman kernel of the first kind the calculated functions P_1, P_2 and P_3 can be substituted into the kernel formula (6)

$$E(z,\zeta,t) = 1 + (z-z_0)P_1\,t^2 + (z-z_0)^2P_2\,t^4 + (z-z_0)^3P_3\,t^6 + \cdots.$$

Considering the case with a vanishing point of reference, i. e. $z_0 = 0$, the first terms of this relation may be put into the form

$$\begin{aligned} E(z,\zeta,t) = 1 &+ 2\left[-(\mu+1)\left[\frac{zt^2}{\eta}\right] + \frac{\eta}{H}\left(\frac{zt^2}{\eta}\right)\right]\lambda + \\ &+ \frac{2}{3}\left[(\mu+2)\left[\frac{zt^2}{\eta}\right]^2 - 2\frac{\eta}{H}\left(\frac{zt^2}{\eta}\right)^2\right](\lambda-1)\lambda(\mu+1) + \\ &+ \frac{4}{15}\left[-\frac{(\mu+3)}{3}\left[\frac{zt^2}{\eta}\right]^3 + \frac{\eta}{H}\left(\frac{zt^2}{\eta}\right)^3\right](\lambda-2)_3(\mu+1)_2 + \cdots. \end{aligned}$$

---

[2] ${}_2F_1(\alpha,\beta;\gamma;z)$ represents the hypergeometric function defined by

$${}_2F_1(\alpha,\beta;\gamma;z) = \sum_{k=0}^{\infty}\frac{(\alpha)_k\,(\beta)_k}{(\gamma)_k}\frac{z^k}{k!}, \qquad |\,z\,| < 1,\quad \alpha,\beta,\gamma \in \mathbb{C},\quad \gamma \neq -m,\quad m \in \mathbb{N}_0,$$

satisfying

$$z(1-z)w'' + [\gamma - (\alpha+\beta+1)z]w' - \alpha\beta w = 0.$$

If one introduces the functions p and q, the kernel series can be assumed in the form

$$E(z,\zeta,t) = p(s) + \frac{\eta}{H}\, q(s) \qquad s = \frac{zt^2}{\eta}. \tag{20}$$

Since $E(z,\zeta,t)$ is a solution of equation (5) we insert it into this equation and obtain

$$\frac{2(\lambda+1)zt}{\eta K}\,\hat{D}_1 + \frac{2}{tH}\,\hat{D}_2 + \frac{2(\mu+1)\eta}{tH^2}\,\hat{D}_3 + \frac{zt}{\eta^2}\,\hat{D}_4\,.$$

The functions $\hat{D}_k$ are defined by

$$\begin{aligned}
\hat{D}_1 &= sp' + \mu q + sq'\,,\\
\hat{D}_2 &= s^2(s-1)q'' + \left[(\mu+1)s^2 + s\right]q' - q + \lambda s^2 p'\,,\\
\hat{D}_3 &= 2s(s-1)q'' + (2\mu s + 1)q + 2\lambda sp\,,\\
\hat{D}_4 &= s(1-s)p'' + \left[\frac{1}{2} - (2-\lambda+\mu)\right]p' + \lambda(\mu+1)p\,.
\end{aligned}$$

Demanding that all $\hat{D}_k$ vanish we again have to solve an overdetermined system as before. But now even four equations are given and two unknown functions have to be calculated. Starting with $\hat{D}_4 = 0$ we have an hypergeometric differential equation which implies the solution

$$p(s) = {}_2F_1(-\lambda, \mu+1; \frac{1}{2}; s)\,. \tag{21}$$

Putting this function into equation $\hat{D}_1 = 0$, the resulting inhomogeneous equation yields the second function

$$q(s) = 2\lambda s\, {}_2F_1(1-\lambda, \mu+1; \frac{3}{2}; s)\,. \tag{22}$$

These two functions (21) and (22) satisfy the remaining equations $\hat{D}_2 = 0$ and $\hat{D}_3 = 0$. Substituting $p(s)$ and $q(s)$ into (20) we have the same result which we had gotten for the kernel by recursions (see (18)).

Proposition 1.

All solutions of the differential equation (11)

$$L_1 v = v_{z\zeta} - \frac{(\mu+1)K}{\eta H}\, v_z - \frac{(\lambda+1)H}{\eta K}\, v_\zeta = 0,$$

$H(z,\zeta) = (\mu+1)\zeta - \lambda z, \quad K(z,\zeta) = (\lambda+1)z - \mu\zeta, \quad \eta = z+\zeta, \quad \lambda, \mu \in \mathbb{C},$

can be represented in the form

$$\begin{aligned} v(z,\zeta;\lambda,\mu) &= \frac{K(z,\zeta)}{\eta^{\mu+1}} \int_{-1 \atop \Gamma}^{1} E(z,\zeta,t)\, f\left(\frac{z}{2}(1-t^2)\right) \frac{dt}{\sqrt{1-t^2}} + \\ &+ \frac{H(z,\zeta)}{\eta^{\lambda+1}} \int_{-1 \atop \Lambda}^{1} E^*(z,\zeta,t)\, g\left(\frac{\zeta}{2}(1-t^2)\right) \frac{dt}{\sqrt{1-t^2}} . \end{aligned} \tag{23}$$

$E(z,\zeta,t;\lambda,\mu)$ *and* $E^*(z,\zeta,t;\lambda,\mu)$ *are Bergman kernels of the first kind and given in* (18) *and* (19). $f(z)$ *and* $g(\zeta)$ *are arbitrary holomorphic functions which are independent of the differential equation.* Γ *and* Λ *are rectifiable arcs in the unit circle of the complex t-plane connecting* -1 *and* 1. *They are usually taken not to pass through the origin. This Bergman representation is valid in every domain* $G = G_1 \times \bar{G}_1 \subset \mathbb{C}^2$, *where* G_1 *is a fundamental domain. Hence it holds for every simply connected domain* G_1 *for which the coefficients of* (11) *are holomorphic.*

Example 2. We next turn to the adjoint differential equation of (11):

$$V_{z\zeta} + \left[\frac{(\mu+1)K}{\eta H} V\right]_z + \left[\frac{(\lambda+1)H}{\eta K} V\right]_\zeta = 0, \tag{24}$$

$H(z,\zeta) = (\mu+1)\zeta - \lambda z, \quad K(z,\zeta) = (\lambda+1)z - \mu\zeta, \quad \eta = z+\zeta, \quad \lambda, \mu \in \mathbb{C}.$

Again we compute the Bergman representation of *all* solutions of (24). Equation (24) may be simplified by the transformation

$$V = \frac{\eta}{HK} \tilde{V} \tag{25}$$

to the form

$$\tilde{V}_{z\zeta} + \frac{\mu H}{\eta K} \tilde{V}_z + \frac{\lambda K}{\eta H} \tilde{V}_\zeta = 0 . \tag{26}$$

Remark:

The factor

$$\frac{\eta}{HK}$$

in relation (25) is a particular solution of differential equation (24).

Equation (26) is similar to the already solved equation (11), but not the same. Comparing these two differential equations, it can be seen, that the functions H and K are changed and for the constant μ resp. λ is written $-\mu-1$ resp. $-\lambda-1$. Looking at the definitions of H and K, we see that H changes to K and K becomes H if

$$\mu \to -\mu - 1 \qquad \text{and} \qquad \lambda \to -\lambda - 1. \tag{27}$$

Application of (27) on equation (26) resp. (11) turns this equation to (11) resp. (26). This implies

$$\tilde{V}(z,\zeta;\lambda,\mu) = v(z,\zeta;-\mu-1,-\lambda-1). \tag{28}$$

Using Proposition 1, transformation (25) and relation (28) we can say:

Proposition 2.

All solutions of the differential equation (24)

$$V_{z\zeta} + \left[\frac{(\mu+1)K}{\eta H}V\right]_z + \left[\frac{(\lambda+1)H}{\eta K}V\right]_\zeta = 0,$$

$$H(z,\zeta) = (\mu+1)\zeta - \lambda z, \quad K(z,\zeta) = (\lambda+1)z - \mu\zeta, \quad \eta = z+\zeta, \quad \lambda,\mu \in \mathbb{C},$$

which is adjoint to (11), *can be represented in the form*

$$\begin{aligned} V(z,\zeta;\lambda,\mu) &= \frac{\eta^{\mu+1}}{K(z,\zeta)} \int_{\substack{-1\\ \Gamma}}^{1} \tilde{E}(z,\zeta,t)\, f\left(\frac{z}{2}(1-t^2)\right) \frac{dt}{\sqrt{1-t^2}} + \\ &+ \frac{\eta^{\lambda+1}}{H(z,\zeta)} \int_{\substack{-1\\ \Lambda}}^{1} \tilde{E}^*(z,\zeta,t)\, g\left(\frac{\zeta}{2}(1-t^2)\right) \frac{dt}{\sqrt{1-t^2}} \end{aligned} \tag{29}$$

The generating functions

$$\tilde{E}(z,\zeta,t;\lambda,\mu) = {}_2F_1(\lambda+1,-\mu;\frac{1}{2};s) - \frac{2(\lambda+1)\eta s}{K}\,{}_2F_1(\lambda+2,-\mu;\frac{3}{2};s) \tag{30}$$

and

$$\tilde{E}^*(z,\zeta,t;\lambda,\mu) = {}_2F_1(\mu+1,-\lambda;\frac{1}{2};S) - \frac{2(\mu+1)\eta s}{H}\,{}_2F_1(\mu+2,-\lambda;\frac{3}{2};S) \tag{31}$$

are Bergman kernels of the first kind with the variable

$$s = \frac{zt^2}{\eta} \qquad \textit{resp.} \qquad S = \frac{\zeta t^2}{\eta}.$$

$f(z)$ and $g(\zeta)$ are arbitrary holomorphic functions which are independent of the differential equation (24). Γ and Λ are rectifiable arcs in the unit circle of the complex t-plane connecting -1 and 1. They are usually taken not to pass through the origin. This Bergman representation is valid in every domain $G = G_1 \times \bar{G}_1 \subset \mathbb{C}^2$, where G_1 is a fundamental domain. Hence it holds for every simple connected domain G_1 for which the coefficients of (24) are holomorphic.

Remark:

By applying the substitution

$$V = \frac{\eta^{\mu+1}}{H}\omega$$

to the given adjoint differential equation (24), this can be simplified to the canonical form without z derivative

$$\omega_{z\zeta} - [\hat{B}(z,\zeta)\,\omega]_\zeta + \hat{C}(z,\zeta)\,\omega = 0. \tag{32}$$

The coefficients are given by

$$\begin{aligned}\hat{B}(z,\zeta;\lambda,\mu) &= \frac{\lambda-\mu}{\eta} - \frac{\lambda+1}{K} - \frac{\lambda}{H},\\ \hat{C}(z,\zeta;\lambda,\mu) &= -\frac{\lambda(\mu+1)}{\eta^2} + \frac{\lambda(\mu+1)}{H^2}.\end{aligned}$$

Looking at (12) we see that these coefficients are quite the same as we have in equation

$$Lw = w_{z\zeta} + B(z,\zeta)w_\zeta + C(z,\zeta)w = 0,$$

which is the canonical form of (11). $\hat{B} \equiv B$ and $\hat{C} \equiv C$ imply that equation (32) is adjoint to $Lw = 0$. Differentiating the second term of equation (32) and comparing its coefficients

$$-\hat{B}(z,\zeta;\lambda,\mu) = \frac{\mu-\lambda}{\eta} + \frac{\lambda+1}{K} + \frac{\lambda}{H}, \tag{33}$$

$$\hat{C}(z,\zeta;\lambda,\mu) - \hat{B}_\zeta(z,\zeta;\lambda,\mu) = -\frac{(\lambda+1)\mu}{\eta^2} + \frac{(\lambda+1)\mu}{K^2}, \tag{34}$$

with these of $Lw = 0$ given in (12), we again see that we get (33) and (34) if we replace in (12) λ by $-\lambda-1$ and μ by $-\mu-1$.

Now we can say:

Let the differential equation $L\,w(z,\zeta;\lambda,\mu)=0$ with the coefficients

$$B(z,\zeta;\lambda,\mu) \equiv \hat{B}(z,\zeta;\lambda,\mu) \qquad \text{and} \qquad C(z,\zeta;\lambda,\mu) \equiv \hat{C}(z,\zeta;\lambda,\mu)$$

be given.
If $w(z,\zeta;-\lambda-1,-\mu-1)$ represents a solution of the equation

$$Lw \;=\; w_{z\zeta} + B(z,\zeta;-\lambda-1,-\mu-1)\,w_\zeta + C(z,\zeta;-\lambda-1,-\mu-1)\,w \;=\; 0,$$

then

$$\omega(z,\zeta;\lambda,\mu) \;=\; w(z,\zeta;-\lambda-1,-\mu-1) \tag{35}$$

is a solution of the adjoint equation to $L\,w(z,\zeta;\lambda,\mu)=0$, that is equation (32)

$$\omega_{z\zeta} \;-\; B(z,\zeta;\lambda,\mu)\,\omega_\zeta \;+\; [C(z,\zeta;\lambda,\mu) - B_\zeta(z,\zeta;\lambda,\mu)]\,\omega \;=\; 0\,.$$

3. The Differential Operator of K. W. Bauer

1964 the German mathematician K. W. Bauer introduced a new kind of representation of solutions of the equation

$$(1\,+\,\varepsilon z\zeta)^2 w_{z\zeta} \,+\, \varepsilon n(n+1)w \;=\; 0\,, \qquad \varepsilon \;=\; \pm 1\,, \quad n \in \mathbb{N}_0\,, \tag{36}$$

often called the *Bauer-Peschl equation.* It is connected with the three dimensional Laplace equation $\triangle_3 U \;=\; 0$ and the wave equation

$$\triangle_3 U \;=\; \frac{\partial^2 U}{\partial t^2}\,.$$

The solutions of the Bauer-Peschl equation are given by

$$w \;=\; \sum_{k=0}^{n} \frac{(2n-k)!}{k!\,(n-k)!}\,\frac{(-\varepsilon)^{n-k}}{(1+\varepsilon z\zeta)^{n-k}}\left[\zeta^{n-k}\,\frac{d^k f(z)}{dz^k} \;+\; z^{n-k}\,\frac{d^k g(\zeta)}{d\zeta^k}\right].$$

All the solutions of (36) can be represented with two arbitrary holomorphic functions $f(z)$ and $g(\zeta)$ and a finite number of their derivatives.

Let us treat the formally hyperbolic differential equation in the canonical form (3), i.e.,

$$Lw \;=\; w_{z\zeta} \,+\, B(z,\zeta)w_\zeta \,+\, C(z,\zeta)w \;=\; 0\,.$$

We want to have solutions of the form

$$w = T_2 f\,, \qquad T_2 \;=\; \sum_{k=0}^{n} b_k(z,\zeta)\,\frac{d^k}{dz^k}\,. \tag{37}$$

$f(z)$ represents arbitrary holomorphic functions which are independent of $Lw = 0$. Inserting $w = T_2 f$ into (3) implies the recursive system

$$f^{(n+1)}(z): \qquad \frac{\partial b_n}{\partial \zeta} = 0, \tag{38}$$

$$f^{(n)}(z): \qquad \frac{\partial b_{n-1}}{\partial \zeta} + C(z,\zeta) b_n = 0, \tag{39}$$

$$f^{(k)}(z): \qquad \frac{\partial b_{k-1}}{\partial \zeta} + L b_k = 0, \qquad k = n-1, \cdots, 1, \tag{40}$$

$$f(z): \qquad L b_0 = b_{0,z\zeta} + B(z,\zeta) b_{0,\zeta} + C(z,\zeta) b_0 = 0. \tag{41}$$

This system possesses $n+2$ equations, but only $n+1$ undetermined functions b_k. So the system is overdetermined. The last equation [3] (41) will not be satisfied for all coefficients $B(z,\zeta)$ and $C(z,\zeta)$. This means, that not every equation $Lw = 0$ possesses a solution in form of a differential operator (37). Which equations $Lw = 0$ can be solved by a differential operator of K. W. Bauer is an unsolved problem till today.

In 1969 the German mathematicians M. Kracht and E. Kreyszig had shown that solutions obtained by the Bergman integral operator with *polynomial kernels*

$$E(z,\zeta,t) = \sum_{k=0}^{N} (z - z_0)^k P_k(z,\zeta) t^{2k}, \qquad P_N \neq 0, \tag{42}$$

can be converted to an *integral-free form*, which is this with a differential operator [9].

Lemma 3. *Integral-free representations of solutions with Bergman operators:*

Let $Lw = 0$ be such an equation that there exists an integral operator T_1 of Bergman (see (4)*) with polynomial kernels* (42), *so that*

$$w = T_1 f = \sum_{k=0}^{N} P_k(z,\zeta)(z - z_0)^k \int_{\Gamma\,-1}^{1} t^{2k} f\left(\frac{z - z_0}{2}(1 - t^2)\right) \frac{dt}{\sqrt{1-t^2}}.$$

That means, that the coefficients $P_k(z,\zeta)$, $k = 1, \cdots, N$, are solution of a finite and overdetermined system

$$\frac{(2k+1)}{2} P_{k+1,\zeta} + L P_k = 0, \qquad k = 0, 1, 2, \cdots, N,$$

[3]Note, that $b_0(z,\zeta)$ is a particular solution of equation (3).

which is defined in $G_1 \times \bar{G}_1$.

Then we have $w = \tilde{w}$, *where*

$$\tilde{w} = \tilde{T}_1 \tilde{f} = \sum_{k=0}^{N} \frac{(2k)!}{2^{2k}\, k!} P_k(z,\zeta) \frac{d^{N-k}\tilde{f}(z)}{dz^{N-k}}.$$

The function $\tilde{f}(z)$ *is the analytic continuation into* G_1 *of the series*

$$\sum_{j=N}^{\infty} \frac{[2(j-N)]!\,\pi}{8^{j-N}\,(j-N)!\,j!}\, b_{j-N}\,(z-z_0)^j.$$

Here, the b_m *are the coefficients of the Taylor series of*

$$f(z) = \sum_{j=0}^{\infty} b_j (z-z_0)^j.$$

Example 3. Looking at Bergman kernel (18)

$$E(z,\zeta,t;\lambda,\mu) = {}_2F_1(-\lambda,\mu+1;\frac{1}{2};s) + \frac{2\lambda\eta s}{H}\,{}_2F_1(1-\lambda,\mu+1;\frac{3}{2};s)$$

of the differential equation (11), we see that the first hypergeometric series breaks off if $\lambda \in \mathbb{N}_0$ or $-\mu \in \mathbb{N}$. The second hypergeometric function becomes a polynomial if $\lambda \in \mathbb{N}$ or $-\mu \in \mathbb{N}$ or it drops if its constant factor λ becomes 0. So we can say, the breaking off condition for the kernel of the first kind (18) is $\lambda \in \mathbb{N}_0$ or $-\mu \in \mathbb{N}$. Since the second kernel E^*, given in (19), was constructed from (18) by changing λ with μ and z with ζ, the breaking off condition for the general Bergman solution (23) in Proposition 1 is given by

$$\{\lambda \in \mathbb{N}_0 \quad \text{or} \quad -\mu \in \mathbb{N}\} \quad \text{and} \quad \{\mu \in \mathbb{N}_0 \quad \text{or} \quad -\lambda \in \mathbb{N}\}.$$

This is eqivalent to

$$\{\lambda \in \mathbb{N}_0 \quad \text{and} \quad \mu \in \mathbb{N}_0\} \quad \text{or} \quad \{-\lambda \in \mathbb{N} \quad \text{and} \quad -\mu \in \mathbb{N}\}.$$

Taking this result, then with the important theorem of M. Kracht and E. Kreyszig (Lemma 3) and by using formula (14) with relation (15) and (16), we can say:

Proposition 3.

a) The differential equation

$$Lw = w_{z\zeta} + \left[\frac{\lambda-\mu}{\eta} - \frac{\lambda+1}{K} - \frac{\lambda}{H}\right] w_\zeta + \lambda(\mu+1)\left[\frac{-1}{\eta^2} + \frac{1}{H^2}\right] w = 0,$$

$$H(z,\zeta) = (\mu+1)\zeta - \lambda z, \quad K(z,\zeta) = (\lambda+1)z - \mu\zeta, \quad \eta = z+\zeta,$$

which represents the canocical form of equation (11), *possesses the integral-free representation of solutions*

$$w = \sum_{k=0}^{\lambda} \frac{(2k)!}{2^{2k}\,k!} P_k(z,\zeta;\lambda,\mu)\,\frac{d^{\lambda-k}\tilde{f}(z)}{dz^{\lambda-k}} + \tag{43}$$

$$+ \frac{K}{H}\eta^{\mu-\lambda}\sum_{k=0}^{\mu} \frac{(2k)!}{2^{2k}\,k!} P_k^*(z,\zeta;\lambda,\mu)\,\frac{d^{\mu-k}\tilde{g}(\zeta)}{d\zeta^{\mu-k}}, \qquad \lambda,\mu \in \mathbb{N}_0,$$

respectively

$$w = \sum_{k=1}^{-\lambda} \frac{(2k)!}{2^{2k}\,k!} P_k(z,\zeta;\lambda,\mu)\,\frac{d^{-\lambda-k}\tilde{f}(z)}{dz^{-\lambda-k}} + \tag{44}$$

$$+ \frac{K}{H}\eta^{\mu-\lambda}\sum_{k=1}^{-\mu} \frac{(2k)!}{2^{2k}\,k!} P_k^*(z,\zeta;\lambda,\mu)\,\frac{d^{-\mu-k}\tilde{g}(\zeta)}{d\zeta^{-\mu-k}}, \qquad -\lambda,-\mu \in \mathbb{N}.$$

Furthermore, we have $P_0 = 1$ *and*

$$P_k(z,\zeta;\lambda,\mu) = \frac{(-\lambda)_k\,(\mu+1)_k}{k!\,\left(\frac{1}{2}\right)_k}\eta^{-k} + 2\lambda\,\frac{(1-\lambda)_{k-1}\,(\mu+1)_{k-1}}{(k-1)!\,\left(\frac{3}{2}\right)_{k-1}}\,\frac{\eta^{-k+1}}{H}$$

where $k \in \mathbb{N}$. *Also* $P_0^* = 1$ *and for* $k \in \mathbb{N}$ *holds*

$$P_k^*(z,\zeta;\lambda,\mu) = \frac{(-\mu)_k\,(\lambda+1)_k}{k!\,\left(\frac{1}{2}\right)_k}\eta^{-k} + 2\mu\,\frac{(1-\mu)_{k-1}\,(\lambda+1)_{k-1}}{(k-1)!\,\left(\frac{3}{2}\right)_{k-1}}\,\frac{\eta^{-k+1}}{K}.$$

$\tilde{f}(z)$ *and* $\tilde{g}(\zeta)$ *are arbitrary holomorphic functions in* G_1 *resp.* $\bar{G}_1$.

b) Conversely, for arbitrary holomorphic functions $\tilde{f}(z)$ *and* $\tilde{g}(\zeta)$ *the differential operator solutions* (43) *and* (44) *represent solutions of* $Lw = 0$ *in* $G_1 \times \bar{G}_1$.

c) The solutions $v(z,\zeta)$ *of the differential equation* (11)

$$L_1 v = v_{z\zeta} - \frac{(\mu+1)K}{\eta H} v_z - \frac{(\lambda+1)H}{\eta K} v_\zeta = 0$$

can be obtained by (43) *and* (44) *with the transformation*

$$v = \frac{K}{\eta^{\mu+1}} w .$$

Now we want to determine the conditions when the Bergman kernels (30) and (31) of equation (24) reduce to polynomial kernels. The first hypergeometric series in (30)

$$\tilde{E}(z,\zeta,t;\lambda,\mu) = {}_2F_1(\lambda+1,-\mu;\frac{1}{2};s) - \frac{2(\lambda+1)\eta s}{K} {}_2F_1(\lambda+2,-\mu;\frac{3}{2};s)$$

and the second with its factor imply the breaking off condition $-\lambda \in \mathbb{N}$ or $\mu \in \mathbb{N}_0$. Using the relation $\tilde{E}^*(z,\zeta,t;\lambda,\mu) = \tilde{E}(z,\zeta,t;\mu,\lambda)$ we have the breaking off condition

$$\{-\lambda \in \mathbb{N} \quad \text{or} \quad \mu \in \mathbb{N}_0\} \quad \text{and} \quad \{-\mu \in \mathbb{N} \quad \text{or} \quad \lambda \in \mathbb{N}_0\}.$$

for the solutions $v(z,\zeta)$ of the equation (24). For this we also can write

$$\{\lambda \in \mathbb{N}_0 \quad \text{and} \quad \mu \in \mathbb{N}_0\} \quad \text{or} \quad \{-\lambda \in \mathbb{N} \quad \text{and} \quad -\mu \in \mathbb{N}\}.$$

By using the theorem of M. Kracht and E. Kreyszig and replacing λ by $-\lambda-1$ and μ by $-\mu-1$ in $P_k(z,\zeta;\lambda,\mu)$ of Proposition 3, one then has:

Proposition 4.

a) The differential equation (32)

$$\omega_{z\zeta} - \left[\left(\frac{\lambda-\mu}{\eta} - \frac{\lambda+1}{K} - \frac{\lambda}{H}\right)\omega\right]_\zeta + \lambda(\mu+1)\left[\frac{-1}{\eta^2} + \frac{1}{H^2}\right]\omega = 0 .$$

$$H(z,\zeta) = (\mu+1)\zeta - \lambda z, \quad K(z,\zeta) = (\lambda+1)z - \mu\zeta, \quad \eta = z+\zeta,$$

which represents the canocical form of the adjoint equation (24) *to* (11), *possesses the integral-free representation of solutions*

$$\begin{aligned}\omega &= \sum_{k=0}^{\lambda} \frac{(2k)!}{2^{2k}\,k!}\,\tilde{P}_k(z,\zeta;\lambda,\mu)\,\frac{d^{\lambda-k}\tilde{f}(z)}{dz^{\lambda-k}} + \\ &+ \frac{H}{K}\eta^{\lambda-\mu}\sum_{k=0}^{\mu} \frac{(2k)!}{2^{2k}\,k!}\,\tilde{P}_k^*(z,\zeta;\lambda,\mu)\,\frac{d^{\mu-k}\tilde{g}(\zeta)}{d\zeta^{\mu-k}}, \qquad \lambda,\mu \in \mathbb{N}_0,\end{aligned} \tag{45}$$

respectively

$$\omega = \sum_{k=1}^{-\lambda} \frac{(2k)!}{2^{2k}\,k!}\,\tilde{P}_k(z,\zeta;\lambda,\mu)\,\frac{d^{-\lambda-k}\tilde{f}(z)}{dz^{-\lambda-k}} + \qquad (46)$$
$$+ \frac{H}{K}\eta^{\lambda-\mu}\sum_{k=1}^{-\mu} \frac{(2k)!}{2^{2k}\,k!}\,\tilde{P}^*_k(z,\zeta;\lambda,\mu)\,\frac{d^{-\mu-k}\tilde{g}(\zeta)}{d\zeta^{-\mu-k}}, \qquad -\lambda, -\mu \in \mathbb{N}.$$

Furthermore, we have $\tilde{P}_0 = 1$ *and for* $k \in \mathbb{N}$ *holds*

$$\tilde{P}_k(z,\zeta;\lambda,\mu) = \frac{(\lambda+1)_k\,(-\mu)_k}{k!\,\left(\frac{1}{2}\right)_k}\,\eta^{-k} - 2(\lambda+1)\,\frac{(\lambda+2)_{k-1}\,(-\mu)_{k-1}}{(k-1)!\,\left(\frac{3}{2}\right)_{k-1}}\,\frac{\eta^{-k+1}}{K}$$

$$\text{and} \qquad \tilde{P}^*_k(z,\zeta;\lambda,\mu) = \tilde{P}_k(\zeta,z;\mu,\lambda) \quad \text{with} \quad \tilde{P}^*_0 = 1\,.$$

$\tilde{f}(z)$ *and* $\tilde{g}(\zeta)$ *are arbitrary holomorphic functions in* G_1 *resp.* $\bar{G}_1$.

b) Conversely, for arbitrary holomorphic functions $\tilde{f}(z)$ *and* $\tilde{g}(\zeta)$ *the differential operator solutions* (45) *and* (46) *represent solutions of equation* (32) *in* $G_1 \times \bar{G}_1$.

c) The solutions $V(z,\zeta)$ *of the differential equation* (24)

$$V_{z\zeta} + \left[\frac{(\mu+1)K}{\eta H}V\right]_z + \left[\frac{(\lambda+1)H}{\eta K}V\right]_\zeta = 0$$

can be obtained by (45) *and* (46) *with the transformation*

$$V = \frac{\eta^{\mu+1}}{H}\,\omega\,.$$

If we want to get *all* solutions of a formally hyperbolic differential equation (2) with the integral operator method of St. Bergman, we must determine two Bergman kernels of the first kind. The problem under which conditions a differential operator solution introduced by K. W. Bauer is representing *all* solutions of such an equation (2), was solved by R. Heersink [8] in 1976:

Lemma 4.
Let

$$T_2 = \sum_{k=0}^{n} b_k(z,\zeta)\,\frac{d^k}{dz^k} \qquad \text{and} \qquad T_2^* = \sum_{k=0}^{m} c_k(z,\zeta)\,\frac{d^k}{d\zeta^k}$$

be the differential operators of K. W. Bauer, for which hold

$$b_n(z,\zeta) \neq 0 \quad \textit{and} \quad c_m(z,\zeta) \neq 0 \quad \textit{for all} \quad (z,\zeta) \in G_1 \times \bar{G}_1 . \tag{47}$$

Furthermore, suppose that solutions of the differential equation (2)

$$L_1 v = v_{z\zeta} + a_1(z,\zeta) v_z + a_2(z,\zeta) v_\zeta + a_3(z,\zeta) v = 0$$

can be represented by differential operators and two arbitrary holomorphic functions $f(z)$ *and* $g(\zeta)$ *in the form*

$$v(z,\zeta) = T_2 f(z) + T_2^* g(\zeta) . \tag{48}$$

Then relation (48) *represents all solutions of equation* $L_1 v = 0$.

Applying this theorem to our solutions in Proposition 3 and 4, we see that in solution (43) and (45) the factors of the highest derivatives only can vanish in the terms including $\tilde{g}(\zeta)$ if η, K or H is equal zero. But then the differential equation is undefined. So we must demand $\eta \cdot H \cdot K \neq 0$. That means, with relation (43) and (45) *all* solutions are given.
In (44) resp. (46) we don't have quite the same situation as before. Now the highest derivative of $\tilde{f}(z)$ or $\tilde{g}(\zeta)$ we get with the summation index $k = 1$ and not $k = 0$. So we don't have the factor $P_0 = 1$ but $P_1 \neq \text{const}$. P_1 resp. $\tilde{P}_1$ is not allowed to become zero. This yields the condition

$$\{[\lambda(\mu+1) + 1]z - \mu(\mu+2)\zeta\}\{[\mu(\lambda+1) + 1]\zeta - \lambda(\lambda+2)z\} \neq 0$$

for solution (44) and the inequality

$$(\lambda+1)\{[(\lambda+1)\mu+1]z - (\mu^2-1)\zeta\} \cdot (\mu+1)\{[(\mu+1)\lambda+1]\zeta - (\lambda^2-1)z\} \neq 0.$$

for solution (46) for obtaining *all* solutions.

These solutions we also can obtain without knowing any Bergman kernels. Then we have to assume the solution in the form (37) and to solve the system (38), (39), (40) and (41).

Further differential equations possessing solutions which can be represented by differential operators are

$$w_{z\zeta} + (n-m)\left[\frac{\phi'(z)}{\phi(z)+\psi(\zeta)}\right] w_\zeta - n(m+1)\left[\frac{\phi'(z)\,\psi'(\zeta)}{[\phi(z)+\psi(\zeta)]^2}\right] w = 0,$$

$n, m \in \mathbb{N}_0$, with any holomorphic functions $\phi(z)$ and $\psi(\zeta)$ and

$$w_{z\zeta} + \left[\frac{-m(m+1)}{(z+\zeta)^2} + \frac{n(n+1)}{(z-\zeta)^2} + \frac{p(p+1)}{(1-z\zeta)^2} + \frac{q(q+1)}{(1+z\zeta)^2}\right] w = 0$$

where $m, n, p, q \in \mathbb{N}_0$, to call only two. They were not solved by the author.

4. The integral operator of I. N. Vekua

There are some other representations of solutions of equation (2)

$$L_1 v = v_{z\zeta} + a_1(z,\zeta)v_z + a_2(z,\zeta)v_\zeta + a_3(z,\zeta)v = 0, \qquad v = v(z,\zeta),$$

that are interesting. The most important of the ones we will give is the integral representation of I. N. Vekua [16]. To represent the solutions of $L_1 v = 0$ with Vekua's integral formula we need the theory of the complex Riemann function.

Now if G_1 is a simply connected fundamental domain (definition is given in the introduction) of equation (2), then $(z,\zeta,t,\tau) \in G \times G = G_1 \times \bar{G}_1 \times G_1 \times \bar{G}_1 \subset \mathbb{C}^4$. The *complex Riemann function* $R = R(z,\zeta;t,\tau)$ (t and τ are complex parameters) of the formally hyperbolic differential equation (2) is defined as the unique solution of the adjoint equation of $L_1 v = 0$:

$$L_1^* V = V_{z\zeta} - [a_1(z,\zeta)V]_z - [a_2(z,\zeta)V]_\zeta + a_3(z,\zeta)V = 0, \tag{49}$$

which satisfies the initial conditions

$$\left[\frac{\partial}{\partial \zeta} - a_1(t,\zeta)\right] R(t,\zeta;t,\tau) = 0, \tag{50}$$

$$\left[\frac{\partial}{\partial z} - a_2(z,\tau)\right] R(z,\tau;t,\tau) = 0, \tag{51}$$

$$R(t,\tau;t,\tau) = 1 \tag{52}$$

in $G \times G$ (cf. [16], [6], [7]).
The characteristic conditions (50) and (51) may be written, using (52), as

$$R(t,\zeta;t,\tau) = \exp\left[-\int_\tau^\zeta a_1(t,\beta)d\beta\right],$$

$$R(z,\tau;t,\tau) = \exp\left[-\int_t^z a_2(\alpha,\tau)d\alpha\right].$$

An equivalent definition for the Riemann function is that $R(z,\zeta;t,\tau)$ represents a solution of the following Volterra integral equation:

$$R(z,\zeta;t,\tau) - \int_\tau^\zeta a_1(z,\beta)R(z,\beta;t,\tau)d\beta - \int_t^z a_2(\alpha,\zeta)R(\alpha,\zeta;t,\tau)d\alpha + \tag{53}$$

$$+ \int_t^z d\alpha \int_\tau^\zeta a_3(\alpha,\beta)R(\alpha,\beta;t,\tau)d\beta = 1.$$

The Georgian mathematician I. N. Vekua had given the following representation theorem [16]:

Lemma 5.

Let $R = R(z,\zeta;t,\tau)$ be the Riemann function of the formally hyperbolic differential equation (2)

$$L_1 v = v_{z\zeta} + a_1(z,\zeta)v_z + a_2(z,\zeta)v_\zeta + a_3(z,\zeta)v = 0, \qquad v = v(z,\zeta).$$

Then holds:

a) For all constants $\alpha_0 \in \mathbb{C}$ and for all holomorphic functions $\varphi(z)$ and $\psi(\zeta)$

$$v(z,\zeta) = \tag{54}$$

$$= \alpha_0 R(z_0,\zeta_0;z,\zeta) + \int_{z_0}^{z} \varphi(t)R(t,\zeta_0;z,\zeta)dt + \int_{\zeta_0}^{\zeta} \psi(\tau)R(z_0,\tau;z,\zeta)d\tau$$

represents a solution of $L_1 v = 0$. (z_0,ζ_0) denotes an arbitrary but fixed point in G. $\varphi(z)$ is holomorphic in G_1 and $\psi(\zeta)$ holomorphic in $\bar{G}_1$.

b) For each solution $v(z,\zeta)$ of (2) *there exist a constant $\alpha_0 \in \mathbb{C}$ and functions $\varphi(z)$, holomorphic in G_1, and $\psi(\zeta)$,holomorphic in $\bar{G}_1$, such that $v(z,\zeta)$ may be represented by* (54) *where α_0, $\varphi(z)$ and $\psi(\zeta)$ are uniquely determined by*

$$\begin{aligned} \alpha_0 &= v(z_0,\zeta_0), \\ \varphi(z) &= v_z(z,\zeta_0) + a_2(z,\zeta_0)v(z,\zeta_0), \\ \psi(\zeta) &= v_\zeta(z_0,\zeta) + a_1(z_0,\zeta)v(z_0,\zeta). \end{aligned}$$

Now we want to use a result of [17] given in Theorem 3. There we reduce the differential equation by transforming away the last term without derivative and replace λ by $\mu+1$ and μ by λ. Then we can say:

Lemma 6.

$$R(z,\zeta;t,\tau) = I\cdot(II + III\cdot IV) \tag{55}$$

where

$$I = \frac{(z+\zeta)^{\mu+1}}{(t+\tau)^{1+\lambda}}(z+\tau)^{\lambda-\mu}\frac{[\lambda z-(\mu+1)\tau]\,[(\lambda+1)t-\mu\zeta]}{[\lambda z-(\mu+1)\zeta]\,[(\lambda+1)z-\mu\zeta]}$$

$$II = {}_2F_1(-\mu,\lambda+1;1;-\Phi)$$

$$III = \lambda(\lambda+1)\mu(\mu+1)\frac{(z-t)(\zeta-\tau)}{[\lambda z-(\mu+1)\tau]\,[(\lambda+1)t-\mu\zeta]}\,.$$

$$IV = \tfrac{\Phi}{2}\,{}_2F_1(1-\mu,\lambda+2;3;-\Phi)$$

and

$$\Phi = \frac{(z-t)(\zeta-\tau)}{(z+\zeta)(t+\tau)}$$

is the Riemann function of the formally hyperbolic differential equation (11)

$$v_{z\zeta} - \frac{\mu+1}{z+\zeta}\frac{(\lambda+1)z-\mu\zeta}{(\mu+1)\zeta-\lambda z}v_z - \frac{\lambda+1}{z+\zeta}\frac{(\mu+1)\zeta-\lambda z}{(\lambda+1)z-\mu\zeta}v_\zeta = 0, \quad \lambda,\mu \in \mathbb{C}.$$

The Riemann function of (24), which is the adjoint equation of (11) easily can be calculated by using the following law:

Lemma 7. *(Symmetry of the Riemann function.)*
The Riemann function $R^*(z,\zeta;t,\tau)$ of the adjoint differential equation (49) *is related to the Riemann function $R(z,\zeta;t,\tau)$ of equation* (2) *by the rule*

$$R^*(z,\zeta;t,\tau) = R(t,\tau;z,\zeta).$$

References

[1] K. W. Bauer, Über die Lösungen der elliptischen Differentialgleichung $(1 \pm z\bar{z})^2 w_{z\bar{z}} + \lambda w = 0$, *J. Reine und Angew. Mathematik*, **221** (1966) Teil I: 48-84; Teil II: 176-196.

[2] K. W. Bauer, Über eine partielle Differentialgleichung 2. Ordnung mit zwei unabhängigen komplexen Variablen, *Monatshefte für Mathematik*, **70/5** (1966) 385-418.

[3] K. W. Bauer, St. Ruscheweyh, Differential operators for Partial Differential Equations and Function Theoretic Applications, *Lecture Notes in Mathematics*, **791** (1980) Springer, Berlin.

[4] P. Berglez, Lineare Differentialoperatoren bei einem Paar adjungierter Differentialgleichungen, *Ber. d. math.-stat. Sektion im Forschungszentrum Graz*, **118**, (1979) 1-80.

[5] S. Bergman, *Integral Operators in the Theory of Linear Partial Differential Equations*, Springer, Berlin, 1971.

[6] R. P. Gilbert, *Function Theoretic Methods in Partial Differential Equations*, Academic Press, New York, 1969.

[7] R. P. Gilbert, *Constructive Methods for Elliptic Equations*, Springer, Berlin, 1974.

[8] R. Heersink, Characterization of Certain Differential Operators in the Solution of Linear Partial Differential Equations, *Glasgow Math. Journal*, Vol. **17**, Part 2, (1976) 83-88.

[9] M. Kracht, E. Kreyszig, Bergman-Operatoren mit Polynomen als Erzeugende, *Manuscripta math*, **1** (1969), 369-376.

[10] M. Kracht, E. Kreyszig, *Methods of Complex Analysis in Partial Differential Equations with Applications*, A Wiley-Interscience Publication, New York · Chichester · Brinsbane · Toronto · Singapore, 1988.

[11] Ursula M. Ch. Stessl, Lösungsoperatoren bei elliptischen Differentialgleichungen, *Complex Variables*, **19** (1992), 49-62.

[12] K. W. Tomantschger, Integral Operators for $W_{z\zeta} - z^{M_1-1}\zeta^{M_2-1}W = 0$, $M_\mu \in \mathbb{Z}$, *Bulletins for Appl. Math.*, **30** (1983), 127-142.

[13] K. W. Tomantschger, Representations for Solutions of Higher Order Partial Differential Equations, *Rostocker Mathematisches Kolloquium*, **31** (1987), 42-50.

[14] K. W. Tomantschger, The Bauer-Peschl Equation - Derivation and Solution of a Partial Differential Equation by Laplace's Method, *Functional-Analytic and Complex Methods, their Interactions, and Applications to Partial Differential Equations*, Proceedings of the International Graz Workshop 2001, Editors: H. Florian, N. Ortner, F. J. Schnitzer, W. Tutschke, 276 – 293.

[15] K. W. Tomantschger, Differential Operator Solutions of the Bauer-Peschl Equation, *Complex Variables*, in press.

[16] I.N. Vekua, *New Methods for Solving Elliptic Equations*, Nort-Holland Publ., Amsterdam, 1968.

[17] H. Wallner, On the Construction of Riemann Functions for Generalized Euler Equations with Rational Coefficients, *Bulletins for Applied Mathematics*, **235** (1984), 35-47.

16

SOME RESULTS ON Q_K-TYPE SPACES

Hasi Wulan *

Department of Mathematics

Shantou University

Shantou, Guangdong 515063, P. R. China

wulan@stu.edu.cn

Abstract: For a nondecreasing function $K : [0, \infty) \to [0, \infty)$, we introduce a Möbius invariant Banach space Q_K of functions analytic in the unit disk in the plane and its meromorphic counterpart $Q_K^\#$. This paper gives a survey of recent research on Q_K and $Q_K^\#$.

AMS-classification: 30D45, 30D50

1. Introduction

Let Δ be the unit disk in the complex plane, and let $dA(z)$ be the Euclidean area element on Δ. $H(\Delta)$ and $M(\Delta)$ denote the classes of functions analytic and meromorphic in Δ, respectively. Green's function in Δ with pole at $a \in \Delta$ is given by $g(z, a) = \log \frac{1}{|\varphi_a(z)|}$, where $\varphi_a(z) = \frac{a-z}{1-\bar{a}z}$ is a Möbius transformation of Δ. For $0 < r < 1$, let $\Delta(a, r) = \{z \in \Delta : |\varphi_a(z)| < r\}$ be the pseudohyperbolic disk with center $a \in \Delta$ and radius r. Let $K : [0, \infty) \to [0, \infty)$ be a nondecreasing function and define the spaces Q_K and $Q_K^\#$ as follows (cf. [7], [9]).

*This research is supported in part by a grant from the NSF of China (No. 10171058).

Definition 1. *$f \in H(\Delta)$ belongs to the space Q_K if*

$$\sup_{a\in\Delta} \iint_{\Delta} |f'(z)|^2 K(g(z,a))\, dA(z) < \infty. \tag{1.1}$$

Definition 2. *$f \in M(\Delta)$ belongs to the class $Q_K^{\#}$ if*

$$\sup_{a\in\Delta} \iint_{\Delta} f^{\#^2}(z) K(g(z,a))\, dA(z) < \infty, \tag{1.2}$$

where $f^{\#}(z) = \frac{|f'(z)|}{1+|f(z)|^2}$ is the spherical derivative of f.

Modulo constants, Q_K is a Banach space under the norm defined in (1.1) but $Q_K^{\#}$ is not since it is not a linear space. It is clear that Q_K and $Q_K^{\#}$ are Möbius invariant. For $0 < p < \infty$, $K(t) = t^p$ gives the space Q_p and the class $Q_p^{\#}$ (cf. [1], [4]). Choosing $K(t) = (1 - e^{-2t})^p$, we obtain M_p and $M_p^{\#}$ (cf. [7], [10]). We know that $Q_p = M_p$ but $Q_p^{\#} \subsetneqq M_p^{\#}$ (cf. [4] and [10]). Choosing $K(t) = 1$, we get the Dirichlet space $\mathcal{D}$ and the spherical Dirichlet class $\mathcal{D}^{\#}$. For a fixed r, $0 < r < 1$, we choose

$$K_0(t) = \begin{cases} 1, & t \geq \log(1/r), \\ 0, & 0 < t < \log(1/r). \end{cases}$$

Then we obtain

$$\iint_{\Delta} |f'(z)|^2 K_0(g(z,a))\, dA(z) = \int\int_{\Delta(a,r)} |f'(z)|^2\, dA(z),$$

and

$$\iint_{\Delta} f^{\#^2}(z) K_0(g(z,a))\, dA(z) = \int\int_{\Delta(a,r)} f^{\#^2}(z)\, dA(z).$$

We conclude that $Q_{K_0} = \mathcal{B}$ (cf. [Ax]) and $Q_{K_0}^{\#} = \mathcal{B}^{\#}$, where $\mathcal{B}^{\#}$ is the class of spherical Bloch functions (cf. [10]). It is easy to see that $\mathcal{N} \subsetneqq \mathcal{B}^{\#}$ (cf. [8] and [10]). This paper is based on joint work with Pengcheng Wu and Matts Essen; see [6], [7] and [9].

2. $\mathbf{Q}_K$-spaces

Let us first give some notations. Writing $f(r) \approx g(r)$ as $r \to r_0$, we mean that there exist positive constants c_1 and c_2 such that

$$c_1 \leq \frac{f(r)}{g(r)} \leq c_2$$

in a neighbourhood of r_0. For a nondecreasing function $K : [0,\infty) \to [0,\infty)$, we say that the space Q_K is trivial if Q_K contains only constant functions.

Theorem 1. *Assume that the integral*

$$\int_0^{1/e} K(\log(1/\rho))\rho\, d\rho = \int_1^{\infty} K(t)e^{-2t}\, dt \tag{2.1}$$

is divergent. Then the space Q_K is trivial.

It is also easy to see that the condition $K(t)e^{-2t} = O(t^{-s})$ for $s > 1$ is sufficient for the integral (2.1) to be convergent. The convergence of (2.1) is related to the growth order of K. The log-order of the function $K(r)$ is defined as

$$\rho = \overline{\lim_{r\to\infty}} \frac{\log^+ \log^+ K(r)}{\log r}, \tag{2.2}$$

where $\log^+ x = \max\{\log x, 0\}$. If $0 < \rho < \infty$, the log-type of the function $K(r)$ is defined as

$$\sigma = \overline{\lim_{r\to\infty}} \frac{\log^+ K(r)}{r^\rho}. \tag{2.3}$$

Theorem 2. *Let ρ and σ be the log-order and the log-type of a nondecreasing function K.*

(i) If $\rho > 1$, then the space Q_K is trivial.

(ii) If $\rho = 1$ and $\sigma > 2$, then the space Q_K is trivial.

In the critical case $\rho = 1$ and $\sigma = 2$, Q_K may be trivial or nontrivial (cf. [9]). From now on, we assume that the function $K : [0,\infty) \to [0,\infty)$ is nondecreasing and that the integral (2.1) is convergent.

Theorem 3. *We assume that $K(1) > 0$ and define*

$$K_1(r) = \inf(K(r), K(1)).$$

Then $Q_K = Q_{K_1}$.

As an immediate consequence of Theorem 3 we have that an analytic function $f \in \mathcal{B}$ if and only if there exists an r, $0 < r < 1$, such that $K(\log(1/r)) > 0$ and

$$\sup_{a\in\Delta} \iint_{\Delta(a,r)} |f'(z)|^2 K(g(z,a))\, dA(z) < \infty. \tag{2.4}$$

In particular, we have $Q_K \subset \mathcal{B}$ (cf. [9]).

Theorem 4. *Let $0 < p < \infty$. Assume that $K(r) \approx r^p$, $r \to 0$. Then $Q_K = Q_p$.*

Theorem 5. *$Q_K = \mathcal{B}$ if and only if*

$$\int_0^1 K(\log(1/r))(1-r^2)^{-2}\, r\, dr < \infty. \tag{2.5}$$

Choosing $K(r) = r^p$, we obtain that (2.5) holds for $p > 1$ and that (2.5) fails for $0 < p \leq 1$, that is, $Q_p = \mathcal{B}$ when $p > 1$ and $Q_p \subsetneqq \mathcal{B}$ when $0 < p \leq 1$ (cf. [1], [4]).

Let us now discuss the space $Q_{K,0}$, defined as

$$\left\{ f \in H(\Delta) : \iint_\Delta |f'(z)|^2 K(g(z,a))\, dA(z) \to 0,\ |a| \to 1 \right\},$$

and the relation between $Q_{K,0}$ and $\mathcal{B}_0$, which is defined as follows:

$$\mathcal{B}_0 = \left\{ f \in H(\Delta) : \lim_{|z|\to 1} |f'(z)|(1-|z|^2) = 0 \right\}.$$

Theorem 6. *$Q_{K,0} = \mathcal{B}_0$ if and only if (2.5) holds.*

Theorem 7. *Let $K_1 \leq K_2$ in $(0,1)$, assume furthermore that $K_1(r)/K_2(r) \to 0$ as $r \to 0$ and that the integral in (2.5) is divergent when $K = K_2$. Then*

$$Q_{K_2} \subsetneqq Q_{K_1}.$$

Theorem 8. *$\mathcal{D} \subset Q_{K,0}$ if and only if $K(0) = 0$; $\mathcal{D} = Q_K$ if and only if $K(0) > 0$.*

An immediate consequence of Theorem 8 is that we have $\mathcal{D} \subset Q_{p,0}$ for all p, $0 < p < \infty$.

Theorem 9. *$\log(1-z) \in Q_K$ if and only if*

$$\int_0^1 (1-r^2)^{-1} K(\log(1/r)) r dr < \infty. \tag{2.6}$$

Choosing $K(t) = t^p$, we obtain that $\log(1-z) \in Q_p$ for all $p \in (0,\infty)$.

Theorem 10. *If (2.6) holds, then $Q_{K,0} \subsetneqq Q_K$.*

3. $Q_K^\#$ classes

It is necessary to know for what functions K the classes $Q_K^\#$ will be trivial. Here, the square of the spherical derivative $f^\#(z)^2$ is not necessarily subharmonic, and the proof of Theorem 1 does not work. The following theorem also answers a question of Wu and Wulan (cf. [[9], Remark 2]).

Theorem 11. *If the integral (2.1) is divergent, then the class $Q_K^\#$ contains only constant functions.*

We are interested in the class of normal functions

$$\mathcal{N} = \{\, f \in M(\Delta) : \sup_{z\in\Delta}(1-|z|^2)f^\#(z) < \infty \,\}$$

and the class of spherical Bloch functions

$$\mathcal{B}^\# = \{\, f \in M(\Delta) : \sup_{a\in\Delta} \iint_{\Delta(a,r)} f^{\#^2}(z)\, dA(z) < \infty \text{ for some } r \in (0,1) \,\}.$$

We clearly have $\mathcal{N} \subset \mathcal{B}^\#$. In the analytic case, we know that the corresponding definitions with $f^\#(z)$ replaced by $|f'(z)|$ both give the space of Bloch functions $\mathcal{B}$. In the meromorphic case, the situation is different. There exists a locally univalent and analytic function $f_0 \in \mathcal{B}^\# \setminus \mathcal{N}$ (cf. [8]). S. Yamashita [11] has proved that there is an essential difference between $\mathcal{N}$ and $\mathcal{B}^\#$:

Theorem 12 (Theorem Y). *A meromorphic function $f \in \mathcal{N}$ if and only if*

$$\sup_{a\in\Delta} \iint_{\Delta(a,r)} f^{\#^2}(z)\, dA(z) < \pi$$

for some $r \in (0,1)$.

What can we say about the meromorphic analogue of (2.4)? Is the condition that there exists $r \in (0,1)$ such that

$$\sup_{a\in\Delta} \iint_{\Delta(a,r)} f^\#(z)^2 K(g(z,a))\, dA(z) < \infty, \tag{3.1}$$

necessary and sufficient to have $f \in \mathcal{B}^\#$?

We note that Theorem 3 and Theorem 4 do not remain valid in the meromorphic case and that it can occur that $Q_K^\# \not\subset \mathcal{N}$. We quote a theorem in [9] which is an immediate consequence of the main result in [3].

Theorem 13. *Let $0 < p < \infty$. Assume that K is bounded and that $K(r) = O(r^p)$ as $r \to 0$. Then there exists a non-normal function $f_0 \in Q_K^\#$.*

What more conditions on K are required for the inclusion $Q_K^\# \subset \mathcal{N}$? When are the classes $Q_{K_1}^\#$ and $Q_{K_2}^\#$ identical for $K_1 \neq K_2$? One result was given in [9], which is stated as follows.

Theorem 14. *Assume that $K(r) \to \infty$ as $r \to \infty$. Then $Q_K^\# \subset \mathcal{N}$.*

If the function K is bounded, Theorem 12 tells us that $Q_K^\# \subset \mathcal{N}$ does not hold in the general case. However, we have

Theorem 15. *Assume that $K(\infty) = 1$. Then $f \in \mathcal{N}$ if and only if*

$$\sup_{a\in\Delta} \int\int_{\Delta(a,r)} f^{\#^2}(z) K(g(z,a))\, dA(z) < \pi, \tag{3.2}$$

for some $r \in (0,1)$.

The meromorphic analogue of Theorem 3 is given by

Theorem 16. *We assume that $K(1) > 0$ and define $K_1(r) = \inf(K(r), K(1))$.*

i) If K is bounded, then $Q_K^\# = Q_{K_1}^\#$.

ii) If K is unbounded, then $Q_K^\# = \mathcal{N} \cap Q_{K_1}^\#$.

Next, we state conditions on K_1 and K_2 which imply that $Q_{K_1}^\# = Q_{K_2}^\#$.

Theorem 17. *Assume that K_1 and K_2 are both either bounded or unbounded and that $K_1(r) \approx K_2(r)$ as $r \to 0$. Then $Q_{K_1}^\# = Q_{K_2}^\#$.*

Theorem 18. *Let $0 < p < \infty$ and assume furthermore that K is bounded and that $K(r) \approx r^p$ as $r \to 0$. Then $Q_K^\# = M_p^\#$.*

Combining Theorems 15, 16 and 17, we obtain (cf. Theorem 2.2.2 in [10]) that if K is unbounded and that $K(r) \approx r^p$ as $r \to 0$. Then $Q_K^\# = Q_p^\# = \mathcal{N} \cap M_p^\#$.

Theorem 19. *i) If K is unbounded and (2.5) holds, then $Q_K^\# = \mathcal{N}$.*

ii) If K is bounded and (2.5) holds, then $Q_K^\# = \mathcal{B}^\#$.

iii) In (i) and (ii), (2.5) is a necessary condition that we have $Q_K^\# = \mathcal{N}$ or $Q_K^\# = \mathcal{B}^\#$, respectively.

It gives the following results (cf. [1] and [10])

Theorem 20. *i)* $Q_p^{\#} = \mathcal{N}$ *when* $p > 1$ *and* $Q_p^{\#} \subsetneqq \mathcal{N}$ *when* $0 < p \leq 1$.
ii) $M_p^{\#} = \mathcal{B}^{\#}$ *when* $p > 1$ *and* $M_p^{\#} \subsetneqq \mathcal{B}^{\#}$ *when* $0 < p \leq 1$.

Finally we consider the following classes:

$$\mathcal{B}_0^{\#} = \{f \in M(\Delta) : \lim_{|a|\to 1} \int\int_{\Delta(a,r)} f^{\#}(z)^2 \, dA(z) = 0 \quad \text{for some } r \in (0,1)\},$$

$$Q_{K,0}^{\#} = \{f \in M(\Delta) : \lim_{|a|\to 1} \iint_{\Delta} f^{\#}(z)^2 K(g(z,a)) dA(z) = 0\},$$

$$\mathcal{N}_0 = \{f \in M(\Delta) : (1-|z|^2) f^{\#}(z) \to 0, \quad |z| \to 1\},$$

and the spherical Dirichlet class

$$\mathcal{D}^{\#} = \{f \in M(\Delta) : \iint_{\Delta} f^{\#}(z)^2 \, dA(z) < \infty\}.$$

Then we have

Theorem 21. $Q_{K,0}^{\#} \subset \mathcal{B}_0^{\#} = \mathcal{N}_0$.

Theorem 22. *If (2.5) holds, then* $Q_{K,0}^{\#} = \mathcal{N}_0$.

Theorem 23. *(i) If* $K(0) > 0$, *then* $\mathcal{D}^{\#} = Q_K^{\#}$.
(ii) $\mathcal{D}^{\#} \subset Q_{K,0}^{\#}$ *if and only if* $K(0) = 0$.
(iii) Assume that $Q_K^{\#} \neq Q_{K,0}^{\#}$. *If* $\mathcal{D}^{\#} = Q_K^{\#}$, *then* $K(0) > 0$.
(iv) If $\mathcal{D}^{\#} = Q_K^{\#} = Q_{K,0}^{\#}$, *then* $K(0) = 0$.

Theorem 24. $\mathcal{D}^{\#} \subset Q_{p,0}^{\#}$ *for all* p, $0 < p < \infty$.

We conclude this paper by two problems related with Q_K and $Q_K^{\#}$.
(1) We don't know whether we have $Q_{K,0} = Q_K$ when (2.6) doesn't hold.
(2) We don't know whether the situation in (iv) of Theorem 23 can ever occur.

References

[1] R. Aulaskari and P. Lappan, *Criteria for an analytic function to be Bloch and a harmonic or meromorphic function to be normal*, Complex analysis and its applications, Pitman Research Notes in Mathematics, 305, Longman Scientific & Technical, Harlow, 1994, 136–146.

[2] R. Aulaskari, D. Stegenga and J. Xiao, *Some subclasses of BMOA and their characterization in terms of Carleson measures*, Rocky Mountain J. Math., Vol. 26, 1996, 485-506.

[3] R. Aulaskari, H. Wulan and R. Zhao, *Carleson measure and some classes of meromorphic functions*, Proc. Amer. Math. Soc., Vol. 128, 2000, 2329-2335.

[4] R. Aulaskari, J. Xiao and R. Zhao, *On subspaces and subsets of* $BMOA$ *and* UBC, Analysis, Vol. 15, 1995, 101–121.

[5] S. Axler, *The Bergman space, the Bloch space, and the commutators of multiplication operators*, Duke Math. J., Vol. 53, 1986, 315-332.

[6] M. Essén, Q_p*-spaces, in "Complex Function Spaces"*, Mekrijärvi 1999 (editor R. Aulaskari), Department of Mathematics, University of Joensuu, Finland, Report No., Vol. 4,, 2001, 9-40.

[7] M. Essén and H. Wulan, *On analytic and meromorphic functions and spaces of* Q_K*-type*, manuscript.

[8] P. Lappan, *A non-normal locally uniformly univalent function*, Bull. London Math. Soc. 5 (1973), 291-294.

[9] H. Wulan and P. Wu, *Characterizations of* Q_T *spaces*, J. Math. Anal. Appl., Vol. 254, 2001, 484-497.

[10] H. Wulan, *On some classes of meromorphic functions*, Ann. Acad. Sci. Fenn. Math. Diss., Vol. 116, 1998, 1–57.

[11] S. Yamashita, *Functions of uniformly bounded characteristic*, Ann. Acad. Sci. Fenn. Ser. A I Math. Vol. 7, 1982, 349–367.

III. ANALYSIS IN HIGHER DIMENSIONS

17

REPRESENTATIONS IN POLYDOMAINS

H. Begehr
Freie Universität Berlin
I. Mathematisches Institut
Arnimallee 3, D – 14195 Berlin, Germany
E-mail: begehr@math.fu-berlin.de

1. Introduction

Iterating Cauchy–Pompeiu operators leads to higher order Cauchy–Pompeiu representation formulas. There are two basic second order representations. One is related to the Laplacian $\partial^2/\partial z\partial\overline{z}$, the other to the Bitsadze operator $\partial^2/\partial\overline{z}^2$. These two formulas applied for different variables and iterated in proper manner lead to different representation formulas of second order for functions of several complex variables in polydomains. They are called of second order as only derivatives up to second order with respect to each variable are involved.

On one hand these representation formulas provide particular solutions to certain inhomogeneous second order overdetermined systems as the inhomogeneous polyharmonic, the inhomogeneous polyanalytic and some mixed kind systems. On the other hand they even can be used to transform more general systems including lower order terms into systems of singular integral equations for some density functions. They determine a particular solution to the inhomogeneous system.

The most general second order representation is given in Theorem 3. For one group of variables the Laplace operator is involved while for the others the Bitsadze operator is. Theorem 1 and Theorem 2 present special results where only one of both operators appears.

The representation formulas (15) and (16) in Theorem 3 correspond to different kinds of systems. While the first formula is related to

$$\begin{aligned} w_{z_\mu \overline{z_\mu}} &= f_\mu \quad \text{for} \quad \mu \in \{\ell_1, \dots, \ell_\lambda\}, \\ w_{\overline{z_\mu}\, \overline{z_\mu}} &= f_\mu \quad \text{for} \quad \mu \in \{\ell_{\lambda+1}, \dots, \ell_n\}, \end{aligned}$$

the latter is related to

$$\begin{aligned} w_{z_\mu \overline{z_\mu}} &= f_{\mu\mu} \quad \text{for} \quad \mu \in \{\ell_1, \dots, \ell_\lambda\}, \\ w_{\overline{z_\mu}\, \overline{z_\mu}} &= f_{\mu\mu} \quad \text{for} \quad \mu \in \{\ell_{\lambda+1}, \dots, \ell_n\}, \\ w_{\overline{z_\mu}\, \overline{z_\nu}} &= f_{\mu\nu} \quad \text{for} \quad 1 \le \mu < \nu \le n. \end{aligned}$$

Here $\{\ell_1, \dots, \ell_\lambda, \ell_{\lambda+1}, \dots, \ell_n\} = \{1, \dots, n\}$ for some λ, $0 \le \lambda \le n$, and proper compatibility conditions have to be satisfied. Dual formulas with the roles of the Laplace and Bitsadze operators interchanged are available, too. For the origin of the Pompeiu operator see [5].

2. Basic representation formulas

The complex forms of the Gauss theorem

$$\int_D w_{\overline{z}} dx\, dy = \frac{1}{2i} \int_{\partial D} w\, dz, \quad \int_D w_z dx\, dy = -\frac{1}{2i} \int_{\partial D} w\, d\overline{z} \tag{1}$$

for $w \in C^1(D; \mathbb{C}) \cap C(\overline{D}; \mathbb{C})$ with bounded smooth domains $D \subset \mathbb{C}$ lead to the Cauchy Pompeiu representations

$$w(z) = \frac{1}{2\pi i} \int_{\partial D} w(\zeta) \frac{d\zeta}{\zeta - z} - \frac{1}{\pi} \int_D w_{\overline{\zeta}}(\zeta) \frac{d\xi\, d\eta}{\zeta - z}, \quad z \in D, \tag{2}$$

$$w(z) = -\frac{1}{2\pi i} \int_{\partial D} w(\zeta) \frac{d\overline{\zeta}}{\overline{\zeta - z}} - \frac{1}{\pi} \int_D w_{\zeta}(\zeta) \frac{d\xi\, d\eta}{\overline{\zeta - z}}, \quad z \in D. \tag{2'}$$

Iterations of these formulas lead to higher order representations in terms of integral operators from a certain hierarchy, see [4]. The simplest are

$$\begin{aligned} w(z) \quad &= \quad \frac{1}{2\pi i} \int_{\partial D} w(\zeta) \frac{d\zeta}{\zeta - z} - \frac{1}{2\pi i} \int_{\partial D} w_{\overline{\zeta}}(\zeta) \frac{\overline{\zeta - z}}{\zeta - z} d\zeta \\ &\quad + \frac{1}{\pi} \int_D w_{\overline{\zeta}\,\overline{\zeta}}(\zeta) \frac{\overline{\zeta - z}}{\zeta - z} d\xi\, d\eta, \quad z \in D, \end{aligned} \tag{3}$$

$$\begin{aligned} w(z) \quad &= \quad \frac{1}{2\pi i} \int_{\partial D} w(\zeta) \frac{d\zeta}{\zeta - z} + \frac{1}{2\pi i} \int_{\partial D} w_{\overline{\zeta}}(\zeta) \log |\zeta - z|^2 d\overline{\zeta} \\ &\quad + \frac{1}{\pi} \int_D w_{\zeta\overline{\zeta}}(\zeta) \log |\zeta - z|^2 d\xi\, d\eta, \quad z \in D, \end{aligned} \tag{3'}$$

for $w \in C^2(D;\mathbb{C})\cap C^1(\overline{D};\mathbb{C})$. As (2′) is a dual representation to (2) there are dual formulas to (3) and (3′) where ζ–derivatives are used instead of $\overline{\zeta}$–derivatives and vice versa. E.g. instead of (3′) one has

$$\begin{aligned} w(z) &= -\frac{1}{2\pi i}\int_{\partial D} w(\zeta)\,\frac{d\overline{\zeta}}{\overline{\zeta}-\overline{z}} - \frac{1}{2\pi i}\int_{\partial D} w_\zeta(\zeta)\,\log|\zeta-z|^2\,d\zeta \\ &\quad +\frac{1}{\pi}\int_D w_{\zeta\overline{\zeta}}(\zeta)\,\log|\zeta-z|^2\,d\xi\,d\eta,\ \ z\in D\,. \end{aligned} \tag{3"}$$

For functions of several complex variables (2) can be used to represent $w \in C^1(D^n;\mathbb{C}) \cap C(\overline{D^n};\mathbb{C})$ for polydomains $D^n = \overset{n}{\underset{k=1}{\times}} D_k \subset \mathbb{C}^n$ with bounded smooth domains $D_k \subset \mathbb{C}$, $1 \le k \le n$, as

$$\begin{aligned} w(z) &= \frac{1}{(2\pi i)^n}\int_{\partial_0 D^n} w(\zeta)\prod_{k=1}^n \frac{d\zeta_k}{\zeta_k - z_k} - \\ &-\sum_{k=1}^n \sum_{1\le\nu_1<\cdots<\nu_k\le n} \frac{1}{\pi^k}\int_{D_{\nu_1}}\cdots\int_{D_{\nu_k}} w_{\overline{\zeta_{\nu_1}}\cdots\overline{\zeta_{\nu_k}}}(\zeta)\prod_{\rho=1}^k \frac{d\xi_{\nu_\rho}d\eta_{\nu_\rho}}{\zeta_{\nu_\rho}-z_{\nu_\rho}}, \\ & z\in D^n. \end{aligned} \tag{4}$$

Here $\partial_0 D^n := \overset{n}{\underset{k=1}{\times}} \partial D_k$ is the distinguished boundary of D so that w has to be continuous just on $D^n \cup \partial_0 D^n$ rather than on the closure $\overline{D^n}$ of D^n. Variables ζ_μ which are not integrated upon are understood to be replaced by z_μ throughout this paper.

In (4) only first order derivatives with respect to each variables z_k, $1 \le k \le n$, are involved as the higher order derivatives occuring only in mixed form. In [3] the iteration of (4) with its dual formula is used to develop a formula where second order derivatives with respect to each variable occur in order to solve the inhomogeneous pluriharmonic system, see also [1, 2].

Here (3) and (3′) will be iterated for creating “second order” representation formulas for functions in several complex variables.

But also first order and second order representations can be combined leading to representations for solutions to mixed systems. There are many combinations possible.

3. Iteration of (3)

Let $w \in C^2(D^2;\mathbb{C})\cap C^1(\overline{D}^2;\mathbb{C})$ for a bounded smooth bidomain $D^2 = D_1 \times D_2 \subset \mathbb{C}^2$. By $w \in C^1(D^2;\mathbb{C})$ we mean a complex–valued function with continuous derivatives $w_{z_1}, w_{z_2}, w_{z_1z_2}, w_{\overline{z_1}}, w_{\overline{z_2}}, w_{\overline{z_1}\,\overline{z_2}}, w_{z_1\overline{z_2}}, w_{\overline{z_1}z_2}$ in D^2, etc.

Applying (3) for D_1 to w, and for D_2 to w and to $w_{\overline{z_1}}$ and inserting the last two into the first gives

$$
\begin{aligned}
w(z) \;=\; & \frac{1}{(2\pi i)^2}\int_{\partial D_1}\int_{\partial D_2} w(\zeta_1,\zeta_2)\,\frac{d\zeta_1}{\zeta_1-z_1}\,\frac{d\zeta_2}{\zeta_2-z_2} \\
& -\frac{1}{(2\pi i)^2}\int_{\partial D_1}\int_{\partial D_2} w_{\overline{\zeta_2}}(\zeta_1,\zeta_2)\,\frac{d\zeta_1}{\zeta_1-z_1}\,\frac{\overline{\zeta_2-z_2}}{\zeta_2-z_2}\,d\zeta_2 \\
& +\frac{1}{2\pi i}\,\frac{1}{\pi}\int_{\partial D_1}\int_{D_2} w_{\overline{\zeta_2}\,\overline{\zeta_2}}(\zeta_1,\zeta_2)\,\frac{d\zeta_1}{\zeta_1-z_1}\,\frac{\overline{\zeta_2-z_2}}{\zeta_2-z_2}\,d\xi_2 d\eta_2 \\
& -\frac{1}{(2\pi i)^2}\int_{\partial D_1}\int_{\partial D_2} w_{\overline{\zeta_1}}(\zeta_1,\zeta_2)\,\frac{\overline{\zeta_1-z_1}}{\zeta_1-z_1}\,d\zeta_1\,\frac{d\zeta_2}{\zeta_2-z_2} \\
& +\frac{1}{(2\pi i)^2}\int_{\partial D_1}\int_{\partial D_2} w_{\overline{\zeta_1}\,\overline{\zeta_2}}\,\frac{\overline{\zeta_1-z_1}}{\zeta_1-z_1}\,d\zeta_1\,\frac{\overline{\zeta_2-z_2}}{\zeta_2-z_2}\,d\zeta_2 \\
& -\frac{1}{2\pi i}\,\frac{1}{\pi}\int_{\partial D_1}\int_{D_2} w_{\overline{\zeta_1}\,\overline{\zeta_2}\,\overline{\zeta_2}}\,\frac{\overline{\zeta_1-z_1}}{\zeta_1-z_1}\,d\zeta_1\,\frac{\overline{\zeta_2-z_2}}{\zeta_2-z_2}\,d\xi_2 d\eta_2 \\
& +\frac{1}{\pi}\int_{D_1} w_{\overline{\zeta_1}\,\overline{\zeta_1}}(\zeta_1,z_2)\,\frac{\overline{\zeta_1-z_1}}{\zeta_1-z_1}\,d\xi_1 d\eta_1\,.
\end{aligned}
$$

Using (3) once more for D_1 to $w_{\overline{z_2}\,\overline{z_2}}$ shows

$$
\begin{aligned}
w(z_1,z_2) = & \frac{1}{(2\pi i)^2}\int_{\partial D_1}\int_{\partial D_2}\left\{\frac{w(\zeta_1,\zeta_2)}{(\zeta_1-z_1)(\zeta_2-z_2)} - \frac{w_{\overline{\zeta_1}}(\zeta_1,\zeta_2)}{\zeta_2-z_2}\,\frac{\overline{\zeta_1-z_1}}{\zeta_1-z_1} -\right. \\
& \left. -\frac{w_{\overline{\zeta_2}}(\zeta_1,\zeta_2)}{\zeta_1-z_1}\,\frac{\overline{\zeta_2-z_2}}{\zeta_2-z_2} + w_{\overline{\zeta_1}\,\overline{\zeta_2}}(\zeta_1,\zeta_2)\frac{\overline{\zeta_1-z_1}}{\zeta_1-z_1}\,\frac{\overline{\zeta_2-z_2}}{\zeta_2-z_2}\right\} d\zeta_1 d\zeta_2 + \\
& +\frac{1}{\pi}\int_{D_1} w_{\overline{\zeta_1}\,\overline{\zeta_1}}(\zeta_1,z_2)\,\frac{\overline{\zeta_1-z_1}}{\zeta_1-z_1}\,d\xi_1 d\eta_1 + \frac{1}{\pi}\int_{D_2} w_{\overline{\zeta_2}\,\overline{\zeta_2}}(z_1,\zeta_2)\,\frac{\overline{\zeta_2-z_2}}{\zeta_2-z_2}\,d\xi_2 d\eta_2 - \\
& -\frac{1}{\pi^2}\int_{D_1}\int_{D_2} w_{\overline{\zeta_1}\,\overline{\zeta_1}\,\overline{\zeta_2}\,\overline{\zeta_2}}(\zeta_1,\zeta_2)\,\frac{\overline{\zeta_1-z_1}}{\zeta_1-z_1}\,\frac{\overline{\zeta_2-z_2}}{\zeta_2-z_2}\,d\xi_1 d\eta_1 d\xi_2 d\eta_2, \qquad (5)
\end{aligned}
$$

$(z_1,z_2)\in D_1, D_2\,.$

In order to eliminate the boundary values of $w_{\overline{z_1}\,\overline{z_2}}$ the Gauß theorem is applied leading to

$$\frac{1}{(2\pi i)^2}\int_{\partial D_1}\int_{\partial D_2} w_{\overline{\zeta_1}\,\overline{\zeta_2}}(\zeta_1,\zeta_2)\,\frac{\overline{\zeta_1 - z_1}}{\zeta_1 - z_1}\,\frac{\overline{\zeta_2 - z_2}}{\zeta_2 - z_2}\,d\zeta_1 d\zeta_2$$

$$= \frac{1}{\pi^2}\int_{D_1}\int_{D_2} \frac{\partial^2}{\partial\overline{\zeta_1}\partial\overline{\zeta_2}}\left[w_{\overline{\zeta_1}\,\overline{\zeta_2}}(\zeta_1,\zeta_2)\,\frac{\overline{\zeta_1 - z_1}}{\zeta_1 - z_1}\,\frac{\overline{\zeta_2 - z_2}}{\zeta_2 - z_2}\right] d\xi_1 d\eta_1 d\xi_2 d\eta_2$$

$$= \frac{1}{\pi^2}\int_{D_1}\int_{D_2}\left\{w_{\overline{\zeta_1}\,\overline{\zeta_1}\,\overline{\zeta_2}\,\overline{\zeta_2}}(\zeta_1,\zeta_2)\,\frac{\overline{\zeta_1 - z_1}}{\zeta_1 - z_1}\,\frac{\overline{\zeta_2 - z_2}}{\zeta_2 - z_2} + \right.$$

$$+ w_{\overline{\zeta_1}\,\overline{\zeta_2}\,\overline{\zeta_2}}(\zeta_1,\zeta_2)\,\frac{\overline{\zeta_1 - z_1}}{\zeta_1 - z_1}\,\frac{1}{\zeta_2 - z_2} + w_{\overline{\zeta_1}\,\overline{\zeta_2}\,\overline{\zeta_2}}(\zeta_1,\zeta_2)\,\frac{1}{\zeta_1 - z_1}\,\frac{\overline{\zeta_2 - z_2}}{\zeta_2 - z_2} +$$

$$\left. + \frac{w_{\overline{\zeta_1}\,\overline{\zeta_2}}(\zeta_1,\zeta_2)}{(\zeta_1 - z_1)(\zeta_2 - z_2)}\right\} d\xi_1 d\eta_1 d\xi_2 d\eta_2\,.$$

Hence, for $(z_1, z_2) \in D_1 \times D_2$

$$w(z) = \frac{1}{(2\pi i)^2}\int_{\partial D_1}\int_{\partial D_2}\left\{\frac{w(\zeta_1,\zeta_2)}{(\zeta_1 - z_2)(\zeta_2 - z_2)} - \frac{w_{\overline{\zeta_1}}(\zeta_1,\zeta_2)}{\zeta_2 - z_2}\,\frac{\overline{\zeta_1 - z_1}}{\zeta_1 - z_1}\right.$$

$$\left. - \frac{w_{\overline{\zeta_2}}(\zeta_1,\zeta_2)}{\zeta_1 - z_1}\,\frac{\overline{\zeta_2 - z_2}}{\zeta_2 - z_2}\right\} d\zeta_1 d\zeta_2$$

$$+ \frac{1}{\pi}\int_{D_1} w_{\overline{\zeta_1}\,\overline{\zeta_1}}(\zeta_1, z_2)\,\frac{\overline{\zeta_1 - z_1}}{\zeta_1 - z_1}\,d\xi_1 d\eta_1 + \frac{1}{\pi}\int_{D_2} w_{\overline{\zeta_2}\,\overline{\zeta_2}}(z_1,\zeta_2)\,\frac{\overline{\zeta_2 - z_2}}{\zeta_2 - z_2}\,d\xi_2 d\eta_2$$

$$+ \frac{1}{\pi^2}\int_{D_1}\int_{D_2} w_{\overline{\zeta_1}\,\overline{\zeta_2}}(\zeta_1,\zeta_2)\,\frac{d\xi_1 d\eta_2}{\zeta_1 - z_1}\,\frac{d\xi_2 d\eta_2}{\zeta_2 - z_2} +$$

$$\frac{1}{\pi^2}\int_{D_1}\int_{D_2}\left\{w_{\overline{\zeta_1}\,\overline{\zeta_1}\,\overline{\zeta_2}}(\zeta_1,\zeta_2)\,\frac{\overline{\zeta_1 - z_1}}{\zeta_1 - z_1}\,\frac{1}{\zeta_2 - z_2}\right. \tag{6}$$

$$hspace*3cm + \left. w_{\overline{\zeta_1}\,\overline{\zeta_2}\,\overline{\zeta_2}}(\zeta_1,\zeta_2)\,\frac{1}{\zeta_1 - z_1}\,\frac{\overline{\zeta_2 - z_2}}{\zeta_2 - z_2}\right\} d\xi_1 d\eta_1 d\xi_2 d\eta_2\,.$$

Generalizing (5) und (6) to more than two variables gives the following result.

Theorem 1 *Let* $w \in C^2(D^n;\mathbb{C}) \cap C^1(\overline{D}^n;\mathbb{C})$ *for a bounded smooth* *polydomain* $D^n := \mathop{\times}\limits_{k=1}^{n} D_k \subset \mathbb{C}^n$. *Then for* $z \in D^n$

$$\begin{aligned} w(z) = \sum_{k=0}^{n}(-1)^k \sum_{1\le\nu_1<\cdots<\nu_k\le n} \frac{1}{(2\pi i)^n}\int_{\partial_0 D^n} w_{\overline{\zeta_{\nu_1}}\,\overline{\zeta_{\nu_2}}\cdots\overline{\zeta_{\nu_k}}}(\zeta)\, \times \\ \times \prod_{\rho=1}^{k}(\overline{\zeta_{\nu_\rho} - z_{\nu_\rho}}) \prod_{\mu=1}^{n} \frac{d\zeta_\mu}{\zeta_\mu - z_\mu} + \\ + \sum_{k=1}^{n}(-1)^{k+1} \sum_{1\le\nu_1<\cdots<\nu_k\le n} \frac{1}{\pi^k}\int_{D_{\nu_1}}\cdots\int_{D_{\nu_k}} w_{\overline{\zeta_{\nu_1}}\,\overline{\zeta_{\nu_1}}\cdots\overline{\zeta_{\nu_k}}\,\overline{\zeta_{\nu_k}}}(\zeta) \\ \times \prod_{\rho=1}^{k} \frac{\overline{\zeta_{\nu_\rho} - z_{\nu_\rho}}}{\zeta_{\nu_\rho} - z_{\nu_\rho}}\, d\xi_{\nu_\rho} d\eta_{\nu_\rho}\,, \end{aligned} \tag{7}$$

and

$$\begin{aligned} w(z) = \sum_{k=0}^{n-1}(-1)^k \sum_{1\le\nu_1<\cdots<\nu_k\le n} \frac{1}{(2\pi i)^n}\int_{\partial_0 D^n} w_{\overline{\zeta_{\nu_1}}\cdots\overline{\zeta_{\nu_k}}}(\zeta)\, \times \\ \times \prod_{\rho=1}^{k}(\overline{\zeta_{\nu_\rho} - z_{\nu_\rho}}) \prod_{\mu=1}^{n} \frac{d\zeta_\mu}{\zeta_\mu - z_\mu} + \\ + \sum_{k=1}^{n-1}(-1)^{k+1} \sum_{1\le\nu_1<\cdots<\nu_k\le n} \frac{1}{\pi^k}\int_{D_{\nu_1}}\cdots\int_{D_{\nu_k}} w_{\overline{\zeta_{\nu_1}}\,\overline{\zeta_{\nu_1}}\cdots\overline{\zeta_{\nu_k}}\,\overline{\zeta_{\nu_k}}}\, \times \\ \times \prod_{\rho=1}^{k} \frac{\overline{\zeta_{\nu_\rho} - z_{\nu_\rho}}}{\zeta_{\nu_\rho} - z_{\nu_\rho}}\, d\xi_{\nu_\rho} d\eta_{\nu_\rho} + \\ + \sum_{k=0}^{n-1} \sum_{\substack{1\le\nu_1<\cdots<\nu_k\le n \\ 1\le\nu_{k+1}<\cdots<\nu_n\le n \\ \{\nu_1,\dots,\nu_n\}=\{1,\dots,n\}}} \frac{(-1)^n}{\pi^n}\int_{D^n} w_{\overline{\zeta_{\nu_1}}\,\overline{\zeta_{\nu_1}}\cdots\overline{\zeta_{\nu_k}}\,\overline{\zeta_{\nu_k}}\,\overline{\zeta_{\nu_{k+1}}}\cdots\overline{\zeta_{\nu_n}}}(\zeta) \\ \times \prod_{\rho=1}^{k}(\overline{\zeta_{\nu_\rho} - z_{\nu_\rho}}) \prod_{\mu=1}^{n} \frac{d\xi_\mu d\eta_\mu}{\zeta_\mu - z_\mu}\,. \end{aligned} \tag{8}$$

Here the variables which are not integrated upon are thought to be z*'s rather than* ζ*'s. Moreover,* $\partial_0 D^n := \mathop{\times}\limits_{k=1}^{n} \partial D_k$ *is the distinguished boundary of* D^n*.*

The proof is by induction and follows the above argumentation in the case $n = 2$.

4. Other Iterations

In an analogue way one can proceed with (3′).

Theorem 2 *Any $w \in C^2(D^n;\mathbb{C}) \cap C^1(\overline{D}^n;\mathbb{C})$ can in D^n be represented as*

$$w(z) = \sum_{k=0}^{n} \sum_{\substack{1\le\nu_1<\cdots<\nu_k\le n \\ 1\le\nu_{k+1}<\cdots<\nu_n\le n \\ \{\nu_1,\dots,\nu_n\}=\{1,\dots,n\}}} \frac{1}{(2\pi i)^n} \int_{\partial_0 D^n} w_{\overline{\zeta_{\nu_1}}\cdots\overline{\zeta_{\nu_k}}}(\zeta)$$
$$\times \prod_{\rho=1}^{k} \log|\zeta_{\nu_\rho} - z_{\nu_\rho}|^2 d\overline{\zeta_{\nu_\rho}} \prod_{\rho=k+1}^{n} \frac{d\zeta_{\nu_\rho}}{\zeta_{\nu_\rho} - z_{\nu_\rho}} + \tag{9}$$
$$+ \sum_{k=1}^{n} (-1)^{k+1} \sum_{1\le\nu_1<\cdots<\nu_k\le n} \frac{1}{\pi^k} \int_{D_{\nu_1}} \cdots \int_{D_{\nu_k}} w_{\zeta_{\nu_1}\overline{\zeta_{\nu_1}}\cdots\zeta_{\nu_k}\overline{\zeta_{\nu_k}}}(\zeta)$$
$$\times \prod_{\rho=1}^{k} \log|\zeta_{\nu_\rho} - z_{\nu_\rho}|^2 d\xi_{\nu_\rho} d\eta_{\nu_\rho}$$

and

$$w(z) = \sum_{k=0}^{n-1} \sum_{\substack{1\le\nu_1<\cdots<\nu_k\le n \\ 1\le\nu_{k+1}<\cdots<\nu_n\le n}} \frac{1}{(2\pi i)^n} \int_{\partial_0 D^n} w_{\overline{\zeta_{\nu_1}}\cdots\overline{\zeta_{\nu_k}}}(\zeta) \times$$
$$\times \prod_{\rho=1}^{k} \log|\zeta_{\nu_\rho} - z_{\nu_\rho}|^2 d\zeta_{\nu_\rho} \prod_{\rho=k+1}^{n} \frac{d\zeta_{\nu_\rho}}{\zeta_{\nu_\rho} - z_{\nu_\rho}} +$$
$$+ \sum_{k=1}^{n-1} (-1)^{k+1} \sum_{1\le\nu_1<\cdots<\nu_k\le n} \frac{1}{\pi^k} \int_{D_{\nu_1}} \cdots \int_{D_{\nu_k}} w_{\zeta_{\nu_1}\overline{\zeta_{\nu_1}}\cdots\zeta_{\nu_k}\overline{\zeta_{\nu_k}}}(\zeta)$$
$$\times \prod_{\rho=1}^{k} \log|\zeta_{\nu_\rho} - z_{\nu_\rho}|^2 d\xi_{\nu_\rho} d\eta_{\nu_\rho} +$$
$$+ \sum_{k=0}^{n-1} \sum_{\substack{1\le\nu_1<\cdots<\nu_k\le n \\ 1\le\nu_{k+1}<\cdots<\nu_n\le n}} \frac{(-1)^n}{\pi^n} \int_{D^n} w_{\zeta_{\nu_1}\overline{\zeta_{\nu_1}}\cdots\zeta_{\nu_k}\overline{\zeta_{\nu_k}}\,\overline{\zeta_{\nu_{k+1}}}\cdots\overline{\zeta_{\nu_n}}}(\zeta) \times \tag{10}$$

$$\times \prod_{\rho=1}^{k} \log|\zeta_{\nu_\rho} - z_{\nu_\rho}|^2 \prod_{\rho=k+1}^{n} \frac{1}{\zeta_{\nu_\rho} - z_{\nu_\rho}} \prod_{\mu=1}^{n} d\xi_\mu d\eta_\mu .$$

Let $(3'')$ applied for D_2 be inserted into $(3')$ used for D_1. Then for $(z_1, z_2) \in D_1 \times D_2$ analogously as before

$$\begin{aligned} w(z_1, z_2) = &-\frac{1}{(2\pi i)^2} \int_{\partial D_1} \int_{\partial D_2} \Big\{ \frac{w(\zeta_1, \zeta_2)}{(\zeta_1 - z_1)(\overline{\zeta_2 - z_2})} d\zeta_1 d\overline{\zeta_2} + \\ &+ \frac{w_{\overline{\zeta}_1}(\zeta_1, \zeta_2)}{\overline{\zeta_2 - z_2}} \log|\zeta_1 - z_1|^2 d\overline{\zeta_1} d\overline{\zeta_2} + \frac{w_{\zeta_2}(\zeta_1, \zeta_2)}{\zeta_1 - z_1} \log|\zeta_2 - z_2|^2 d\zeta_1 d\zeta_2 + \\ &+ w_{\overline{\zeta_1}\zeta_2}(\zeta_1, \zeta_2) \log|\zeta_1 - z_1|^2 \log|\zeta_2 - z_2|^2 d\overline{\zeta_1} d\zeta_2 \Big\} + \\ &+ \frac{1}{\pi} \int_{D_1} w_{\zeta_1\overline{\zeta_1}}(\zeta_1, z_2) \log|\zeta_1 - z_1|^2 d\xi_1 d\eta_1 + \\ &+ \frac{1}{\pi} \int_{D_2} w_{\zeta_2\overline{\zeta_2}}(z_1, \zeta_2) \log|\zeta_2 - z_2|^2 d\xi_2 d\eta_2 - \\ &- \frac{1}{\pi^2} \int_{D_1} \int_{D_2} w_{\zeta_1\overline{\zeta_1}\zeta_2\overline{\zeta_2}}(\zeta_1, \zeta_2) \log|\zeta_1 - z_1|^2 \log|\zeta_2 - z_2|^2 d\xi_1 d\eta_1 d\xi_2 d\eta_2 . \end{aligned} \tag{11}$$

This can be reformulated as

$$\begin{aligned} &w(z_1, z_2) \\ &= -\frac{1}{(2\pi i)^2} \int_{\partial D_1} \int_{\partial D_2} \Big\{ \frac{w(\zeta_1, \zeta_2)}{(\zeta_1 - z_1)(\overline{\zeta_2 - z_2})} d\zeta_1 d\overline{\zeta_2} + \\ &+ \frac{w_{\overline{\zeta_1}}(\zeta_1, \zeta_2)}{\overline{\zeta_2 - z_2}} \log|\zeta_1 - z_1|^2 d\overline{\zeta_1} d\overline{\zeta_2} + \frac{w_{\zeta_2}(\zeta_1, \zeta_2)}{\zeta_1 - z_1} \log|\zeta_2 - z_2|^2 d\zeta_1 d\zeta_2 \Big\} \\ &+ \frac{1}{\pi} \int_{D_1} w_{\zeta_1\overline{\zeta_1}}(\zeta_1, z_2) \log|\zeta_1 - z_1|^2 d\xi_1 d\eta_1 + \\ &+ \frac{1}{\pi} \int_{D_2} w_{\zeta_2\overline{\zeta_2}}(z_1, \zeta_2) \log|\zeta_2 - z_2|^2 d\xi_2 d\eta_2 + \\ &+ \frac{1}{\pi^2} \int_{D_1} \int_{D_2} \Big\{ \frac{w_{\overline{\zeta_1}\zeta_2}(\zeta_1, \zeta_2)}{(\zeta_1 - z_1)(\overline{\zeta_2 - z_2})} + \frac{w_{\zeta_1\overline{\zeta_1}\zeta_2}(\zeta_1, \zeta_2)}{\overline{\zeta_2 - z_2}} \log|\zeta_1 - z_1|^2 \\ &\qquad + \frac{w_{\overline{\zeta_1}\zeta_2\overline{\zeta_2}}(\zeta_1, \zeta_2)}{\zeta_1 - z_1} \log|\zeta_2 - z_2|^2 \Big\} d\xi_1 d\eta_1 d\xi_2 d\eta_2 . \end{aligned} \tag{12}$$

Generalizations of (11) and (12) are not yet available.

5. Iteration of (3) with (3')

Inserting (3′) for D_2 into (3) applied for D_1 gives for $(z_1, z_2) \in D_1 \times D_2$

$$
\begin{aligned}
w(z_1,z_2) = & \qquad (13)\\
= & \frac{1}{(2\pi i)^2}\int_{\partial D_1}\int_{\partial D_2} w(\zeta_1,\zeta_2)\frac{d\zeta_1}{\zeta_1 - z_1}\frac{d\zeta_2}{\zeta_2 - z_2}\\
& -\frac{1}{(2\pi i)^2}\int_{\partial D_1}\int_{\partial D_2} w_{\overline{\zeta_2}}(\zeta_1,\zeta_2)\frac{d\zeta_1}{\zeta_1 - z_1}\frac{\overline{\zeta_2 - z_2}}{\zeta_2 - z_2}\,d\zeta_2\\
& +\frac{1}{2\pi i}\frac{1}{\pi}\int_{\partial D_1}\int_{D_2} w_{\overline{\zeta_2}\,\overline{\zeta_2}}(\zeta_1,\zeta_2)\frac{d\zeta_1}{\zeta_1 - z_1}\frac{\overline{\zeta_2 - z_2}}{\zeta_2 - z_2}d\xi_2 d\eta_2\\
& +\frac{1}{(2\pi i)^2}\int_{\partial D_1}\int_{\partial D_2} w_{\overline{\zeta_1}}(\zeta_1,\zeta_2)\log|\zeta_1 - z_1|^2 d\overline{\zeta_1}\frac{d\zeta_2}{\zeta_2 - z_2}\\
& -\frac{1}{(2\pi i)^2}\int_{\partial D_1}\int_{\partial D_2} w_{\overline{\zeta_1}\,\overline{\zeta_2}}(\zeta_1,\zeta_2)\log|\zeta_1 - z_1|^2 d\overline{\zeta_1}\frac{\overline{\zeta_2 - z_2}}{\zeta_2 - z_2}\,d\zeta_2\\
& +\frac{1}{2\pi i}\frac{1}{\pi}\int_{\partial D_1}\int_{D_2} w_{\overline{\zeta_1}\,\overline{\zeta_2}\,\overline{\zeta_2}}(\zeta_1,\zeta_2)\log|\zeta_1 - z_1|^2 d\overline{\zeta_1}\frac{\overline{\zeta_2 - z_2}}{\zeta_2 - z_2}d\xi_2 d\eta_2\\
& +\frac{1}{\pi}\int_{D_1} w_{\zeta_1\overline{\zeta_1}}(\zeta_1,z_2)\log|\zeta_1 - z_1|^2 d\xi_1 d\eta_1\\
= & \frac{1}{(2\pi i)^2}\int_{\partial D_1}\int_{\partial D_2}\Bigg\{\frac{w(\zeta_1,\zeta_2)}{(\zeta_1 - z_1)(\zeta_2 - z_2)}d\zeta_1 d\zeta_2\\
& +\frac{w_{\overline{\zeta_1}}(\zeta_1,\zeta_2)}{\zeta_2 - z_2}\log|\zeta_1 - z_1|^2 d\overline{\zeta_1}d\zeta_2 - \frac{w_{\overline{\zeta_2}}(\zeta_1,\zeta_2)}{\zeta_1 - z_1}\frac{\overline{\zeta_2 - z_2}}{\zeta_2 - z_2}\,d\zeta_1 d\zeta_2\\
& \qquad -w_{\overline{\zeta_1}\,\overline{\zeta_2}}(\zeta_1,\zeta_2)\log|\zeta_1 - z_1|^2\frac{\overline{\zeta_2 - z_2}}{\zeta_2 - z_2}\,d\overline{\zeta_1}d\zeta_2\Bigg\}\\
& +\frac{1}{\pi}\int_{D_1} w_{\zeta_1\overline{\zeta_1}}(\zeta_1,z_2)\log|\zeta_1 - z_1|^2 d\xi_1 d\eta_1\\
& +\frac{1}{\pi}\int_{D_2} w_{\overline{\zeta_2}\,\overline{\zeta_2}}(z_1,\zeta_2)\frac{\overline{\zeta_2 - z_2}}{\zeta_2 - z_2}d\xi_2 d\eta_2\\
& -\frac{1}{\pi^2}\int_{D_1}\int_{D_2} w_{\zeta_1\overline{\zeta_1}\,\overline{\zeta_2}\,\overline{\zeta_2}}(\zeta_1,\zeta_2)\log|\zeta_1 - z_1|^2\frac{\overline{\zeta_2 - z_2}}{\zeta_2 - z_2}d\xi_1 d\eta_1 d\xi_2 d\eta_2
\end{aligned}
$$

where again (3′) is used to get rid of the mixed integrals. The Gauß theorem

$$
\frac{1}{(2\pi i)^2}\int_{\partial D_1}\int_{\partial D_2} w_{\overline{\zeta_1}\,\overline{\zeta_2}}(\zeta_1,\zeta_2)\log|\zeta_1 - z_1|^2\frac{\overline{\zeta_2 - z_2}}{\zeta_2 - z_2}d\overline{\zeta_1}d\zeta_2 =
$$

$$= -\frac{1}{\pi^2}\int_{D_1}\int_{D_2} \frac{\partial^2}{\partial\zeta_1\partial\overline{\zeta_2}}\Big[w_{\overline{\zeta_1}\,\overline{\zeta_2}}(\zeta_1,\zeta_2)\log|\zeta_1 - z_1|^2 \frac{\overline{\zeta_2 - z_2}}{\zeta_2 - z_2}\Big] d\xi_1 d\eta_1 d\xi_2 d\eta_2$$

$$= -\frac{1}{\pi^2}\int_{D_1}\int_{D_2}\Big\{w_{\zeta_1\overline{\zeta_1}\,\overline{\zeta_2}\,\overline{\zeta_2}}(\zeta_1,\zeta_2)\log|\zeta_1 - z_1|^2\frac{\overline{\zeta_2 - z_2}}{\zeta_2 - z_2} + w_{\zeta_1\overline{\zeta_1}\,\overline{\zeta_2}}(\zeta_1,\zeta_2)\times$$

$$\times\frac{\log|\zeta_1 - z_1|^2}{\zeta_2 - z_2} + w_{\overline{\zeta_1}\,\overline{\zeta_2}\,\overline{\zeta_2}}(\zeta_1,\zeta_2)\,\frac{1}{\zeta_1 - z_1}\,\frac{\overline{\zeta_2 - z_2}}{\zeta_2 - z_2} +$$

$$+ w_{\overline{\zeta_1}\,\overline{\zeta_2}}(\zeta_1,\zeta_2)\frac{1}{(\zeta_1 - z_1)(\zeta_2 - z_2)}\Big\} d\xi_1 d\eta_1 d\xi_2 d\eta_2$$

gives then for $(z_1, z_2) \in D_1 \times D_2$

$$\begin{aligned}
w(z) = {} & \frac{1}{(2\pi i)^2}\int_{\partial D_1}\int_{\partial D_2}\Big\{w(\zeta_1,\zeta_2)\frac{d\zeta_1}{\zeta_1 - z_2}\,\frac{d\zeta_2}{\zeta_2 - z_2} + \frac{w_{\overline{\zeta_1}}(\zeta_1,\zeta_2)}{\zeta_2 - z_2} \\
& \times\log|\zeta_1 - z_1|^2 d\overline{\zeta_1}d\zeta_2 - \frac{w_{\overline{\zeta_2}}(\zeta_1,\zeta_2)}{\zeta_1 - z_1}\,\frac{\overline{\zeta_2 - z_2}}{\zeta_2 - z_2}\,d\zeta_1 d\zeta_2\Big\} + \\
& + \frac{1}{\pi}\int_{D_1} w_{\zeta_1\overline{\zeta_1}}(\zeta_1, z_2)\log|\zeta_1 - z_1|^2 d\xi_1 d\eta_1 + \\
& + \frac{1}{\pi}\int_{D_2} w_{\overline{\zeta_2}\,\overline{\zeta_2}}(z_1,\zeta_2)\frac{\overline{\zeta_2 - z_2}}{\zeta_2 - z_2}\,d\xi_2 d\eta_2 \\
& + \frac{1}{\pi^2}\int_{D_1}\int_{D_2} w_{\overline{\zeta_1}\,\overline{\zeta_2}}(\zeta_1,\zeta_2)\frac{d\xi_1 d\eta_1}{\zeta_1 - z_1}\,\frac{d\xi_2 d\eta_2}{\zeta_2 - z_2} \\
& + \frac{1}{\pi^2}\int_{D_1}\int_{D_2}\Big\{w_{\zeta_1\overline{\zeta_1}\,\overline{\zeta_2}}(\zeta_1,\zeta_2)\frac{\log|\zeta_1 - z_1|^2}{\zeta_2 - z_2} + \\
& + w_{\overline{\zeta_1}\,\overline{\zeta_2}\,\overline{\zeta_2}}(\zeta_1,\zeta_2)\frac{1}{\zeta_1 - z_1}\,\frac{\overline{\zeta_2 - z_2}}{\zeta_2 - z_2}\Big\} d\xi_1 d\eta_1 d\xi_2 d\eta_2\,.
\end{aligned} \tag{14}$$

These representations can be generalized by induction.

Theorem 3 *If $w \in C^2(D^n;\mathbb{C}) \cap C^1(\overline{D}^n;\mathbb{C})$ then for $z \in D^n$ and any grouping $\{\ell_1,\dots,\ell_\lambda\}$, $\{\ell_{\lambda+1},\dots,\ell_n\}$, $0 \le \lambda \le n$, with $\{\ell_1,\dots,\ell_n\} = \{1,\dots,n\}$, $1 \le \ell_1 < \dots < \ell_\lambda \le n$, $1 \le \ell_{\lambda+1} < \dots < \ell_n \le n$,*

$$\begin{aligned}
w(z) = {} & \sum_{k=0}^{n}\sum_{1\le\nu_1<\dots<\nu_k\le n}\frac{1}{(2\pi i)^n}\int_{\partial_0 D^n} w_{\overline{\zeta_{\nu_1}}\cdots\overline{\zeta_{\nu_k}}}(\zeta)\times \\
& \times\prod_{\nu_\rho\in\{\ell_1,\dots,\ell_\lambda\}}\log|\zeta_{\nu_\rho} - z_{\nu_\rho}|^2 d\overline{\zeta_{\nu_\rho}}\times
\end{aligned} \tag{15}$$

$$\times \prod_{\nu_\rho\in\{\ell_{\lambda+1},\dots,\ell_n\}} \frac{\overline{z_{\nu_\rho}-\zeta_{\nu_\rho}}}{\zeta_{\nu_\rho}-z_{\nu_\rho}} d\zeta_{\nu_\rho} \prod_{\mu\notin\{\nu_1,\dots,\nu_k\}} \frac{d\zeta_\mu}{\zeta_\mu - z_\mu} +$$

$$+ \sum_{k=1}^{n} (-1)^{k+1} \sum_{1\le\nu_1<\dots<\nu_k\le n} \frac{1}{\pi^k} \int_{D_{\nu_1}} \cdots \int_{D_{\nu_k}} \prod_{\nu_\rho\in\{\ell_1,\dots,\ell_\lambda\}} \frac{\partial^2}{\partial\zeta_{\nu_\rho}\partial\overline{\zeta_{\nu_\rho}}} \times$$

$$\times \prod_{\nu_\rho\in\{\ell_{\lambda+1},\dots,\ell_n\}} \frac{\partial^2}{\partial\overline{\zeta_{\nu_\rho}}^2} w(\zeta) \prod_{\nu_\rho\in\{\ell_1,\dots,\ell_\lambda\}} \log|\zeta_{\nu_\rho}-z_{\nu_\rho}|^2 \times$$

$$\times \prod_{\nu_\rho\in\{\ell_{\lambda+1},\dots,\ell_n\}} \frac{\overline{\zeta_{\nu_\rho}-z_{\nu_\rho}}}{\zeta_{\nu_\rho}-z_{\nu_\rho}} \prod_{\rho=1}^{k} d\xi_{\nu_\rho} d\eta_{\nu_\rho}$$

and

$$w(z) = \sum_{k=0}^{n-1} \sum_{1\le\nu_1<\dots<\nu_k\le n} \frac{1}{(2\pi i)^n} \int_{\partial_0 D^n} w_{\overline{\zeta_1}\cdots\overline{\zeta_{\nu_k}}}(\zeta) \times$$

$$\times \prod_{\nu_\rho\in\{\ell_1,\dots,\ell_\lambda\}} \log|\zeta_{\nu_\rho}-z_{\nu_\rho}|^2 d\overline{\zeta_{\nu_\rho}} \times$$

$$\times \prod_{\nu_\rho\in\{\ell_{\lambda+1},\dots,\ell_n\}} \frac{\overline{z_{\nu_\rho}-\zeta_{\nu_\rho}}}{\zeta_{\nu_\rho}-z_{\nu_\rho}} d\zeta_{\nu_\rho} \prod_{\mu\notin\{\nu_1,\dots,\nu_k\}} \frac{d\zeta_\mu}{\zeta_\mu - z_\mu} +$$

$$+ \sum_{k=1}^{n-1} (-1)^{k+1} \sum_{1\le\nu_1<\dots<\nu_k\le n} \frac{1}{\pi^k} \int_{D_{\nu_1}} \cdots \int_{D_{\nu_k}} \prod_{\nu_\rho\in\{\ell_1,\dots,\ell_\lambda\}} \frac{\partial^2}{\partial\zeta_{\nu_\rho}\partial\overline{\zeta_{\nu_\rho}}} \times$$

$$\times \prod_{\nu_\rho\in\{\ell_{\lambda+1},\dots,\ell_n\}} \frac{\partial^2}{\partial\overline{\zeta_{\nu_\rho}}^2} w(\zeta) \prod_{\nu_\rho\in\{\ell_1,\dots,\ell_\lambda\}} \log|\zeta_{\nu_\rho}-z_{\nu_\rho}|^2 \times \tag{16}$$

$$\times \prod_{\nu_\rho\in\{\ell_{\lambda+1},\dots,\ell_n\}} \frac{\overline{\zeta_{\nu_\rho}-z_{\nu_\rho}}}{\zeta_{\nu_\rho}-z_{\nu_\rho}} \prod_{\rho=1}^{k} d\xi_{\nu_\rho} d\eta_{\nu_\rho} +$$

$$+ \sum_{\rho=0}^{\lambda-1} \sum_{\sigma=\lambda}^{n-1} \sum_{\substack{1\le\nu_1<\dots<\nu_\rho\le\lambda \\ 1\le\nu_{\rho+1}<\dots<\nu_\lambda\le\lambda}} \sum_{\substack{\lambda+1\le\nu_{\lambda+1}<\dots<\nu_\sigma\le n \\ \lambda+1\le\nu_{\sigma+1}<\dots<\nu_n\le n}} \frac{(-1)^n}{\pi^n} \int_{D^n}$$

$$\times \prod_{\tau=1}^{\rho} \frac{\partial^2}{\partial\zeta_{\ell_{\nu_\tau}}\partial\overline{\zeta_{\ell_{\nu_\tau}}}} \prod_{\iota=\lambda+1}^{\sigma} \frac{\partial^2}{\partial\overline{\zeta_{\ell_{\nu_\iota}}}^2} \prod_{\mu\notin\{\ell_1,\dots,\ell_{\nu_\rho},\ell_{\nu_{\lambda+1}},\dots,\ell_{\nu_\sigma}\}} \frac{\partial}{\partial\overline{\zeta_\mu}} w(\zeta)$$

$$\times \prod_{\tau=1}^{\rho} \log|\zeta_{\ell_{\nu_\tau}} - z_{\ell_{\nu_\tau}}|^2 \prod_{\iota=\lambda+1}^{\sigma} \frac{\overline{\zeta_{\ell_{\nu_\iota}} - z_{\ell_{\nu_\iota}}}}{\zeta_{\ell_{\nu_\iota}} - z_{\ell_{\nu_\iota}}}$$

$$\times \prod_{\mu \notin \{\ell_{\nu_1}, \dots, \ell_{\nu_\rho}, \ell_{\nu_{\lambda+1}}, \dots, \ell_{\nu_\sigma}\}} \frac{1}{\zeta_\mu - z_\mu} \prod_{k=1}^{n} d\xi_k d\eta_k \,.$$

As (2) and (2′) are related to one another by complex conjugation i.e. by interchanging the roles of z and $\overline{z}$ similarly there are dual formulas for (3) and (3′) and hence to the representations given in the theorems above. These formulas are second order representations in some sense as they express the function w through its derivatives up to second order with respect to each single variable. They can be used to solve overdetermined second order systems. E.g. the Bitsadze system

$$w_{\overline{z_k}\,\overline{z_\ell}} = f_{k\ell}, \quad 1 \le k, \ell \le n, \quad \text{in } D^n,$$

satisfying $f_{k\ell} = f_{\ell k}$ and the compatibility conditions

$$f_{k\ell\overline{z_j}} = f_{kj\overline{z_\ell}}, \quad 1 \le k, \ell, j \le n$$

can be solved via (8). Let on $\partial_0 D^n$

$$w_{\overline{z_{\nu_1}} \cdots \overline{z_{\nu_k}}}(z) = \varphi_{\nu_1 \cdots \nu_k}(z), \quad 1 \le \nu_1 < \cdots < \nu_k \le n, \ 0 \le k \le n-1 \,.$$

Then in D^n

$$w(z) = \sum_{k=0}^{n-1} (-1)^k \sum_{1 \le \nu_1 < \cdots < \nu_k \le n} \frac{1}{(2\pi i)^n} \int_{\partial_0 D^n} \varphi_{\nu_1 \cdots \nu_k}(\zeta) \times \tag{17}$$

$$\times \prod_{\rho=1}^{k} (\overline{\zeta_{\nu_\rho} - z_{\nu_\rho}}) \prod_{\mu=1}^{n} \frac{d\zeta_\mu}{\zeta_\mu - z_\mu}$$

$$+ \sum_{k=1}^{n-1} (-1)^{k+1} \sum_{1 \le \nu_1 < \cdots < \nu_k \le n} \frac{1}{\pi^k} \int_{D_{\nu_1}} \cdots \int_{D_{\nu_k}} f_{\nu_1 \nu_1 \overline{\zeta_{\nu_2}}\,\overline{\zeta_{\nu_2}} \cdots \overline{\zeta_{\nu_k}}\,\overline{\zeta_{\nu_k}}}(\zeta)$$

$$\times \prod_{\rho=1}^{k} \frac{\overline{\zeta_{\nu_\rho} - z_{\nu_\rho}}}{\zeta_{\nu_\rho} - z_{\nu_\rho}} d\xi_{\nu_\rho} d\eta_{\nu_\rho}$$

$$+ \sum_{k=0}^{n-1} \sum_{\substack{1 \le \nu_1 < \cdots < \nu_k \le n \\ 1 \le \nu_{k+1} < \cdots < \nu_n \le n}} \frac{(-1)^n}{\pi^n} \int_{D^n} f_{\nu_1 \nu_{k+1} \overline{\zeta_{\nu_1}}\,\overline{\zeta_{\nu_2}}\,\overline{\zeta_{\nu_2}} \cdots \overline{\zeta_{\nu_k}}\,\overline{\zeta_{\nu_k}}\,\overline{\zeta_{\nu_{k+2}}} \cdots \overline{\zeta_{\nu_n}}}(\zeta)$$

$$\times \prod_{\rho=1}^{k} (\overline{\zeta_{\nu_\rho} - z_{\nu_\rho}}) \prod_{\mu=1}^{n} \frac{d\xi_\mu d\eta_\mu}{\zeta_\mu - z_\mu} \,.$$

It is easily checked that the boundary integrals in (17) form a solution to the homogeneous system

$$w_{\overline{z_k}\,\overline{z_\ell}} = 0\,,$$

i.e. are pluriholomorphic functions while the area integrals altogether form a particular solution to the inhomogeneous Bitsadze system. Similar remarks hold for (10) and (16). It should be mentioned that (17) does not give a solution to some boundary value problem. Only if w is the solution to the above problem it can be represented by (17). To solve this problem solvability conditions have to be found first and then the solution has to be determined. For examples see [3].

Similarly from (7) a solution to the system

$$w_{\overline{z_k}\,\overline{z_k}} = f_k\,, \quad f_{k\overline{z_\ell}\,\overline{z_\ell}} = f_{\ell\overline{z_k}\,\overline{z_k}}, \quad 1 \le k, \ell \le n, \text{ in } D$$

with

$$w_{\overline{z_{\nu_1}}\,\overline{z_{\nu_2}}\cdots\overline{z_{\nu_k}}}(z) = \varphi_{\nu_1\nu_2\cdots\nu_k}(z)\,, \quad 1 \le \nu_1 < \nu_2 < \cdots < \nu_k \le n,\ 0 \le k \le n,$$

on $\partial_0 D^n$ is seen to be given as

$$\begin{aligned} w(z) \;=\; & \sum_{k=0}^{n}(-1)^k \sum_{1\le\nu_1<\cdots<\nu_k\le n} \frac{1}{(2\pi i)^n}\int_{\partial_0 D^n} \varphi_{\nu_1\nu_2\cdots\nu_k}(\zeta)\,\times \\ & \times \prod_{\rho=1}^{k}(\overline{\zeta_{\nu_\rho} - z_{\nu_\rho}}) \prod_{\mu=1}^{n} \frac{d\zeta_\mu}{\zeta_\mu - z_\mu} \\ & + \sum_{k=1}^{n}(-1)^{k+1} \sum_{1\le\nu_1<\cdots<\nu_k\le n} \frac{1}{\pi^k}\int_{D_{\nu_1}}\cdots\int_{D_{\nu_k}} f_{\nu_1\overline{\zeta_{\nu_2}}\,\overline{\zeta_{\nu_2}}\cdots\overline{\zeta_{\nu_k}}\,\overline{\zeta_{\nu_k}}}(\zeta) \\ & \times \prod_{\rho=1}^{k} \frac{\overline{\zeta_{\nu_\rho} - z_{\nu_\rho}}}{\zeta_{\nu_\rho} - z_{\nu_\rho}}\, d\xi_{\nu_\rho} d\eta_{\nu_\rho}\,. \end{aligned}$$

Similar remarks can be made for (9) and (15). Moreover, higher order representation formulas are available in an analoguous way starting from higher order Cauchy Pompeiu representations with higher order operators from the hierarchy in [4].

References

[1] H. BEGEHR: *Complex analytic methods for partial differential equations.* ZAMM 76 (1996), Suppl. 2, 21–24.

[2] H. BEGEHR: *Boundary value problems in $\mathbb{C}$ and $\mathbb{C}^n$.* Acta Math. Vietnamica 22 (1997), 407–425.

[3] H. BEGEHR, A. DZHURAEV: *An introduction to several complex variables and partial differential equations.* Addison Wesley Longman, Harlow, 1997.

[4] H. BEGEHR, G.N. HILE: *A hierarchy of integral operators.* Rocky Mountain J. Math. 27 (1997), 669–706.

[5] M. MITREA, F. SABAC: *Pompeiu's integral representation formula.* History and mathematics. Revue Romaine Pure Appl. 43 (1998), 211–226.

18

ON B^Q SPACES OF HYPERHOLOMORPHIC FUNCTIONS AND THE BLOCH SPACE IN R^3

K. Gürlebeck
Bauhaus-Universität Weimar
Institut für Mathematik und Physik
Coudray-Str. 13 B, D-99423 Weimar, Germany
E-mail: guerlebe@fossi.uni-weimar.de

A. El-Sayed Ahmed
South Valley University
Department of Mathematics, Sohag, Egypt
E-mail: ahsayed80@hotmail.com

Abstract: Recently a definition of $\mathbf{B^q}$ spaces of hyperholomorphic functions of three real variables has been presented by the same authors (see [15]). This definition allows to characterize the hypercomplex Bloch space equivalently by the integral norms of $\mathbf{B^q}$. The aim of this paper is to describe a "wider" scale of $\mathbf{B^q}$ spaces of hyperholomorphic functions by using another weight function. A part of the scale still characterizes the Bloch space but the whole scale is not longer equivalent to the Bloch space. Besides, some important basic properties of these weighted $\mathbf{B^q}$ spaces are also considered.

Keywords: quaternionic analysis, $\mathbf{B^q}$ spaces, monogenic Bloch functions.

AMS-classification: 30G35, 46E15.

1. Introduction

We start here with some terminology, notation and the definition of various classes of analytic functions defined on the open unit disk $\Delta =$

$\{z : |z| < 1\}$ in the complex plane $\mathbb{C}$ (see e.g. [2], [3], [4], [6], [8], [20], [21], [28], [31] and [32]) and the corresponding classes of hyperholomorphic functions defined and studied on the unit ball $B_1(0)$ by several authors (see e.g. [15], [16] and [17]).

Recall that the well known Bloch space (see e.g. [1], [6] and [12]) is defined as follows:

$$\mathcal{B} = \{f : f \text{ analytic in } \Delta \text{ and } \sup_{a\in\Delta}(1-|z|^2)|f'(z)| < \infty\} \tag{1}$$

and the Dirichlet space (see e.g. [1] and [32]) is given by

$$\mathcal{D} = \{f : f \text{ analytic in } \Delta \text{ and } \int_\Delta |f'(z)|^2 d\sigma_z < \infty\}, \tag{2}$$

where $d\sigma_z$ is the Euclidean area element $dxdy$.

Let $0 < q < \infty$. Then the Besov-type spaces

$$\mathbf{B^q} = \{f : f \text{ analytic in } \Delta \text{ and } \sup_{a\in\Delta}\int_\Delta |f'(z)|^q(1-|z|^2)^{q-2}(1-|\varphi_a(z)|^2)^2 d\sigma_z < \infty\} \tag{3}$$

are introduced and studied intensively (see [28]). From [28] it is known that the $\mathbf{B^q}$ spaces defined by (3) can be used to describe the Bloch space $\mathcal{B}$ equivalently by the integral norms of $\mathbf{B^q}$. On the other hand there are some papers employing the weight function $(1-|\varphi_a(z)|^2)$ instead of $(1-|\varphi_a(z)|^2)^2$ (see [7] and [21]). Also, if the exponent of $(1-|\varphi_a(z)|)$ is equal to zero, then we will get the Besov paces $\mathbf{B_p}$, $1 < p < \infty$ which were studied by many authors (see e.g. [5], [23], [29], [32] and others). Here, φ_a always stands for the Möbius transformation $\varphi_a(z) = \frac{a-z}{1-\bar{a}z}$.

In 1994, Aulaskari and Lappan [6] introduced a new class of holomorphic functions, the so called $\mathbf{Q_p}$-spaces as follows:

$$\mathbf{Q_p} = \left\{f : f \text{ analytic in } \Delta \text{ and } \sup_{a\in\Delta}\int_\Delta |f'(z)|^2 g^p(z,a) d\sigma_z < \infty\right\}, \tag{4}$$

where the weight function $g(z,a) = \ln\left|\frac{1-\bar{a}z}{a-z}\right|$ is defined as the composition of the Möbius transformation φ_a and the fundamental solution of the two-dimensional real Laplacian. One idea of this work was to "close" the gap between the Dirichlet space and the Bloch space. Main results are

$\mathcal{D} \subset \mathbf{Q_p} \subset \mathbf{Q_q} \subset$ BMOA, $\quad 0 < p < q < 1$ (see [8]) where, BMOA is the space of analytic functions of bounded mean oscillations,

$\mathbf{Q_1}$ = BMOA (see [6]),

$\mathbf{Q_p} = \mathcal{B}$, for $p > 1$ (see [6]).

For more information about the study of $\mathbf{Q_p}$ spaces of analytic functions we refer to [6], [14], and [25].

It should be mentioned here also that several authors (see e.g. [10], [24], [25]] and [28]) tried to generalize the idea of these spaces to higher dimensions in the unit ball of $\mathbf{C}^n$. Essen et al. [13] studied also $\mathbf{Q_p}$ spaces in $\mathbf{R}^n$.

In 1999 Gürlebeck et al. [16] defined $\mathbf{Q_p}$ spaces of hyperholomorphic functions instead of analytic functions.

To introduce the meaning of hyperholomorphic functions let Δ be the set of real quaternions. This means that elements of Δ are of the form:

$$a = \sum_{k=0}^{3} a_k e_k, \quad \text{where} \quad \{a_k | k \in \mathbf{N}_3^0 := \mathbf{N}_3 \cup \{0\}; \mathbf{N}_3 := \{1,2,3\}\} \subset \mathbf{R};$$

$e_0 = 1$ the unit; e_1, e_2, e_3 are called imaginary units, and they define arithmetic rules in Δ; by definition $e_k^2 = -e_0$, $k \in \mathbf{N}_3$; $e_1 e_2 = -e_2 e_1 = e_3$; $e_2 e_3 = -e_3 e_2 = e_1$; $e_3 e_1 = -e_1 e_3 = e_2$.

Natural operations of addition and multiplication in Δ turn it into a skew-field. The main involution in Δ, the quaternionic conjugation, is defined by

$$\bar{e_0} := e_0; \quad \bar{e_k} := -e_k; \quad for \quad k \in \mathbf{N}_3,$$

and it extends onto Δ by $\mathbf{R}$-linearity, i.e., for $a \in \Delta$

$$\bar{a} := \sum_{k=0}^{3} \bar{a_k e_k} = \sum_{k=0}^{3} a_k \bar{e_k} = a_0 - \sum_{k=1}^{3} a_k e_k.$$

Note that

$$\bar{a} a = a \bar{a} = \sum_{k=0}^{3} a_k^2 = |a|^2_{\mathbf{R}^4} =: |a|^2_{\Delta}$$

Therefore, for $a \in \Delta \setminus \{0\}$ the quaternion

$$a^{-1} := \frac{1}{|a|^2} \bar{a}$$

is an inverse to a. Whereas the above mentioned properties are analogous to the complex one-dimensional case we have for the quaternionic conjugation that for any $a, b \in \Delta$

$$\bar{ab} = \bar{b}\, \bar{a}.$$

Let Ω be a domain in $\mathbf{R}^3$, then we shall consider Δ-valued functions defined in Ω (depending on $x = (x_0, x_1, x_2)$):

$$f : \Omega \longrightarrow \Delta.$$

The notation $C^p(\Omega;\Delta), p \in \mathbf{N}\cup\{0\}$, has the usual component-wise meaning.
On $C^1(\Omega;\Delta)$ we define a generalized Cauchy-Riemann operator D by

$$D(f) := \sum_{k=0}^{2} e_k \frac{\partial f}{\partial x_k} =: \sum_{k=0}^{2} e_k\, \partial_k f.$$

D is a right-linear operator with respect to scalars from Δ. The operator $\overline{D}$

$$\overline{D}(f) := \sum_{k=0}^{2} \bar{e}_k \frac{\partial f}{\partial x_k} =: \sum_{k=0}^{2} \bar{e}_k\, \partial_k f$$

is the adjoint Cauchy-Riemann operator. The solutions of $Df = 0,\ x \in \Omega$ are called (left) hyperholomorphic (or monogenic) functions and generalize the class of holomorphic functions from the one-dimensional complex function theory. Let $\triangle$ be the three-dimensional Laplace operator $\triangle := \sum_{k=0}^{2} \partial_k^2$. Then on $C^2(\Omega;\Delta)$ analogously to the complex case the following equalities hold:

$$\triangle = D\overline{D} = \overline{D}D.$$

Using the adjoint generalized Cauchy-Riemann operator $\overline{D}$ instead of the derivative $f'(z)$, the quaternionic Möbius transformation $\varphi_a(x) = (a - x)(1 - \bar{a}x)^{-1}$, and the modified fundamental solution $g(x) = \frac{1}{4\pi}\left(\frac{1}{|x|} - 1\right)$ of the real Laplacian in [16] generalized $\mathbf{Q_p}$-spaces are defined by

$$\mathbf{Q_p} = \left\{ f \in \ker D : \sup_{a \in B_1(0)} \int_{B_1(0)} \left|\overline{D}f(x)\right|^2 (g(\varphi_a(x)))^p dB_x < \infty \right\}, \quad (5)$$

where $B_1(0)$ stands for the unit ball in $\mathbf{R}^3$ also, some times we use the expression $g(x,a)$ instead of $g(\varphi_a(x))$. Here, the generalizations of the Green function and of the higher dimensional Möbius transformation seem to be naturally; that $-\frac{1}{2}\overline{D}$ plays the role of a derivative is shown in [17] for arbitrary dimensions and in [22] and [30] for dimension four.

From the consideration of $\mathbf{Q_p}$ as $p \to \infty$ in [16] is introduced the following definition of the Bloch norm in three dimensional case:

$$\mathcal{B}(f) = \sup_{x \in B_1(0)} (1 - |x|^2)^{\frac{3}{2}} |\overline{D}f(x)|.$$

Based on these definitions it is proved in [16] that

$$\mathcal{D} \subset \mathbf{Q_p} \subset \mathbf{Q_q} \subset \mathcal{B} \text{ for } 0 < p < q \leq 2 \quad \text{and} \quad \mathbf{Q_q} = \mathcal{B} \text{ for } q > 2,$$

where $\mathcal{D}$ be the hyperholomorphic Dirichlet space and is given by (see [16]):

$$\mathcal{D}=\left\{f: f\in kerD \quad \text{and} \quad \int_{B_1(0)} |\overline{D}f(x)|^2 dB_x<\infty\right\}.$$

For more information about the study of $\mathbf{Q_p}$ spaces of hyperholomorphic functions, we refer to [11], [16] and [17].

Recently, $\mathbf{B^q}$ spaces of quaternion-valued functions were defined in [15] as follows:

$$\mathbf{B^q}=\left\{f\in\ker D:\right.$$
$$\left.\sup_{a\in B_1(0)}\int_{B_1(0)} |\overline{D}f(x)|^q(1-|x|^2)^{\frac{3}{2}q-3}(1-|\varphi_a(x)|^2)^3 dB_x<\infty\right\}, \quad (6)$$

where $0<q<\infty$ and the exponent 3 is related to the real space dimension.

We will study in the following $\mathbf{B^q}$ spaces of quaternion-valued functions by employing the weight function $(1-|\varphi_a(x)|^2)^2$ in lieu of $(1-|\varphi_a(x)|^2)^3$ as follows:

$$\mathbf{B^q}=\left\{f\in\ker D:\right.$$
$$\left.\sup_{a\in B_1(0)}\int_{B_1(0)} |\overline{D}f(x)|^q(1-|x|^2)^{\frac{3}{2}q-3}(1-|\varphi_a(x)|^2)^2 dB_x<\infty\right\}, \quad (7)$$

where, $0<q<\infty$. The main aim of this paper is to study these $\mathbf{B^q}$ spaces and their relations to the above mentioned quaternionic Bloch space. It will be shown that this exponent $2=3-1(=n-1)$ generates a new scale of spaces, not equivalent to the Bloch space for the whole range of q. This behaviour corresponds to complex results proved in [28]. The concept may be generalized in the context of Clifford analysis to arbitrary real dimensions. We will restrict us for simplicity to $\mathbb{R}^3$ and quaternion-valued functions as (the lowest non-commutative case) a model case. For more details about quaternionic analysis and general Clifford analysis, we refer to [9], [18], [19], [27] and [30].

We will need the following lemmas in the sequel:

Lemma 1. [17] *Let $0<q\leq 2$, $|a|<1$, $r\leq 1$. Then*

$$\int_{\partial B_1(0)} \frac{1}{|1-\bar{a}ry|^{2q}} d\Gamma_y \leq \lambda \frac{1}{(1-|a|r)^q},$$

where λ is a constant not depending on a.

Lemma 2. [15] *Let $1 \leq q < \infty$ and $0 < R < 1$, then $\forall f \in kerD$, we have that*

$$\frac{4\pi}{3} R^3 |\overline{D} f(0)|^q \leq \int_{B_R} |\overline{D} f(x)|^q dB_x.$$

2. Some basic properties of $\mathbf{B^q}$ spaces of quaternion valued functions

We will consider now the inclusions of $\mathbf{B^q}$ spaces of quaternion-valued functions as basic scale properties and we will also throw some light in the relations between the norms of $\mathbf{B^q}$ spaces of quaternion-valued functions and the norms of $\mathbf{Q_p}$ spaces of quaternion-valued functions.

Proposition 1. *Let f be a hyperholomorphic function in $B_1(0)$ and let $f \in \mathcal{B}$. Then, for*
$0 < p < q < \infty$, *we will get that*

$$\mathcal{B} \cap \mathbf{B^p} \subset \mathcal{B} \cap \mathbf{B^q}. \tag{8}$$

Proof. Let $f \in \mathbf{B^p}$, for any $0 < p < \infty$. Then for any $0 < p < q < \infty$, we obtain

$$\int_{B_1(0)} |\overline{D} f(x)|^q (1 - |x|^2)^{\frac{3}{2}q-3} (1 - |\varphi_a(x)|^2)^2 dB_x$$
$$= \int_{B_1(0)} |\overline{D} f(x)|^{q-p} (1 - |x|^2)^{\frac{3}{2}(q-p)} |\overline{D} f(x)|^p (1 - |x|^2)^{\frac{3}{2}p-3} \times$$
$$\times (1 - |\varphi_a(x)|^2)^2 dB_x.$$

Since, $f \in \mathcal{B}$ and

$$|\overline{D} f(x)| (1 - |x|^2)^{\frac{3}{2}} \leq \mathcal{B}(f),$$

we get

$$\int_{B_1(0)} |\bar{D} f(x)|^q (1 - |x|^2)^{\frac{3}{2}q-3} (1 - |\varphi_a(x)|^2)^2 dB_x$$
$$\leq \mathcal{B}^{q-p}(f) \int_{B_1(0)} |\overline{D} f(x)|^p \times$$
$$\times (1 - |x|^2)^{\frac{3}{2}p-3} (1 - |\varphi_a(x)|^2)^2 dB_x < \infty.$$

Thus $f \in \mathbf{B^q}$ and $f \in \mathcal{B}$ for any q, $0 < p < q < \infty$ and the proof of our proposition is finished.
Later in this paper we will study under which conditions the additional assumption $f \in \mathcal{B}$ can be removed.

Lemma 3. *Let f be a hyperholomorphic function in $B_1(0)$. Then,*

$$\sup_{a\in B_1(0)} (1-|a|^2)^{\frac{1}{2}} \int_0^1 (M_2^2(\overline{D}f,r))^{\frac{1}{2}}(1-r^2)^{\frac{1}{2}} r^2 dr < \infty \Longrightarrow f \in \mathbf{B^1},$$

with

$$M_2^2(\overline{D}f,r) = \int_0^{\pi}\int_0^{2\pi} \left|h(r)\overline{D}f(r,\theta_1,\theta_2)\right|^2 \sin\theta_1 d\theta_2 d\theta_1,$$

where, $h(r)$ stands for $\frac{1}{|1-\bar{a}x|^2}$ in spherical coordinates.

Proof. Suppose that,

$$\sup_{a\in B_1(0)} (1-|a|^2)^{\frac{1}{2}} \int_0^1 (M_2^2(\overline{D}f,r))^{\frac{1}{2}}(1-r^2)^{\frac{1}{2}} r^2 dr < \infty.$$

Then,

$$\int_{B_1(0)} |\overline{D}f(x)|(1-|x|^2)^{-\frac{3}{2}}(1-|\varphi_a(x)|^2)^2 dB_x$$
$$= \int_{B_1(0)} |\overline{D}f(\varphi_a(x))|(1-|\varphi_a(x)|^2)^{-\frac{3}{2}}(1-|x|^2)^2 \frac{(1-|a|^2)^3}{|1-\bar{a}x|^6} dB_x.$$

Here, we used that the Jacobian determinant is

$$\frac{(1-|a|^2)^3}{|1-\bar{a}x|^6}.$$

Now, using the equality

$$(1-|\varphi_a(x)|^2) = \frac{(1-|a|^2)(1-|x|^2)}{|1-\bar{a}x|^2}, \tag{9}$$

we obtain that,

$$\int_{B_1(0)} |\overline{D}f(x)|(1-|x|^2)^{-\frac{3}{2}}(1-|\varphi_a(x)|^2)^2 dB_x$$

$$= \int_{B_1(0)} |\overline{D}f(\varphi_a(x))|(1-|x|^2)^{\frac{1}{2}}\frac{(1-|a|^2)^{\frac{3}{2}}}{|1-\bar{a}x|^3} dB_x$$

$$= \int_{B_1(0)} \left|\frac{1}{|1-\bar{a}x|^2}\overline{D}f(\varphi_a(x))\right|(1-|x|^2)^{\frac{1}{2}}\frac{(1-|a|^2)^{\frac{3}{2}}}{|1-\bar{a}x|} dB_x$$

$$\leq 2^{\frac{3}{2}}(1-|a|^2)^{\frac{1}{2}}\int_0^1 (M_2^2(\overline{D}f,r))^{\frac{1}{2}}(1-r^2)^{\frac{1}{2}}r^2 dr.$$

Taking $\sup\limits_{a\in B_1(0)}$ in both sides of the above inequality, we deduce that $f \in \mathbf{B}^1$.

Proposition 2. *Let f be a hyperholomorphic function in $B_1(0)$ and $f \in \mathcal{B}$; satisfying the condition*

$$J(a,r) = (1-|a|^2)^{\frac{1}{2}}\int_0^1 (M_2^2(\overline{D}f,r))^{\frac{1}{2}}(1-r^2)^{\frac{1}{2}}r^2 dr < \infty. \tag{10}$$

Then for $1 < q < \infty$, we have that

$$\int_{B_1(0)} |\overline{D}f(x)|^q(1-|x|^2)^{\frac{3}{2}q-3}(1-|\varphi_a(x)|^2)^2 dB_x \leq 2^{\frac{3}{2}}\mathcal{B}^{q-1}(f)J(a,r).$$

Proof. Using equality (9) in (7), we obtain that

$$\int_{B_1(0)} |\overline{D}f(x)|^q(1-|x|^2)^{\frac{3}{2}q-3}(1-|\varphi_a(x)|^2)^2 dB_x$$

$$\leq \mathcal{B}^{q-1}(f)\int_{B_1(0)} |\overline{D}f(x)|(1-|x|^2)^{-\frac{3}{2}}(1-|\varphi_a(x)|^2)^2 dB_x$$

$$= \mathcal{B}^{q-1}(f)\int_{B_1(0)} |\overline{D}f(\varphi_a(x))|(1-|\varphi_a(x)|^2)^{-\frac{3}{2}}(1-|x|^2)^2\frac{(1-|a|^2)^3}{|1-\bar{a}x|^6} dB_x$$

$$= \mathcal{B}^{q-1}(f)\int_{B_1(0)} |\overline{D}f(\varphi_a(x))|(1-|x|^2)^{\frac{1}{2}}\frac{(1-|a|^2)^{\frac{3}{2}}}{|1-\bar{a}x|^3} dB_x$$

$$= \mathcal{B}^{q-1}(f)\int_{B_1(0)} \left|\frac{1}{|1-\bar{a}x|^2}\overline{D}f(\varphi_a(x))\right|(1-|x|^2)^{\frac{1}{2}}\frac{(1-|a|^2)^{\frac{3}{2}}}{|1-\bar{a}x|} dB_x$$

$$\leq 2^{\frac{3}{2}}\sqrt{\pi}\mathcal{B}^{q-1}(f)(1-|a|^2)^{\frac{1}{2}}\int\limits_0^1 (M_2^2(\overline{D}f,r))^{\frac{1}{2}}(1-r^2)^{\frac{1}{2}}r^2dr$$

and our proposition is proved.

Remark 2.1. Proposition 2.2 implies that each hyperholomorphic function $f \in \mathcal{B}(B_1(0))$ with the additional property $\sup_{a\in B_1(0)} J(a,r) < \infty$ belongs to $\mathbf{B^q}$ for all $1 < q < \infty$.

Theorem 1. *Let f be a hyperholomorphic function in $B_1(0)$. Then, for $0 < q \leq 2$, we have that*

$$f \in \mathbf{B^q} \Longleftrightarrow \sup_{a\in B_1(0)} \int_{B_1(0)} |\overline{D}f(x)|^q(1-|x|^2)^{\frac{3}{2}q-3}(g(x,a))^2 dB_x < \infty.$$

Proof. At first we suppose that

$$\sup_{a\in B_1(0)} \int_{B_1(0)} |\overline{D}f(x)|^q(1-|x|^2)^{\frac{3}{2}q-3}(g(x,a))^2 dB_x < \infty.$$

Since,

$$(1-|\varphi_a(x)|^2)^2 \leq (8\pi g(x,a))^2.$$

Then the assertion " $\Longleftarrow$ " follows. i.e., $f \in \mathbf{B^q}$.

Secondly, we assume that $f \in \mathbf{B^q}$. Now our task is to prove that

$$\begin{aligned} I_1 &= \int_{B_1(0)} |\overline{D}f(x)|^q(1-|x|^2)^{\frac{3}{2}q-3}(g(x,a))^2 dB_x \\ &\leq \mu^2 \int_{B_1(0)} |\overline{D}f(x)|^q(1-|x|^2)^{\frac{3}{2}q-3}(1-|\varphi_a(x)|^2)^2 dB_x < \infty, \end{aligned}$$

where, μ^2 is any constant not depending on a. Since

$$I = \int_{B_1(0)} |\overline{D}f(x)|^q(1-|x|^2)^{\frac{3}{2}q-3}(g^2(x,a)-\mu^2(1-|\varphi_a(x)|^2)^2)dB_x$$

$$= \int_{B_1(0)} |\overline{D}f(\varphi_a(x))|^q(1-|\varphi_a(x)|^2)^{\frac{3}{2}q-3}\frac{(1-|a|^2)^3}{|1-\bar{a}x|^6}(1-|x|)^2\Psi(x)dB_x$$

$$= \int_{B_1(0)} \left|\frac{1}{|1-\bar{a}x|^2}\overline{D}f(\varphi_a(x))\right|^q(1-|\varphi_a(x)|^2)^{\frac{3}{2}q-3}\frac{(1-|a|^2)^3}{|1-\bar{a}x|^{6-2q}} \times$$

$$\times\ (1-|x|)^2\Psi(x)dB_x$$

where,

$$\Psi(x) = \frac{1}{4\pi}\left(\frac{1}{|x|} - 4\pi\mu(1+|x|)\right)\left(\frac{1}{|x|} + 4\pi\mu(1+|x|)\right).$$

Using equality (9), we obtain that

$$I = \zeta \int_{B_1(0)} \left| \frac{1}{|1-\bar{a}x|^2} \overline{D} f(\varphi_a(x)) \right|^q (1-|x|^2)^{\frac{3}{2}q-3} (1-|x|)^2 \frac{\Psi(x)}{|1-\bar{a}x|^q} dB_x$$
$$= \zeta \int_{B_{\frac{1}{10}}(0)} \left| \frac{1}{|1-\bar{a}x|^2} \overline{D} f(\varphi_a(x)) \right|^q (1-|x|^2)^{\frac{3}{2}q-3} (1-|x|)^2 \frac{\Psi(x)}{|1-\bar{a}x|^q} dB_x$$
$$- \zeta \int_{B_1(0)\setminus B_{\frac{1}{10}}(0)} \left| \frac{1}{|1-\bar{a}x|^2} \overline{D} f(\varphi_a(x)) \right|^q (1-|x|^2)^{\frac{3}{2}q-3} (1-|x|)^2 \frac{-\Psi(x)}{|1-\bar{a}x|^q} dB_x$$

where , $\zeta = (1-|a|^2)^{\frac{3}{2}q}$. Since μ is an arbitrary constant we can assume that $\mu = \frac{100}{44\pi}$, then $\Psi(x) \le 0$; $\forall |x| \in (\frac{1}{10}, 1]$. Also, we have

$$\frac{1}{(1+|x|)^q} \le \frac{1}{|1-\bar{a}x|^q} \le \frac{1}{(1-|x|)^q}$$

Using the above relation in the above equality, then we obtain

$$I \le \frac{1}{4\pi}(1-|a|^2)^{\frac{3}{2}q} \int_0^{\frac{1}{10}} (M_q(\overline{D}f,r))^q (1-r)^{\frac{q}{2}-1} \times$$
$$\times \left(\frac{1}{r} + \frac{100}{11}(1+r)\right)\left(\frac{1}{r} - \frac{100}{11}(1+r)\right) r^2 dr +$$
$$+ \frac{(2)^{\frac{q}{2}-3}}{4\pi}(1-|a|^2)^{\frac{3}{2}q} \int_{\frac{1}{10}}^{1} (M_q(\overline{D}f,r))^q \times$$
$$\times (1-r)^{\frac{3}{2}q-1}\left(\frac{1}{r} + \frac{100}{11}(1+r)\right)\left(\frac{1}{r} - \frac{100}{11}(1+r)\right) r^2 dr$$
$$= \frac{1}{4\pi}(1-|a|^2)^{\frac{3}{2}q} \int_0^{\frac{1}{10}} (M_q(\overline{D}f,r))^q (1-r)^{\frac{q}{2}-1} \times$$
$$\times \left(1 + \frac{100}{11} r(1+r)\right)\left(1 - \frac{100}{11} r(1+r)\right) dr +$$
$$+ \frac{(2)^{\frac{q}{2}-3}}{4\pi}(1-|a|^2)^{\frac{3}{2}q} \int_{\frac{1}{10}}^{1} (M_q(\overline{D}f,r))^q \times$$
$$\times (1-r)^{\frac{3}{2}q-1}\left(1 + \frac{100}{11} r(1+r)\right)\left(1 - \frac{100}{11} r(1+r)\right) dr$$

where,

$$\left(M_q(\overline{D}f,r)\right)^q = \int_0^{\pi}\int_0^{2\pi} \left|h(r)\overline{D}f(r,\theta_1,\theta_2)\right|^q \sin\theta_1 d\theta_2 d\theta_1,$$

since, $M_q(\overline{D}f,r) \geq 0;\ \forall r \in [0,1]$ and $1 - \frac{100}{11}r(1+r) \leq 0;\ \forall r \notin [0,\frac{1}{10}]$.

Now we want to compare the integral $\int_0^{\frac{1}{10}} \left(M_q(\overline{D}f,r)\right)^q (1-r)^{\frac{q}{2}-1}\big(1+\frac{100}{11}r(1+r)\big)\big(1-\frac{100}{11}r(1+r)\big)dr$ and the integral $(2)^{\frac{q}{2}-3}\int_{\frac{5}{10}}^{\frac{6}{10}} \left(M_q(\overline{D}f,r)\right)^q (1-r)^{\frac{3}{2}q-1}\big(1+\frac{100}{11}r(1+r)\big)\big(\frac{100}{11}r(1+r)-1\big)dr$.

Then, after simple calculation we can obtain that

$$\int_0^{\frac{1}{10}} \left(M_q(\overline{D}f,r)\right)^q (1-r)^{\frac{q}{2}-1}\Big(1+\frac{100}{11}r(1+r)\Big)\Big(1-\frac{100}{11}r(1+r)\Big)dr$$

$$< (2)^{\frac{q}{2}-3}\int_{\frac{5}{10}}^{\frac{6}{10}} \left(M_q(\overline{D}f,r)\right)^q (1-r)^{\frac{3}{2}q-1} \times$$

$$\times \Big(1+\frac{100}{11}r(1+r)\Big)\Big(\frac{100}{11}r(1+r)-1\Big)dr.$$

In particular we have $M_q(\overline{D}f,r)$ is a nondecreasing function, this because $\overline{D}f$ is harmonic in $B_1(0)$ and belongs to $L_q(B_1(0))$; $\forall 0 \leq r < 1$. This gives our statement. Hence, the assertion " $\Longrightarrow$ " follows.

By the help of Theorem 2.1, we see that $\mathbf{B^2} = \mathbf{Q_2}$ and this property is analogous to the complex one-dimensional case.

Theorem 2. *Let $0 < p < 2$, and $0 < q < 2$. Then, we have that*

$$\cup\mathbf{Q_p} \subset \cap\mathbf{B^q}.$$

Proof. Let $f \in \mathbf{Q_p}$, for any fixed p, $0 < p < 2$. Then by using Hölder's inequality for $0 < q < 2$, we can get that

$$\begin{aligned}
&\int_{B_1(0)} |\overline{D}f(x)|^q (1-|x|^2)^{\frac{3}{2}q-3} (1-|\varphi_a(x)|^2)^2 dB_x \\
\leq & \left\{\int_{B_1(0)} \left[|\overline{D}f(x)|^q (1-|\varphi_a(x)|^2)^{\frac{qp}{2}}\right]^{\frac{2}{q}} dB_x\right\}^{\frac{q}{2}} \\
\cdot & \left\{\int_{B_1(0)} \left[(1-|x|^2)^{\frac{3}{2}q-3} (1-|\varphi_a(x)|^2)^{2-\frac{qp}{2}}\right]^{\frac{2}{2-q}} dB_x\right\}^{\frac{2-q}{2}} \\
= & \left\{\int_{B_1(0)} |\overline{D}f(x)|^2 (1-|\varphi_a(x)|^2)^p dB_x\right\}^{\frac{q}{2}} \\
\cdot & \left\{\int_{B_1(0)} (1-|x|^2)^{\frac{3q-6}{2-q}} (1-|\varphi_a(x)|^2)^{\frac{4-qp}{2-q}} dB_x\right\}^{\frac{2-q}{2}}
\end{aligned}$$

Since, we have from [16] for any monogenic function f that

$$f \in \mathbf{Q_p} \Longleftrightarrow \sup_{a\in B_1(0)} \int_{B_1(0)} |\overline{D}f(x)|^2 (1-|\varphi_a(x)|^2)^p dB_x < \infty,$$

Then, the last inequality will take the following formula

$$\begin{aligned}
&\int_{B_1(0)} |\overline{D}f(x)|^q (1-|x|^2)^{\frac{3}{2}q-3} (1-|\varphi_a(x)|^2)^2 dB_x \\
\leq & \left\{\int_{B_1(0)} |\overline{D}f(x)|^2 (g(x,a))^p dB_x\right\}^{\frac{q}{2}} \\
\cdot & \left\{\int_{B_1(0)} (1-|x|^2)^{\frac{3q-6}{2-q}} (1-|\varphi_a(x)|^2)^{\frac{4-qp}{2-q}} dB_x\right\}^{\frac{2-q}{2}} \\
\leq & \mathcal{L}^{\frac{q}{2}} \left\{(1-|a|^2)^{\frac{3q-6}{2-q}+3} . \int_{B_1(0)} \frac{(1-|x|^2)^{\frac{3q-qp-2}{2-q}}}{|1-\bar{a}x|^{2\left(\frac{3q-6}{2-q}+3\right)}} dB_x\right\}^{\frac{2-q}{2}} \\
= & \mathcal{L}^{\frac{q}{2}} \left\{(1-|a|^2)^{\frac{3q-6}{2-q}+3}\right. \\
\cdot & \left.\int_0^1 r^2(1-r^2)^{\frac{3q-qp-2}{2-q}} dr \int_{\partial B_1(0)} \frac{1}{|1-\bar{a}ry|^{2\left(\frac{3q-6}{2-q}+3\right)}} d\Gamma_y\right\}^{\frac{2-q}{2}}
\end{aligned}$$

where,

$$\mathcal{L} = \int_{B_1(0)} |\overline{D}f(x)|^2 (g(x,a))^p dB_x.$$

Applying lemma 1.1, in the last inequality, we obtain that

$$\int_{B_1(0)} |\overline{D}f(x)|^q (1-|x|^2)^{\frac{3}{2}q-3}(1-|\varphi_a(x)|^2)^2 dB_x \leq \lambda_1 \mathcal{L}^{\frac{q}{2}}$$

where λ_1 is a constant not depending on a. Then, taking $\sup\limits_{a\in B_1(0)}$, we obtain that

$$\| f \|_{\mathbf{B^q}} \leq \| f \|_{\mathbf{Q_p}} < \infty.$$

Thus $f \in \mathbf{B^q}$ for any q, $0 < q < 2$ and our theorem is proved.

3. Relation between monogenic Bloch space and $\mathbf{B^q}$ spaces of monogenic functions

Proposition 3. *Let f be a hyperholomorphic function in the unit ball $B_1(0)$ and $1 \leq q < \infty$, then*

$$(1-|a|^2)^{\frac{3q}{2}} |\overline{D}f(a)|^q \leq \frac{1}{\eta^*(R)} \int_{B_1(0)} |\overline{D}f(x)|^q (1-|x|^2)^{\frac{3}{2}q-3}(1-|\varphi_a(x)|^2)^2 dB_x$$

where,

$$\eta^*(R) = \frac{k}{3\sqrt{\pi}2^{3q+1}}(1-R^2)^{\frac{3}{2}q+5} \max\left\{(1-R)^{2q-6}, (1+R)^{2q-6}\right\};$$

k is a constant and $0 < R < 1$.

Proof. Let $U(a,R) = \{x : |\varphi_a(x)| < R\}$ be the pseudo hyperbolic ball with radius R, where $0 < R < 1$. Analogously to the complex case (see [28]), for a point $a \in \Delta$ and $0 < R < 1$, we can get that $U(a,R)$ with pseudo hyperbolic center a and pseudo hyperbolic radius R is a Euclidean disc: its Euclidean center and Euclidean radius are $\frac{(1-R^2)a}{1-R^2|a|^2}$ and $\frac{(1-|a|^2)R}{1-R^2|a|^2}$, respectively. Then,

$$\int_{B_1(0)} |\overline{D}f(x)|^q (1-|x|^2)^{\frac{3}{2}q-3}(1-|\varphi_a(x)|^2)^2 dB_x$$
$$\geq \int_{U(a,R)} |\overline{D}f(x)|^q (1-|x|^2)^{\frac{3}{2}q-3}(1-|\varphi_a(x)|^2)^2 dB_x.$$

Now, we will estimate the following:

$$\int_{U(a,R)} |\overline{D}f(x)|^q (1-|x|^2)^{\frac{3}{2}q-3} dB_x$$
$$= \int_{U(a,R)} |\overline{D}f(x)|^q \left\{\frac{(1-|\varphi_a(x)|^2)(|1-\bar{a}x|^2)}{(1-|a|^2)}\right\}^{\frac{3}{2}q-3} dB_x.$$

Since,

$$\left(1-|x|^2\right)^3 \approx \left|U(a,R)\right|^{\frac{3}{2}}, \quad \text{whenever} \quad x \in U(a,R) \quad \text{and}$$

where, $\left|U(a,R)\right|$ stands for the surface area of the pseudo hyperbolic ball $U(a,R)$ given as below.
Then,

$$\begin{aligned}
&\int_{U(a,R)} \left|\overline{D}f(x)\right|^q \left(1-|x|^2\right)^{\frac{3}{2}q-3} dB_x \\
\geq\ & \frac{k}{\left|U(a,R)\right|^{\frac{3}{2}}} \int_{U(a,R)} \left|\overline{D}f(x)\right|^q \left(1-|x|^2\right)^{\frac{3}{2}q} dB_x \\
\geq\ & \frac{k\left(1-|a|\right)^{2(\frac{3}{2}q)}\left(1-R^2\right)^{\frac{3}{2}q}}{\left|U(a,R)\right|^{\frac{3}{2}}\left(1-|a|^2\right)^{\frac{3}{2}q}} \int_{U(a,R)} \left|\overline{D}f(x)\right|^q dB_x \\
=\ & \frac{k\left(1-|a|^2\right)^{\frac{3}{2}q}\left(1-R^2\right)^{\frac{3}{2}q}}{\left|U(a,R)\right|^{\frac{3}{2}}\left(1+|a|\right)^{3q}} \int_{U(a,R)} \left|\overline{D}f(x)\right|^q dB_x \\
=\ & \frac{k\left(1-|a|^2\right)^{\frac{3}{2}q}\left(1-R^2\right)^{\frac{3}{2}q}}{\left|U(a,R)\right|^{\frac{3}{2}}\left(1+|a|\right)^{3q}} \int_{B_R} \left|\overline{D}f(\varphi_a(x))\right|^q \frac{\left(1-|a|^2\right)^3}{\left|1-\bar{a}x\right|^6} dB_x.
\end{aligned}$$

The problem here is that, $\overline{D}f(x)$ is hyperholomorphic but after the change of variables $\overline{D}f(\varphi_a(x))$ is not hyperholomorphic. But we know from [26] that $\frac{1-\bar{x}a}{|1-\bar{a}x|^3}\overline{D}f(\varphi_a(x))$ is again hyperholomorphic.
We also refer to [30] who studied this problem for the four-dimensional case already in 1979.
Therefore,

$$\begin{aligned}
&\int_{U(a,R)} \left|\overline{D}f(x)\right|^q \left(1-|x|^2\right)^{\frac{3}{2}q-3} dB_x \\
\geq\ & \frac{k\left(1-|a|^2\right)^{\frac{3}{2}q}\left(1-R^2\right)^{\frac{3}{2}q}\left(1-|a|^2\right)^3}{\left|U(a,R)\right|^{\frac{3}{2}}\left(1+|a|\right)^{3q}} \times \\
& \times \int_{B_R} \left|\frac{1-\bar{x}a}{\left|1-\bar{a}x\right|^3}\overline{D}f(\varphi_a(x))\right|^q \frac{\left|1-\bar{a}x\right|^{3q}}{\left|1-\bar{a}x\right|^{q+6}} dB_x \\
\geq\ & \frac{k\left(1-|a|^2\right)^{\frac{3}{2}q}\left(1-R^2\right)^{\frac{3}{2}q}\left(1-|a|^2\right)^3}{2^{3q}\left|U(a,R)\right|^{\frac{3}{2}}} \times \\
& \times \int_{B_R} \left|\frac{1-\bar{x}a}{\left|1-\bar{a}x\right|^3}\overline{D}f(\varphi_a(x))\right|^q \frac{\left|1-\bar{a}x\right|^{3q}}{\left|1-\bar{a}x\right|^{q+6}} dB_x
\end{aligned}$$

where k is a constant depending on R but not on a. Now, since

$$1-R \leq |1-\bar{a}x| \leq 1+R$$

and

$$\left|U(a,R)\right| = 4\pi \frac{\left(1-|a|^2\right)^2}{\left(1-R^2|a|^2\right)^2} R^2$$

by using lemma 1.2, one can get

$$\begin{aligned}
&\int_{U(a,R)} |\overline{D}f(x)|^q \left(1-|x|^2\right)^{\frac{3}{2}q-3} dB_x \\
&\geq \left(1-|a|^2\right)^{\frac{3}{2}q} \eta(R) \int_{B_R} \left| \frac{1-\bar{x}a}{|1-\bar{a}x|^3} \overline{D}f(\varphi_a(x)) \right|^q dB_x \\
&\geq \frac{4\pi}{3} R^3 \left(1-|a|^2\right)^{\frac{3}{2}q} \eta(R) |\overline{D}f(a)|^q.
\end{aligned}$$

where,

$$\begin{aligned}
\eta(R) &= \frac{1}{8\pi\sqrt{\pi}} \frac{k}{2^{3q}} \frac{\left(1-|a|^2R^2\right)^3 \left(1-R^2\right)^{\frac{3}{2}q}}{R^3} \max\left\{(1-R)^{2q-6}, (1+R)^{2q-6}\right\} \\
&\geq \frac{1}{8\pi\sqrt{\pi}} \frac{k}{2^{3q}} \frac{\left(1-R^2\right)^{\frac{3}{2}q+3}}{R^3} \max\left\{(1-R)^{2q-6}, (1+R)^{2q-6}\right\} = \eta_1(R).
\end{aligned}$$

Therefore,

$$\begin{aligned}
&\int_{B_1(0)} |\overline{D}f(x)|^q \left(1-|x|^2\right)^{\frac{3}{2}q-3} \left(1-|\varphi_a(x)|^2\right)^2 dB_x \\
&\geq \int_{U(a,R)} |\overline{D}f(x)|^q \left(1-|x|^2\right)^{\frac{3}{2}q-3} \left(1-|\varphi_a(x)|^2\right)^2 dB_x \\
&\geq \frac{4\pi}{3} R^3 \left(1-R^2\right)^2 \left(1-|a|^2\right)^{\frac{3}{2}q} \eta_1(R) |\overline{D}f(a)|^q \\
&= \eta^*(R) \left(1-|a|^2\right)^{\frac{3q}{2}} |\overline{D}f(a)|^q,
\end{aligned}$$

where

$$\eta^*(R) = \frac{4\pi}{3} R^3 \left(1-R^2\right)^2 \eta_1(R).$$

Corollary 1. *From proposition 3.1, we get for $1 \leq q < \infty$ that*

$$\mathbf{B^q} \subset \mathcal{B}.$$

Proposition 4. *Let f be a hyperholomorphic function in $B_1(0)$. Then, for $1 \leq p < q < \infty$, we have that*

$$\mathbf{B^p} \subset \mathbf{B^q}.$$

Proof. We can obtain the proof of this proposition directly by Proposition 2.1 and Corollary 3.1 .

From $\mathbf{B^p} \subset \mathbf{B^q}$, for $1 \leq p < q < \infty$, $\mathbf{B^q} \subset \mathcal{B}$ and $\mathbf{B^2} = \mathbf{Q_2} \subset \mathcal{B}$, we get that $\mathbf{B^q} \neq \mathcal{B}$ for $q \leq 2$.

Remark 3.1. It is still an open problem if we can get for some $q > 2$ that $\mathbf{B^q} = \mathcal{B}$ without any restrictions.

References

[1] A. Aleman, *Hilbert spaces of analytic functions between the Hardy and the Dirichlet spaces*, Proc. Amer. Math. Soc. 115(1992), 97-104.

[2] J. Anderson, J. Clunie and Ch. Pommerenke, *On Bloch functions and normal functions*, J. Reine Angew. Math. 270 (1974), 12-37.

[3] J. Arazy and S. Fisher, *Some aspects of the minimal , Möbius invariant space of analytic functions in the unit disk*, Springer Lecture Notes Math., 1070(1984), 24-44.

[4] J. Arazy , S. Fisher and J. Peetre, *Möbius invariant function spaces*, J. reine angew. Math., 363(1985), 110-145.

[5] R. Aulaskari, N. Danikas and R. Zhao, *The Algebra property of the integrals of some analytic functions in the unit disk*, Ann. Acad. Sci. Fenn., Math., 24 (1999), 343-351.

[6] R. Aulaskari and P. Lappan, *Criteria for an analytic function to be Bloch and a harmonic or meromorphic function to be normal, Complex analysis and its applications*, Pitman Research Notes in Math. 305, Longman Scientific and Technical Harlow (1994), 136-146.

[7] R. Aulaskari and L.M. Tovar, *On the function spaces* $\mathbf{B^q}$ *and* $\mathbf{Q_p}$, Bull. Hong Kong Math. Soc., 1(1997), 203-208.

[8] R. Aulaskari, J. Xiao and R. Zhao, *On subspaces and subsets of BMOA and UBC*, Analysis, 15 (1995), 101-121.

[9] F. Brackx, R. Delanghe and F. Sommen, *Clifford analysis*, Pitman Research Notes in Math. Boston, London, Melbourne, 1982.

[10] J.S. Choa, H.O. Kim and Y.Y. Park, *A Bergman-Carleson measure characterization of Bloch functions in the unit ball of* $\mathbb{C}^n$, Bull. Korean Math. Soc. 29 (1992), 285-293.

[11] J. Cnops and R. Delange, *Möbius invariant spaces in the unit ball*, Appl. Anal., 73(2000), 45-64.

[12] N. Danikas, *Some Banach spaces of analytic functions, Function spaces and complex analysis*, Summer school Ilomantsi, Finland, August 25-29, Joensuu 1997, Department of Mathematics. Report Series. 2., (1999), 9-35.

[13] M. Essen, S. Janson, L. Peng and J. Xiao, **Q** *spaces of several real variables*, Indiana Univ. Math. J., 49 No.2(2000), 575-615.

[14] M. Essen and J. Xiao, $\mathbf{Q}_\mathbf{p}$ *spaces-a survey, Complex function spaces*, Proceedings of the Summer School, Mekrijrvi, Finland, August 30-September 3, 1999, Rep. Ser. Univ. Joensuu. 4(2001), 11-60.

[15] K. Gürlebeck and A. El-Sayed Ahmed, *Integral norms for hyperholomorphic Bloch-functions in the unit ball of* $\mathbf{R}^3$, to appear in the Proceedings of the 3rd ISAAC Congress, held in Berlin, 2001.

[16] K. Gürlebeck, U. Kähler, M. Shapiro and L.M. Tovar, *On* $\mathbf{Q}_\mathbf{p}$ *spaces of quaternion-valued functions*, Complex Variables, 39(1999), 115-135.

[17] K. Gürlebeck and H.R. Malonek, *On strict inclusions of weighted Dirichlet spaces of monogenic functions*, Bull. Austral. Math. Soc., Vol. 64(2001), 33-50.

[18] K. Gürlebeck and W. Sprößig, *Quaternionc analysis and elliptic boundary value problems*, Int. Ser. Num. Math. 89 Birkhäuser Verlag, Basel, 1990.

[19] K. Gürlebeck and W. Sprößig, *Quaternionc and Clifford calculus for engineers and physicists*, John Wiley and Sons, Chichester, 1997.

[20] H. Leutwiler, *Harmonic functions of bounded mean oscillation*, Math. Ann. 244 (1979), 167-183.

[21] J. Miao, *A property of Analytic functions with Hadamard gaps*, Bull. Austral. Math. Soc., 45(1992), 105-112.

[22] I.M. Mitelman and M.V. Shapiro, *Differentiation of the Martinelli-Bochner integrals and the notion of hyperderivability*, Math. Nachr., 172 (1995), 211-238.

[23] A. Nicolau, *The corona property for bounded analytic functions in some Besov spaces*, Proc. Am. Math. Soc. 110, No.1(1990), 135-140.

[24] C. Ouyang, W. Yang and R. Zhao, *Characterizations of Bergman spaces and Bloch space in the unit ball of* $\mathbf{C}^n$, Transactions of the American Math. Soc., (11) 347 (1995), 4301-4313.

[25] C. Ouyang, W. Yang and R. Zhao, *Möbius invariant* $\mathbf{Q}_\mathbf{p}$ *spaces associated with the Green's function on the unit ball of* $\mathbf{C}^n$, Pac. J. Math., (1) 182 (1998), 69-99.

[26] J. Ryan, *Conformally covariant operators in Clifford analysis*, ZAA 14(1995), 677-704.

[27] J. Ryan and W. Sprößig, *Clifford Algebras and their applications in Mathematical physics*, Birkhäuser Boston, 2000.

[28] K. Stroethoff, *Besov-type characterizations for the Bloch space*, Bull. Austral. Math. Soc. 39(1989), 405-420.

[29] K. Stroethoff, *The Bloch space and Besov spaces of analytic functions*, Bull. Aust. Math. Soc. 54, No.2(1996), 211-219.

[30] A. Sudbery, *Quaternionic analysis*, Math. Proc. Cambridge Philos. Soc., 85(1979), 199-225.

[31] H. Wulan, *On some classes of meromorphic functions*, Annales Academiae Scientiarum Fennicae. Series A I. Mathematica. Dissertationes. 116. Helsinki: Suomalainen Tiedeakatemia, (1998), 1-57.

[32] K. Zhu, *Operator theory in function spaces*, Pure and Applied Math., No. 139 Marcel Dekker, New York, 1990.

19

REGENERATION IN COMPLEX, QUATERNION AND CLIFFORD ANALYSIS

Joji Kajiwara
Faculty of Mathematics(Emeritus)
Kyushu University 33, Fukuoka 812-8581, Japan
E-mail: kajiwara@math.kyushu-u.ac.jp

Xiao Dong Li
Department of Applied Mathematics
University of Electronic Science and
Technology of China, Chengdu, 610054 Sichuan, China
E-mail: xli@uestc.edu.cn

Kwang Ho Shon
Department of Mathematics
Pusan National University
Jangjon-Dong, Pusan 609-735, Korea
E-mail: khshon@hyowon.pusan.ac.kr

Abstract: For any complex valued harmonic function f_1, which satisfies the condition of integrability in a pseudoconvex domain D of $\mathbf{C}^4$, there exists a complex valued harmonic function f_2 in D such that the quaternion valued function $f_1 + f_2 j$ is hyperholomorphic in D.

Keywords: regeneration, quaternion analysis, hyperholomorphic function, Cousin domain, domain of holomorphy.

AMS-classification: Primary 32F20, Secondary 30G35.

1. Introduction

It is a great pleasure for us to submit the present paper under Professor LE Hung Son who[22] solved the Cousin Problem in Clifford Analysis, because we([18], [20], [21] and [23]) also investigated the Cousin problem in Complex Analysis of several and infinite variables for a long time.

In Complex Analysis of several variables, Matsugu[25] gave a necessary and sufficient condition for that a real valued pluriharmonic function f_1 in a Riemann domain D over a Stein manifold is a real part of a holomorphic function $f_1 + if_2$ on D. Fukushima[11] obtained similar results for a Riemann domain over complex projective space, Fukushima-Watanabe[14] and Adachi-Fukushima-Watanabe[1] over a Grassmann manifold and Fukushima[12] for domains over infinite dimensional spaces.

In Quaternion Analysis, M. Naser[26] represented one quaternion variable $z = x_1 + ix_2 + jx_3 + kx_4$ by a pair $z = z_1 + z_2 j$ of two complex variables $z_1 = x_1 + ix_2$ and $z_2 = x_3 + ix_4$, and established the identification $\mathcal{H} \cong \mathbf{C}^2$ between the quaternion field $\mathcal{H}$ and the space $\mathbf{C}^2$. He[26] introduced the notion of hyperholomorphy of quaternion valued functions $f = f_1 + f_2 j$ of a quaternion variables $z = z_1 + z_2 j$ in $\mathbf{C}^2$ and proved that any complex valued harmonic function f_1 in a domain of holomorphy D in $\mathbf{C}^2$ has a hyperconjugate harmonic function f_2 such that the quaternion valued function $f = f_1 + f_2 j$ is hyperholomorphic in D. K. Nôno[28] showed the validity of the converse proving that, if any complex valued harmonic function f_1 in a domain D in $\mathbf{C}^2$ has a hyperconjugate harmonic function f_2, then D is a domain of holomorphy in $\mathbf{C}^2$. M. S. Marinov[24] established the regeneration theory on Quaternion Analysis. Y. Fukushima[13] gave a theorem on regeneration in Clifford Analysis, as is promptly skeched in the last section. The main purpose of the present paper is to give a regeneration theorem in Quaternion Analysis in view of Complex Analysis.

2. Notations on Quaternion Analysis

The field $\mathcal{H}$ of quaternions

$$z = x_1 + ix_2 + jx_3 + kx_4, \qquad x_1, x_2, x_3, x_4 \in \mathbf{R} \tag{1}$$

is a four dimensional non-commutative $\mathbf{R}$-field generated by four base elements 1, i, j and k with the following non commutative multiplication rule:

$$i^2 = j^2 = k^2 = -1, ij = -ji = k, jk = -kj = i, ki = -ik = j. \tag{2}$$

We associate two complex numbers

$$z_1 := x_1 + ix_2, \quad z_2 := x_3 + ix_4 \in \mathbf{C} \tag{3}$$

to (1), regarding as

$$z = z_1 + z_2 j \in \mathcal{H}. \tag{4}$$

Thus, we identify $\mathcal{H}$ with $\mathbf{C}^2 \cong \mathbf{R}^4$ and speak of the topology of $\mathcal{H}$. We define the non commutative multiplication of two quaternions $z = z_1 + z_2 j, w = w_1 + w_2 j \in \mathcal{H}$ by

$$zw := (z_1 w_1 - z_2 \overline{w_2}) + (z_1 w_2 + z_2 \overline{w_1}) j \in \mathcal{H}, \qquad \overline{z_1} := x_1 - i x_2. \tag{5}$$

On the contrary, the quaternion conjugate z^* of $z = z_1 + z_2 j \in \mathcal{H}$ is defined by

$$z^* := \overline{z_1} - z_2 j. \tag{6}$$

The absolute value

$$|z| := \sqrt{|z_1|^2 + |z_2|^2} \tag{7}$$

coincides with the usual norm of $z \in \mathbf{C}^2 \cong \mathbf{R}^4$.

In the present paper, we use the following quaternion differential operators:

$$\frac{\partial}{\partial z} := \frac{\partial}{\partial z_1} - j \frac{\partial}{\partial \overline{z_2}}, \qquad \frac{\partial}{\partial z^*} := \frac{\partial}{\partial \overline{z_1}} + j \frac{\partial}{\partial \overline{z_2}}. \tag{8}$$

where $\frac{\partial}{\partial z_1}, \frac{\partial}{\partial \overline{z_1}}, \frac{\partial}{\partial z_2}$ and $\frac{\partial}{\partial \overline{z_2}}$ are usual differential operators used in Complex Analysis, e. g., [17]:

$$\frac{\partial}{\partial z_1} := \frac{1}{2}(\frac{\partial}{\partial x_1} - i \frac{\partial}{\partial x_2}), \qquad \frac{\partial}{\partial \overline{z_1}} := \frac{1}{2}(\frac{\partial}{\partial x_1} + i \frac{\partial}{\partial x_2}). \tag{9}$$

So, we have

$$\begin{aligned} \frac{\partial}{\partial z_1} j &= \frac{1}{2}(\frac{\partial}{\partial x_1} - i \frac{\partial}{\partial x_2}) j = \frac{1}{2}(j \frac{\partial}{\partial x_1} - ij \frac{\partial}{j \partial x_2}) \\ &= \frac{1}{2}(j \frac{\partial}{\partial x_1} + ji \frac{\partial}{\partial x_2}) = j \frac{\partial}{\partial \overline{z_1}} \end{aligned} \tag{10}$$

and

$$\begin{aligned} \frac{\partial}{\partial \overline{z_1}} j &= \frac{1}{2}(\frac{\partial}{\partial x_1} + i \frac{\partial}{\partial x_2}) j = \frac{1}{2}(j \frac{\partial}{\partial x_1} + ij \frac{\partial}{\partial x_2}) \\ &= \frac{1}{2}(j \frac{\partial}{\partial x_1} - ji \frac{\partial}{\partial x_2}) = j \frac{\partial}{\partial z_1}. \end{aligned} \tag{11}$$

The operator

$$\frac{\partial^2}{\partial z \partial z^*} = (\frac{\partial}{\partial z_1} - j\frac{\partial}{\partial \overline{z_2}})(\frac{\partial}{\partial \overline{z_1}} + j\frac{\partial}{\partial \overline{z_2}}) =$$
$$= \frac{\partial}{\partial z_1}\frac{\partial}{\partial \overline{z_1}} - j\frac{\partial}{\partial \overline{z_2}}\frac{\partial}{\partial \overline{z_1}} + \frac{\partial}{\partial z_1}j\frac{\partial}{\partial \overline{z_2}} - j\frac{\partial}{\partial \overline{z_2}}j\frac{\partial}{\partial \overline{z_2}} =$$
$$= \frac{\partial}{\partial z_1}\frac{\partial}{\partial \overline{z_1}} - j\frac{\partial}{\partial \overline{z_2}}\frac{\partial}{\partial \overline{z_1}} + j\frac{\partial}{\partial \overline{z_1}}\frac{\partial}{\partial \overline{z_2}} - jj\frac{\partial}{\partial z_2}\frac{\partial}{\partial \overline{z_2}}$$
$$= \frac{1}{4}(\frac{\partial^2}{\partial x_1^2} + \frac{\partial^2}{\partial x_2^2} + \frac{\partial^2}{\partial x_3^2} + \frac{\partial^2}{\partial x_4^2}) \tag{12}$$

is the usual real Laplacian Δ_z, which is a typical elliptic differential operator with constant coefficients on $\mathbf{R}^4$ with respect to the variables x_1, x_2, x_3, x_4.

In the space $\mathcal{H}^2 \cong \mathbf{C}^4$ of two quaternion variables $z = z_1 + z_2 j, w = w_1 + w_2 j, z_1 = x_1 + ix_2, z_2 = x_3 + ix_4, w_1 = y_1 + iy_2, w_2 = y_3 + iy_4$. we use the quaternion differential operators

$$\frac{\partial}{\partial w} := \frac{\partial}{\partial w_1} - j\frac{\partial}{\partial \overline{w_2}}, \qquad \frac{\partial}{\partial w^*} := \frac{\partial}{\partial \overline{w_1}} + j\frac{\partial}{\partial \overline{w_2}} \tag{13}$$

too.

Let Ω be an open set in $\mathcal{H}^2 \cong \mathbf{C}^4$ and $f(z,w) = f_1(z,w) + f_2(z,w)j$ be a C^1-function on $\Omega \subset \mathcal{H}^2 \cong \mathbf{R}^8$. f is said to be *hyperholomorphic* in Ω if f satisfies the equations

$$\frac{\partial}{\partial z^*}f = 0, \qquad f\frac{\partial}{\partial w^*} = 0 \tag{14}$$

in Ω. Since, we have

$$f\frac{\partial}{\partial w^*} = (f_1 + f_2 j)(\frac{\partial}{\partial \overline{w_1}} + j\frac{\partial}{\partial \overline{w_2}}) =$$
$$= f_1\frac{\partial}{\partial \overline{w_1}} + f_2 j\frac{\partial}{\partial \overline{w_1}} + f_1 j\frac{\partial}{\partial \overline{w_2}} + f_2 jj\frac{\partial}{\partial \overline{w_2}} =$$
$$= f_1\frac{\partial}{\partial \overline{w_1}} + f_2\frac{\partial}{\partial w_1}j + f_1\frac{\partial}{\partial w_2}j - f_2\frac{\partial}{\partial \overline{w_2}} =$$
$$= \frac{\partial f_1}{\partial \overline{w_1}} + \frac{\partial f_2}{\partial w_1}j + \frac{\partial f_1}{\partial w_2}j - \frac{\partial f_2}{\partial \overline{w_2}}, \tag{15}$$

(14) is equivalent to

$$\frac{\partial f_1}{\partial \overline{z_1}} = \frac{\partial \overline{f_2}}{\partial z_2}, \quad \frac{\partial \overline{f_1}}{\partial z_2} = -\frac{\partial f_2}{\partial \overline{z_1}}, \quad \frac{\partial f_1}{\partial \overline{w_1}} = \frac{\partial f_2}{\partial \overline{w_2}}, \quad \frac{\partial f_1}{\partial w_2} = -\frac{\partial f_2}{\partial w_1}. \tag{16}$$

By (16) we have

$$\frac{\partial}{\partial z_1}\frac{\partial f_1}{\partial \overline{z_1}} + \frac{\partial}{\partial z_2}\frac{\partial f_1}{\partial \overline{z_2}} = \frac{\partial}{\partial z_1}\frac{\partial \overline{f_2}}{\partial z_2} - \frac{\partial}{\partial z_2}\frac{\partial \overline{f_2}}{\partial z_1} = 0, \tag{17}$$

$$\frac{\partial}{\partial w_1}\frac{\partial f_1}{\partial \overline{w_1}} + \frac{\partial}{\partial \overline{w_2}}\frac{\partial f_1}{\partial w_2} = \frac{\partial}{\partial w_1}\frac{\partial f_2}{\partial \overline{w_2}} - \frac{\partial}{\partial \overline{w_2}}\frac{\partial f_2}{\partial w_1} = 0. \tag{18}$$

3. Regeneration in Quaternion Analysis

In this way, any hyperholomorphic function $f_1 + f_2 j$ satisfies $\Delta_z f_1 = \Delta_z f_2 = 0$, $\Delta_w f_1 = \Delta_w f_2 = 0$ and, consequently, $\Delta_{z,w} f_1 = \Delta_{z,w} f_2 = 0$ in the sense of distribution, where the operator $\Delta_{z,w}$ stands for the usual Laplacian operator in 8 real variables $x_1, x_2, x_3, x_4, y_1, y_2, y_3, y_4$ in $\mathbf{R}^8$. By a theorem of K. O. Friedrichs[8], f_1 and f_2, defined firstly as functions of class C^1, are of class C^∞ in 8 real variables $x_1, x_2, x_3, x_4, y_1, y_2, y_3, y_4$ in $\mathbf{R}^8$.

By (16), we have

$$\begin{aligned} -\frac{\partial^2 f_1}{\partial \overline{z_1}\partial w_2} + \frac{\partial^2 \overline{f_1}}{\partial w_1 \partial z_2} &= -\frac{\partial}{\partial \overline{z_1}}\frac{\partial f_1}{\partial w_2} + \frac{\partial}{\partial w_1}\frac{\partial \overline{f_1}}{\partial z_2} = \\ &= \frac{\partial}{\partial \overline{z_1}}\frac{\partial f_2}{\partial w_1} - \frac{\partial}{\partial w_1}\frac{\partial f_2}{\partial \overline{z_1}} = 0, \end{aligned} \tag{19}$$

$$\begin{aligned} \frac{\partial^2 f_1}{\partial \overline{z_1}\partial \overline{w_1}} + \frac{\partial^2 \overline{f_1}}{\partial \overline{w_2}\partial z_2} &= \frac{\partial}{\partial \overline{z_1}}\frac{\partial f_1}{\partial \overline{w_1}} + \frac{\partial}{\partial \overline{w_2}}\frac{\partial \overline{f_1}}{\partial z_2} = \\ &= \frac{\partial}{\partial \overline{z_1}}\frac{\partial f_2}{\partial \overline{w_2}} - \frac{\partial}{\partial \overline{w_2}}\frac{\partial f_2}{\partial \overline{z_1}} = 0, \end{aligned} \tag{20}$$

$$\begin{aligned} \frac{\partial^2 f_1}{\partial \overline{z_1}\partial \overline{w_1}} + \frac{\partial^2 \overline{f_1}}{\partial \overline{w_2}\partial z_2} &= \frac{\partial}{\partial \overline{z_1}}\frac{\partial f_1}{\partial \overline{w_1}} + \frac{\partial}{\partial \overline{w_2}}\frac{\partial \overline{f_1}}{\partial z_2} = \\ &= \frac{\partial}{\partial \overline{z_1}}\frac{\partial f_2}{\partial \overline{w_2}} - \frac{\partial}{\partial \overline{w_2}}\frac{\partial f_2}{\partial \overline{z_1}} = 0, \end{aligned} \tag{21}$$

$$\begin{aligned} \frac{\partial^2 f_1}{\partial \overline{z_2}\partial w_2} + \frac{\partial^2 \overline{f_1}}{\partial w_1 \partial z_1} &= \frac{\partial}{\partial \overline{z_2}}\frac{\partial f_1}{\partial w_2} + \frac{\partial}{\partial w_1}\frac{\partial \overline{f_1}}{\partial z_1} = \\ &= -\frac{\partial}{\partial \overline{z_2}}\frac{\partial f_2}{\partial w_1} + \frac{\partial}{\partial w_1}\frac{\partial f_2}{\partial \overline{z_2}} = 0, \end{aligned} \tag{22}$$

$$\begin{aligned}\frac{\partial^2 f_1}{\partial \overline{z_1} \partial \overline{w_1}} + \frac{\partial^2 \overline{f_1}}{\partial \overline{w_2} \partial z_2} &= \frac{\partial}{\partial \overline{z_1}} \frac{\partial f_1}{\partial \overline{w_1}} + \frac{\partial}{\partial \overline{w_2}} \frac{\partial \overline{f_1}}{\partial z_2} = \\ &= \frac{\partial}{\partial \overline{z_1}} \frac{\partial f_2}{\partial \overline{w_2}} - \frac{\partial}{\partial \overline{w_2}} \frac{\partial f_2}{\partial \overline{z_1}} = 0,\end{aligned} \tag{23}$$

$$\begin{aligned}\frac{\partial^2 f_1}{\partial \overline{z_2} \partial \overline{w_1}} - \frac{\partial^2 \overline{f_1}}{\partial \overline{w_2} \partial z_1} &= \frac{\partial}{\partial \overline{z_2}} \frac{\partial f_1}{\partial \overline{w_1}} - \frac{\partial}{\partial \overline{w_2}} \frac{\partial \overline{f_1}}{\partial z_1} = \\ &= \frac{\partial}{\partial \overline{z_2}} \frac{\partial f_2}{\partial \overline{w_2}} - \frac{\partial}{\partial \overline{w_2}} \frac{\partial f_2}{\partial \overline{z_2}} = 0,\end{aligned} \tag{24}$$

$$\begin{aligned}\frac{\partial^2 f_1}{\partial w_1 \partial \overline{w_1}} + \frac{\partial^2 f_1}{\partial \overline{w_2} \partial w_2} &= \frac{\partial}{\partial w_1} \frac{\partial f_1}{\partial \overline{w_1}} + \frac{\partial}{\partial \overline{w_2}} \frac{\partial f_1}{\partial w_2} = \\ &= \frac{\partial}{\partial w_1} \frac{\partial f_2}{\partial \overline{w_2}} - \frac{\partial}{\partial \overline{w_2}} \frac{\partial f_2}{\partial w_1} = 0\end{aligned} \tag{25}$$

and so on. Among them, the system of the six equations

$$\begin{aligned}&\frac{\partial^2 f_1}{\partial z_1 \partial \overline{z_1}} + \frac{\partial^2 f_1}{\partial z_2 \partial \overline{z_2}} = 0, && -\frac{\partial^2 f_1}{\partial \overline{z_1} \partial w_2} + \frac{\partial^2 \overline{f_1}}{\partial w_1 \partial z_2} = 0, \\ &\frac{\partial^2 f_1}{\partial \overline{z_1} \partial \overline{w_1}} + \frac{\partial^2 \overline{f_1}}{\partial \overline{w_2} \partial z_2} = 0, && \frac{\partial^2 f_1}{\partial \overline{z_2} \partial w_2} + \frac{\partial^2 \overline{f_1}}{\partial w_1 \partial z_2} = 0 \\ &\frac{\partial^2 f_1}{\partial \overline{z_2} \partial \overline{w_1}} - \frac{\partial^2 \overline{f_1}}{\partial \overline{w_2} \partial z_1} = 0, && \frac{\partial^2 f_1}{\partial w_1 \partial \overline{w_1}} + \frac{\partial^2 f_1}{\partial w_2 \partial \overline{w_2}} = 0\end{aligned} \tag{26}$$

is called the *condition of integrability.*

We consider an automorphism

$$(z_1, z_2, \omega_1, w_2) = \kappa(z_1, z_2, w_1, w_2) := (z_1, z_2, \overline{w_1}, w_2) \tag{27}$$

of $\mathbf{C}^4$. A domain D in $\mathbf{C}^4 \cong \mathcal{H}^4$ is said to be *pseudoconvex* with respect to the complex variables $z_1, z_2, \overline{w_1}, w_2$, if $\kappa(D)$ is a pseudoconvex domain of the space $\mathbf{C}^4$ of 4 complex variables z_1, z_2, ω_1, w_2 in the sense of Complex Analysis, e. g., L. Hörmander[17].

Theorem. *Let D be a domain in $\mathbf{C}^4 \cong \mathcal{H}^2$, which is a pseudoconvex domain with respect to the complex variables $z_1, z_2, \overline{w_1}, w_2$ and f_1 be a complex valued function of class C^2 on D satisfying the condition of integrability (26). Then, there exists a complex valued function f_2 of class C^2 on D such that $f = f_1 + f_2 j$ is a hyperholomorphic function on D.*

Proof of Theorem. We consider the 1-form

$$\Omega := -\frac{\partial \overline{f_1}}{\partial z_2} d\overline{z_1} + \frac{\partial \overline{f_1}}{\partial z_1} d\overline{z_2} - \frac{\partial f_1}{\partial w_2} dw_1 + \frac{\partial f_1}{\partial \overline{w_1}} d\overline{w_2} \tag{28}$$

and the differential operator

$$\delta = \frac{\partial}{\partial \overline{z_1}} d\overline{z_1} + \frac{\partial}{\partial \overline{z_2}} d\overline{z_2} + \frac{\partial}{\partial w_1} dw_1 + \frac{\partial}{\partial \overline{w_2}} d\overline{w_2}. \tag{29}$$

The operator $\delta \circ \kappa$ is nothing but the $\bar{\partial}$ operator in the space $\mathbf{C}^4$ of 4 complex variables z_1, z_2, ω_1, w_2 and the strong weapon of the theory of L. Hörmander[17] is applicable. We operate the operator δ from left handside for the 1-form Ω:

$$\delta\Omega := (\frac{\partial}{\partial \overline{z_1}} d\overline{z_1} + \frac{\partial}{\partial \overline{z_2}} d\overline{z_2} + \frac{\partial}{\partial w_1} dw_1 + \frac{\partial}{\partial \overline{w_2}} d\overline{w_2}) \wedge$$

$$(-\frac{\partial \overline{f_1}}{\partial z_2} d\overline{z_1} + \frac{\partial \overline{f_1}}{\partial z_1} d\overline{z_2} - \frac{\partial f_1}{\partial w_2} dw_1 + \frac{\partial f_1}{\partial \overline{w_1}} d\overline{w_2}) \tag{30}$$

and arrange it as follows:

$$\delta\Omega = (\frac{\partial^2 \overline{f_1}}{\partial \overline{z_1} \partial z_1} + \frac{\partial^2 \overline{f_1}}{\partial \overline{z_2} \partial z_2}) d\overline{z_1} \wedge d\overline{z_2} + (-\frac{\partial^2 f_1}{\partial \overline{z_1} \partial w_2} + \frac{\partial^2 \overline{f_1}}{\partial w_1 \partial z_2}) d\overline{z_1} \wedge dw_1 +$$

$$(\frac{\partial^2 f_1}{\partial \overline{z_1} \partial \overline{w_1}} + \frac{\partial^2 \overline{f_1}}{\partial \overline{w_2} \partial z_2}) d\overline{z_1} \wedge d\overline{w_2} + (-\frac{\partial^2 f_1}{\partial \overline{z_2} \partial w_2} - \frac{\partial^2 \overline{f_1}}{\partial w_1 \partial z_1}) d\overline{z_2} \wedge dw_1 +$$

$$(\frac{\partial^2 f_1}{\partial \overline{z_2} \partial \overline{w_1}} - \frac{\partial^2 \overline{f_1}}{\partial \overline{w_2} \partial z_1}) d\overline{z_2} \wedge d\overline{w_2} + (\frac{\partial^2 f_1}{\partial w_1 \partial \overline{w_1}} + \frac{\partial^2 f_1}{\partial \overline{w_2} \partial w_2}) dw_1 \wedge d\overline{w_2}. \tag{31}$$

By the condition (26) of integrability, all coefficients vanish. And the $\bar{\partial}$-closed (0,1)-form $\Omega \circ \kappa$ on the pseudoconvex domain $\kappa^{-1}(D) = \kappa(D)$ of four complex variables z_1, z_2, ω_1, w_2 is a $\bar{\partial}$-axact (0,1) form . Therefore, Ω is δ-exact and there exists a complex valued function f_2 of class C^∞ on D such that $\delta f_2 = \Omega$. By the definition (29) of δ, we have (16). Hence $f_1 + f_2 j$ is a hyperholomorphic function on D. q. e. d.

4. Regeneration in Clifford Analysis

We use notations in Le[22] on Clifford Analysis without explanation.
Theorem of Fukushima[13]. *Let m and k be positive integers with $1 < m, k \leq n, m + k \leq n$. Let $P = (h_1, h_2, \cdots, h_r)$ be a permutation with $1 \leq h_1 < h_2 < \cdots < h_r$. Let Ω be an open subset of $\mathbf{R}^m \times \mathbf{R}^k$ and*

$$f : \Omega \to \mathcal{A} \tag{32}$$

$$\Omega \ni (x,y) = (x_1, x_2, \cdots, x_m, y_1, y_2, \cdots, y_k) \rightsquigarrow f(x,y) = \sum_{A \neq P} f_A(x,y) e_A. \tag{33}$$

be a function of class C^∞ on Ω. If there exits a function f_P of class C^∞ on Ω such that the $\mathcal{A}$ valued function $f = \sum_{A \neq P} f_A(x,y)e_A + f_P(x,y)e_P$ is a biregular function on Ω, i.e., satisfies (36)-(37) below, the real valued functions $f_A (A \neq P)$ satisfies the integrability condition

$$\mathrm{d}\omega = 0 \tag{34}$$

on Ω, where the differential form ω of degree 1 is given by

$$\omega = -\sum_1^m \Big(\sum_{\{i\}' \cup A = \{i\} \cup P, A \neq P} \epsilon_{iP}^{i'A} \frac{\partial f_A}{\partial x_{i'}} \Big) \mathrm{d}x_i - \\ - \sum_1^k \Big(\sum_{A \cup \{m+j'\} = B, A \neq P} \epsilon_{Pm+j'}^{Am+j'} \frac{\partial f_A}{\partial y_{j'}} \Big) \mathrm{d}y_j. \tag{35}$$

Conversely, if $f_A (A \neq P)$ satisfies the integrability condition (34)-(35) on Ω and if the domain Ω satisfies $\mathrm{H}^1(\Omega, \mathbf{Z}) = 0$ for the ring $\mathbf{Z}$ of integers, then there exits a function f_P of class C^∞ on Ω such that the $\mathcal{A}$ valued function $f = \sum_{A \neq P} f_A(x,y)e_A + f_P(x,y)e_P$ is a biregular function on Ω.

Proof. If there exits a function f_P of class C^∞ on Ω such that the $\mathcal{A}$ valued function $f = \sum_{A \neq P} f_A(x,y)e_A + f_P(x,y)e_P$ is a biregular function on Ω, by the definition, there hold

$$0 = D_x f = \sum_B \Big(\sum_{\{i\} \cup A = B} \epsilon_B^{iA} \frac{\partial f_A}{\partial x_i} \Big) e_B, \tag{36}$$

$$0 = f \mathrm{D}_y = \sum_B \Big(\sum_{A \cup \{m+j\} = B} \epsilon_B^{Am+j} \frac{\partial f_A}{\partial y_j} \Big) e_B. \tag{37}$$

Hence, there hold the following relations between the derivatives of f_A's:

$$\sum_{\{i\} \cup A = B} \epsilon_B^{iA} \frac{\partial f_A}{\partial x_i} = 0, \tag{38}$$

$$\sum_{A\cup\{m+j\}=B} \epsilon_B^{Am+j} \frac{\partial f_A}{\partial y_j} = 0 \tag{39}$$

for any permutation B with order from small to large. Then the derivatives of the function f_P satisfy

$$\sum_{\{i'\}\cup A=B, A\neq P} \epsilon_B^{i'A} \frac{\partial f_A}{\partial x_{i'}} + \sum_{\{i\}\cup P=B} \epsilon_B^{iP} \frac{\partial f_P}{\partial x_i} = 0 \tag{40}$$

and

$$\sum_{A\cup\{m+j'\}=B, A\neq P} \epsilon_B^{Am+j'} \frac{\partial f_A}{\partial y_{j'}} + \sum_{P\cup\{m+j\}=B} \epsilon_B^{Pm+j} \frac{\partial f_P}{\partial y_j} = 0 \tag{41}$$

for any permutation B with order from small to large. The derivatives of f_P are given by

$$\frac{\partial f_P}{\partial x_i} = - \sum_{\{i\}'\cup A=\{i\}\cup P, A\neq P} \epsilon_{\{i\}P}^{i'A} \frac{\partial f_A}{\partial x_{i'}}. \tag{42}$$

Then, the differential ω of the function f_P

$$\omega = \sum_1^m \frac{\partial f_P}{\partial x_i} \mathrm{d}x_i + \sum_1^k \frac{\partial f_P}{\partial y_j} \mathrm{d}y_j \tag{43}$$

is represented by other $f_A (A \neq P)$ as follows:

$$\omega = -\sum_1^m \Big(- \sum_{\{i\}'\cup A=\{i\}\cup P, A\neq P} \epsilon_{\{i\}P}^{i'A} \frac{\partial f_A}{\partial x_{i'}}\Big) \mathrm{d}x_i - \\ - \sum_1^k \Big(\sum_{A\cup\{m+j'\}=B, A\neq P} \epsilon_{Pm+j'}^{Am+j'} \frac{\partial f_A}{\partial y_{j'}} \Big) \mathrm{d}y_j. \tag{44}$$

Of course, the integrability condition $\mathrm{d}\omega = 0$ holds, since ω is the differential of the function f_P.

Let p be a non negative integer, $\mathbf{R}$ be the constant sheaf of real numbers over Ω, $\mathcal{E}^p$ be the sheaf of germs of differential forms of degree p with coefficients of class C^∞ over the domain $\Omega \subset \mathbf{R}^m \times \mathbf{R}^k$, d be the usual differential operator $\mathrm{d}^p : \mathcal{E}^p \to \mathcal{E}^{p+1}$ and $\iota : \mathbf{R} \to \mathcal{E}^0$ be the canonical injection. Then, by the lemma of Poincaré, the above operators give a fine resolution

$$0 \to \mathbf{R} \to \mathcal{E}^0 \to \mathcal{E}^1 \to \cdots \to \mathcal{E}^p \to \mathcal{E}^{p+1} \cdots . \tag{45}$$

of the constant sheaf $\mathbf{R}$ over Ω. By the theorem of Dolbeault[5], we have the following Dolbeault's isomorphism

$$\mathrm{H}^p(\Omega, \mathbf{R}) \cong \mathrm{H}^p(\Omega, (\mathrm{d}^p)^{-1}(0))/\mathrm{d}^{p-1}(\mathrm{H}^0(\Omega, \mathcal{E}^{p-1})) \tag{46}$$

for any positive integer p. By the universal coefficient theorem[35], we have $\mathrm{H}^p(\Omega, \mathbf{R}) \cong \mathrm{H}^p(\Omega, \mathbf{Z}) \bigotimes \mathbf{R}$ and, hence, $\mathrm{H}^p(\Omega, \mathbf{R}) = 0$ if and only if $\mathrm{H}^p(\Omega, \mathbf{Z}) = 0$, for any positive integer p. Therefore, from the assumptions $\mathrm{H}^1(\Omega, \mathbf{Z}) = 0$ and (34)-(35), we have $\omega \in \mathrm{H}^0(\Omega, (\mathrm{d}^1)^{-1}(0)) = \mathrm{d}^0(\mathrm{H}^0(\Omega, \mathcal{E}^0))$ and there exists $f_P \in \mathrm{H}^0(\Omega, \mathcal{E}^0)$ such that $\omega = \mathrm{d}^0 f_P$. The $\mathcal{A}$ valued function $f := f(x, y) = \sum_A f_A(x, y)e_A$ of class C^∞ on Ω satisfies $D_x f = 0$ by (34)-(35) and $\omega = \mathrm{d}^0 f_P$, and $f D_y = 0$ by (34)-(35) and $\omega = \mathrm{d}^0 f_4$. Hence the function f is t he desired biregular function on Ω with f_P as e_P component for other components f_A given. q.e.d.

Fukushima[13] is the first fifth of his thesis, Studies on regeneration of holomorphic functions and biregular functions, for the doctorate of mathematics from Kyushu University.

References

[1] K. Adachi, Y. Fukushima and K. Watanabe, *Pluriharmonic functions on a domain over a Grassman manifolds*, Sci. Bull. Educ. Nagasaki Univ. **37**(1986), 151-157.

[2] F. Brackx, *On (k)-monogenic functions of a quaternion variable*, Research Notes in Math. **8**(1976), 22-44.

[3] F. Brackx, *Non (k)-monogenic points of functions of a quaternion variable*, Lecture Notes in Math. **561**(1976), 138-149.

[4] F. Brackx and W. Pincket, *Domains of biregularity in Clifford analysis*, Rend. Circ. Mat. Palermo (2) suppl. **9**(1985), 21-35(1986).

[5] C. A. Deavours, *The quaternion calculus*, Amer. Math. Monthly **80**(1973), 995-1008.

[6] R. Delanghe, F. Brackx and F. Sommen, Clifford analysis, Research Notes in Math. **76**, Pitman Books, London, 1982.

[7] F. Gürsey and H. C. Tze, *Complex and Quaternion Analyticity in Chiral and Gauge Theories-I*, Ann. of Physics **128**(1980), 29-130.

[8] K. O. Friedrichs, *On differentiability of the solutions of linear elliptic differential equations*, Comm. Pure and Appl. Math. **6**(1953), 299-325.

[9] R. Fueter, *Die Funktionentheorie der Differentialgleichungen* $\Delta u = 0$ *und* $\Delta\Delta u = 0$ *mit vier reellen Variablen*, Comment. Math. Helv. **7**(1934), 307-330.

[10] R. Fueter, *Über die Analysche Darstellung der regulären Funktionen einer Quaternionenvariablen*, Comment. Math. Helv. **8**(1935), 371-378.

[11] Y. Fukushima, *On the relation between pluriharmonic functions and holomorphic functions*, Fukuoka Univ. Rep. **15**(1983), 33-37.

[12] Y. Fukushima, *Pluriharmonic functions on a topological vector space with finite open topology*, Fukuoka Univ. Rep. **163**(1994), 31-35.

[13] Y. Fukushima, *Regeneration in Clifford Analysis*, Comptes rendus de l'Académie bulgare des Sciences **54**(2001), 7-12.

[14] Y. Fukushima and K. Watanabe, *Pluriharmonic functions on a domain over a Grassman manifolds*, Fukuoka Univ. Sci. Rep. **15**(1985), 1-4.

[15] R. C. Gunning and H. Rossi, Analytic functions of several complex variables, Prentice-Hall N.J.(1965).

[16] L. Hörmander, Linear partial differential operators, Springer-Verlag and Academic Press, Inc., New York, 1963.

[17] L. Hörmander, An introduction to complex analysis in several variables, Van Nostrand, Princeton, N. J., 1966.

[18] J. Kajiwara, *La réciproque du théorème d'annulation et de finitude de cohomology dans L'espace produit d'une famille dénombrable de sphère de Riemann*, Bull. Soc. Math. France **103**(1975), 129-139.

[19] J. Kajiwara and H. Kazama, *Two dimensional complex manifold with vanishing cohomology set*, Math. Ann. **204**(1973),1-12.

[20] J. Kajiwara and K. H. Shon, *Continuation and vanishing theorem for cohomology of infinite dimensional spaces*, Pusan Kyongnam Math. J. **9**-1(1993), 65-73.

[21] J. Kajiwara, Y. Fukushima, L. Li and X. D. Li, *Characterizations of Holomorphy of Domains through Validity of Theorem A, B or Oka's Principle*, Finite or infinite Dimensional Complex Analysis, Marcel Dekker Inc.(New York/Basel)2000, 179-199.

[22] H. S. Le, *Cousin Problem for biregular functions with values in a Clifford Algebra*, Complex Variables **20**(1992), 255-263.

[23] X. D. Li and J. Kajiwara, *Global Generation and Holomorphy of Domains*, Proceedings of the Eighth International Colloquium on

Finite or Infinite Dimensional Complex Analysis, Shangdong Science and Technology Press(2000) 105-109 .

[24] M. S. Marinov, *Regeneration of regular quaternion functions*, 20th Summer School "Applications of Mathematics in Engeneering" Varna(1994), 85-101.

[25] Y. Matsugu, *Pluriharmonic functions as the real parts of holomorphic functions*, Mem. Fac. Sci. Kyushu Univ., Ser. A **36**(1982), 157-163.

[26] M. Naser, *Hyperholomorphic functions*, Siberian Math. J. **12**(1971), 959-968.

[27] K. Nôno, *Hyperholomorphic Functions of a Quaternion Variable*, Bull. Fukuoka Univ. of Educ. **32**(1983), 21-37.

[28] K. Nôno, *Characterization of domains of holomorphy by the existence of hyper-conjugate harmonic functions*, Revue Roumaine de math. pures et appl. **31**-2(1986), 159-161.

[29] K. Nôno, *Runge's Theorem for complex valued harmonic and quaternion valued hyperholomorphic functions*, Revue Roumaine de math. pures et appl. **32**-2(1987), 155-158.

[30] K. Nôno, *Domains of Hyperholomorphy in* $\mathbf{C}^2 \times \mathbf{C}^2$, Bull. Fukuoka Univ. of Educ. **36**(1987), 1-9.

[31] K. Oka, *Sur les fonctions analytiques de plusieurs variables: II Domaine d'holomorphie*, J. Sci. Hiroshima Univ. **17**(1937), 115-130.

[32] K. Oka, *Sur les fonctions analytiques de plusieurs variables: III Deuxième problème de Cousin*, J. Sci. Hiroshima Univ. **9**(1938), 7-19.

[33] K. Oka, *Sur les fonctions analytiques de plusieurs variables: VII Sur quelques notions arithmétiques*, Bull. Soc. Math. France **78**(1950), 1-27.

[34] K. Oka, *Sur les fonctions analytiques de plusieurs variables: IX Domaines finis sans point critique intérieur*, Jap. J. Math. **23**(1953), 1-27.

[35] E. H. Spinier, *Algebraic topology*, MacGraw-Hill, New York, 1966.

20

A NOTE ON SESQUILINEAR FORMS AND THE GENERALIZED UNCERTAINTY RELATIONS IN *-ALGEBRAS

Ken Kuriyama
Yamaguchi University
Ube city, Yamaguchi, 755-8611, Japan

Shigeru Furuichi
Science University of Tokyo in Yamaguchi
Onoda city, Yamaguchi, 756-0884, Japan

Kenjiro Yanagi
Yamaguchi University
Ube city, Yamaguchi, 755-8611, Japan

Abstract: Sesquilinear forms induced by a positive linear form on a *-algebra are introduced. Some relations among them are investigated and then the inequality related to the uncertainty principle is shown in general situation. As a by-product, GNS construction theorem by the right regular representation is constructed by the *-antihomomorphism.

Keywords: sesquilinear form, positive linear form, uncertainty principle, Operator algebra methods.

AMS-classification: 81R15

1. Introduction

In quantum measurement theory, Holevo introduced the notion of measurement with values in U, and showed the mathematically generalized uncertainty relation [1, 2]. Let $B(\mathcal{H})$ be the set of all bounded linear operators on a Hilbert space $\mathcal{H}$ and $\mathfrak{S}(\mathcal{H})$ the convex set of all density operators on $\mathcal{H}$. Let U is a set in R^n. A mapping $S \to \mu_S$ from $\mathfrak{S}(\mathcal{H})$ to the convex set of probabitity measures on measurable space is called a measurement with values in U (U-measurement), if the condition

$$\mu_{\sum_{i=1}^{n} \alpha_i S_i} = \sum_{i=1}^{n} \alpha_i \mu_{S_i}$$

is satisfied, where $S_i \in \mathfrak{S}(\mathcal{H}), \alpha_i \geq 0, \sum_{i=1}^{n} \alpha_i = 1$. The following relation between a U-measurement and a resolution of identity (positive operator-valued measure) holds. For a U-measurement μ, there exists a resolution of identity $\{M(B) : B \in \mathcal{A}(U)\}$ such that $\mu_S(B) = \mathrm{tr} SM(B)$ for $S \in \mathfrak{S}(\mathcal{H})$ and $B \in \mathcal{A}(\mathcal{H})$. Conversely for a resolution of identity $\{M(B) : B \in \mathcal{A}(U)\}$, if we set $\mu_S(B) = \mathrm{tr} SM(B)$ for $S \in \mathfrak{S}(\mathcal{H})$ and $B \in \mathcal{A}(U)$, then the mapping μ is a U-measurement.

The expected value $E_S\{M\}$ and the variance $D_S\{M\}$ of the real-valued measurement M with respect to $S \in \mathfrak{S}(\mathcal{H})$ are defined by

$$\begin{aligned} E_S\{M\} &= \int \lambda d\mu_S(\lambda), \\ D_S\{M\} &= \int (\lambda - E_S(\lambda))^2 d\mu_S(\lambda). \end{aligned}$$

For two real-valued measurements M_1 and M_2, Holevo proved the *generalized uncertainty relation*

$$D_S\{M_1\} D_S\{M_2\} \geq \frac{1}{4} |[X_{M_1}, X_{M_2}]_S|^2. \tag{1}$$

The key points of the proof are as follows;

(1) An element X_M of a certain Hilbert space $\mathcal{L}^2(S)$ is defined for the real-valued measurement M.

(2) The following inequality is proved

$$\langle X_{M_1}, X_{M_1} \rangle_S \langle X_{M_2}, X_{M_2} \rangle_S \geq \frac{1}{4} |[X_{M_1}, X_{M_2}]_S|^2, \quad for\ X_{M_1}, X_{M_2} \in \mathcal{L}^2(S). \tag{2}$$

The main purpose in the paper is to prove the inequality (2) in a general situation. We use a positive linear form on a *-algebra in stead of a normal state on a von Neumann algebra.

2. GNS construction theorem by right regular representation

It is well known the Gelfand-Naimark-Segal(GNS) construction theorem by left regular representation[3]. In this section, we construct the GNS construction theorem by right regular representation.

Let ω be a linear form on *-algebras A. ω is a positive linear form if and only if $\omega(x^*x) \geq 0$ for any $x \in A$.

Proposition 1. *For a positive linear form ω on *-algegras A (we assume $1 \in A$ for simplicity) we have*

(1) $\omega(y^*x) = \overline{\omega(x^*y)}$,

(2) $\omega(x^*) = \overline{\omega(x)}$,

(3) $|\omega(y^*x)|^2 \leq \omega(x^*x)\omega(y^*y)$,

(4) $|\omega(x)|^2 \leq \omega(1)\omega(x^*x)$.

(1)-(4) of the above proposition are applied to the GNS construction theorem. For a proof of this proposition, the reader may refer to a standerd textbook over operator algebras[4, 5]. As an example for quantum theory, the following is well known. Let S be a density operator on Hilbert space $\mathcal{H}$ (i.e., S is a positive trace-class operator and $TrS = 1$). Then defining $\omega(A) \equiv TrSA$ for any $A \in B(\mathcal{H})$, ω is a positive linear form on the set of all bounded linear operators on Hilbert spape $B(\mathcal{H})$.

We construct the cyclic representation GNS theorem by the above positive linear form on *-algebras. Firstly, we state the so-called standard GNS construction theorem by means of the left regular representation. Let A and ω be *-algebras and a positive linear form on A, respectively. We now set $\langle x, y\rangle_\omega^+ = \omega(y^*x)$. Then $\langle x, y\rangle_\omega^+$ is a pre-inner product (positive sesquilinear form) on the comlpex vector space A. Defining

$$N_\omega^+ \equiv \{x \in A : \omega(x^*x) = \langle x, x\rangle_\omega^+ = 0\},$$

we find the following set is the left ideal of A:

$$N_\omega^+ \equiv \left\{x \in A : \omega(y^*x) = 0, \ {}^\forall y \in A\right\}.$$

Since $\langle , \rangle_\omega^+$ is a pre-inner product, it does not satisfy one of the conditions for an inner product. In order to settle this, we consider the quotient space A/N_ω^+. That is, we define the canonical map:

$$\eta_\omega^+ : A \to A/N_\omega^+.$$

Thus we have $\langle \eta_\omega^+(x), \eta_\omega^+(y)\rangle_\omega^+ = \langle x, y\rangle_\omega^+ = \omega(y^*x)$ on A/N_ω^+. This is well-defined and the inner-product on A/N_ω^+. Let $\mathcal{H}_\omega^+$ be the completion for A/N_ω^+ by this inner-product. (Thus we construct Hilbert space.) For each $x \in A$, setting

$$\pi_\omega^+(x) : A/N_\omega^+ \ni \eta_\omega^+(y) \to \eta_\omega^+(xy) \in A/N_\omega^+$$

$\pi_\omega^+(x)$ is a linear operator on A/N_ω^+. Moreover $\pi_\omega^+(x)$ can be uniquely extended to the linear operator on $\mathcal{H}_\omega^+$. Then the following map from A to the set of all linear operators $\mathcal{L}(A/N_\omega^+)$ on A/N_ω^+

$$\pi_\omega^+ : A \ni x \to \pi_\omega^+(x) \in \mathcal{L}(A/N_\omega^+)$$

is a *-homomorphism, i.e., it satisfies

$$\begin{aligned}
&\pi_\omega^+(x+y) = \pi_\omega^+(x) + \pi_\omega^+(y),\\
&\pi_\omega^+(\lambda x) = \lambda \pi_\omega^+(x),\\
&\pi_\omega^+(xy) = \pi_\omega^+(x)\pi_\omega^+(y),\\
&\pi_\omega^+(x^*) = \pi_\omega^+(x)^*|_{A/N_\omega^+}.
\end{aligned}$$

Theorem 1. *[3]***(GNS construction theorem by left regular representation)**
The mapping $\pi_\omega^+ : A \ni x \to \pi_\omega^+(x) \in \mathcal{L}(A/N_\omega^+)$ is *-homomorphism.

Next, we construct GNS construction theorem by right regular representation as by-product of the present work. Let A and ω be *-algebras and a positive linear form on A, respectively. We now set $\langle x, y\rangle_\omega^- = \omega(y^*x)$. Then $\langle x, y\rangle_\omega^-$ is a pre-inner product (positive sesquilinear form) on the complex vector space A. Defining

$$N_\omega^- \equiv \left\{x \in A : \omega(x^*x) = \langle x, x\rangle_\omega^- = 0\right\},$$

we find the following set is the right ideal of A:

$$N_\omega^- \equiv \left\{x \in A : \omega(y^*x) = 0,\ {}^\forall y \in A\right\}.$$

Since $\langle , \rangle_\omega^-$ is a pre-inner product, it does not satisfy one of the conditions for inner product. In order to settle this, we consider the quotient space A/N_ω^-. That is, we define the canonical map:

$$\eta_\omega^- : A \to A/N_\omega^-$$

Thus we have $\langle \eta_\omega^-(x), \eta_\omega^-(y)\rangle_\omega^- = \langle x, y\rangle_\omega^- = \omega(y^*x)$ on A/N_ω^-. This is well-defined and the inner-product on A/N_ω^-. Let $\mathcal{H}_\omega^-$ be the completion for A/N_ω^- by this inner-product. (Thus we construct Hilbert space.) For each $x \in A$, setting

$$\pi_\omega^-(x) : A/N_\omega^- \ni \eta_\omega^-(y) \to \eta_\omega^-(xy) \in A/N_\omega^-,$$

$\pi_\omega^-(x)$ is a linear operator on A/N_ω^-. Moreover $\pi_\omega^-(x)$ can be uniquely extended to the linear operator on $\mathcal{H}_\omega^-$. Then the following map A to the set of all linear operators $\mathcal{L}(A/N_\omega^-)$ on A/N_ω^-

$$\pi_\omega^- : A \ni x \to \pi_\omega^-(x) \in \mathcal{L}(A/N_\omega^-)$$

is a *-antihomeomorphism, i.e., satisfies

$$\begin{aligned}
\pi_\omega^-(x+y) &= \pi_\omega^-(x) + \pi_\omega^-(y),\\
\pi_\omega^-(\lambda x) &= \lambda\pi_\omega^-(x),\\
\pi_\omega^-(xy) &= \pi_\omega^-(x)\pi_\omega^-(y),\\
\pi_\omega^-(x^*) &= \pi_\omega^-(x)^*|_{A/N_\omega^-}.
\end{aligned}$$

Thus we have the following theorem in terms of the above *-antihomomorphi

Theorem 2. (GNS construction theorem by right regular representation)
*The mapping $\pi_\omega^- : A \ni x \to \pi_\omega^-(x) \in B(\mathcal{H}_\omega^-)$ is *-antihomomorphism.*

3. Sesquilinear form induced by a positive linear form

We should note that the relation among $\mathcal{H}_\omega, \mathcal{H}_\omega^+$ and $\mathcal{H}_\omega^-$. Let A and ω be *-algebras and a positive linear form on A, respectively. Setting

$$\langle x, y\rangle_\omega = \frac{1}{2}\left\{\langle x, y\rangle_\omega^+ + \langle x, y\rangle_\omega^-\right\} = \frac{1}{2}\left\{\omega(y^*x) + \omega(xy^*)\right\}$$

we find that $\langle , \rangle_\omega$ is a pre-inner product (positive sesquilinear form) on the complex vector space A. Defining

$$N_\omega \equiv \{x \in A : \langle x, x\rangle_\omega = 0\},$$

$N_\omega = N_\omega^+ \cap N_\omega^-$ is *-subalgebra on A, not ideal on A. We also set the canonical map such that:

$$\eta_\omega : A \to A/N_\omega.$$

Thus we have

$$\langle \eta_\omega(x), \eta_\omega(y)\rangle_\omega = \langle x, y\rangle_\omega = \frac{1}{2}\left\{\omega(y^*x) + \omega(x^*y)\right\}$$

on A/N_ω. This is well-defined and is an inner-product on A/N_ω. Let $\mathcal{H}_\omega$ be the completion for A/N_ω by this inner-product. For the identity map $i : A \to A$, we have $i(N_\omega) \subset N_\omega^+$. Therefore the following map is induced:

$$i : A/N_\omega \ni \eta_\omega(x) \to \eta_\omega^+(i(x)) = \eta_\omega^+(x) \in A/N_\omega^+.$$

Since this map is a bounded linear operator, it can be uniquely extended to $i : \mathcal{H}_\omega \to \mathcal{H}_\omega^+$ as a bounded linear operator.

Up to now we discussed about $\langle , \rangle_\omega^+, \langle , \rangle_\omega^-$ and $\langle , \rangle_\omega$. Form now on we discuss about the following commutator and its relations :

$$[x, y]_\omega = i\omega([x, y^*]) = i\{\omega(xy^*) - \omega(y^*x)\}.$$

If we take as $A_h = \{x \in A : x^* = x\}$, then it is a real vector space of A.

Proposition 2. *Then we have the following for $x, y \in A_h$.*

(1) $\langle x, y\rangle_\omega = remark\langle x, y\rangle_\omega^+ = remark\langle x, y\rangle_\omega^-$.

(2) $\langle x, x\rangle_\omega = \langle x, x\rangle_\omega^+ = \langle x, x\rangle_\omega^-$.

(3) $[x, y]_\omega = -2Im\langle x, y\rangle_\omega^- = -2Im\omega(xy) \in R$. *In particular, if* $x \in A_h$, *then* $[x, x]_\omega = 0, [x, y]_\omega = -[y, x]_\omega$.

Proposition 3. $[x, y]_\omega$ *is a skew-Hermitian sesquilinear form on A. That is, we have*

(1) $[x + y, z]_\omega = [x, z]_\omega + [y, z]_\omega$,

(2) $[x, y + z]_\omega = [x, y]_\omega + [x, z]_\omega$,

(3) $[\lambda x, y]_\omega = \lambda[x, y]_\omega$,

(4) $[x, \lambda y]_\omega = \overline{\lambda}[x, y]_\omega$,

(5) $[x, y]_\omega = -\overline{[y, x]_\omega}$.

Note that (1)-(4) are the properties of the sesquilinear form and (5) is a property of the skew-Hermitian corresponding to purely imagnary number.

Lemma 1. *(1)* $|[x, y]_\omega| \leq \langle x, x\rangle_\omega^{\frac{1}{2}}\langle y, y\rangle_\omega^{\frac{1}{2}}$ *for* $^\forall x, y \in A$.

(2) $x, y \in N_\omega$, *i.e.,* $\langle x, x\rangle_\omega = \langle y, y\rangle_\omega = 0 \Rightarrow [x, y]_\omega = 0$.

Note that the above (1) shows a kind of continuousness and (2) does the condition of well-definedness. For any $\eta_\omega(x), \eta_\omega(y) \in A/N_\omega$ of canonical map $\eta_\omega : A \to A/N_\omega$, if we set

$$[\eta_\omega(x), \eta_\omega(y)]_\omega = [x, y]_\omega,$$

then we have the following.

Theorem 3.

(1) $[,]_\omega$ *is a bounded skew-Hermitian sesquilinear form* A/N_ω.

(2) $|[\eta_\omega(x), \eta_\omega(y)]_\omega| \leq 2\|\eta_\omega(x)\|\|\eta_\omega(y)\|$ *for* $^\forall\eta_\omega(x), \eta_\omega(y) \in A/N_\omega$.

(3) $[,]_\omega$ *can be uniquely extended to a bounded skew-Hermitian sesquilinear form on* $\mathcal{H}_\omega = \overline{A/N_\omega}$.

As similarly, for any $\eta_\omega(x), \eta_\omega(y) \in A/N_\omega$, if we set,

$$\langle\eta_\omega(x), \eta_\omega(y)\rangle_\omega^+ = \langle x, y\rangle_\omega^+ = \omega(y^*x),$$
$$\langle\eta_\omega(x), \eta_\omega(y)\rangle_\omega^- = \langle x, y\rangle_\omega^- = \omega(xy^*),$$

then we have the following.

Theorem 4.

(1) $|[\eta_\omega(x), \eta_\omega(y)]_\omega^+| \leq 2\|\eta_\omega(x)\|\|\eta_\omega(y)\|$ *for* $^\forall\eta_\omega(x), \eta_\omega(y) \in A/N_\omega$.

(2) $|[\eta_\omega(x), \eta_\omega(y)]_\omega^-| \leq 2\|\eta_\omega(x)\|\|\eta_\omega(y)\|$ *for* $^\forall\eta_\omega(x), \eta_\omega(y) \in A/N_\omega$.

Therefore $\langle,\rangle_\omega^+, \langle,\rangle_\omega^-$ and $[,]_\omega$ can be defined in $\mathcal{H}_\omega$. Thus we have the following proposition by getting them together.

Proposition 4. *Three bounded sesquilinear forms are defined on Hilbert space and we have*

(1) $\langle,\rangle_\omega^+, \langle,\rangle_\omega^-$ *are bounded positive sesquilinear forms,*

(2) $[,]_\omega$ *is bounded skew-Hermitian sesquilinear form.*

(3) On the densely subspace A/N_ω *of* $\mathcal{H}_\omega$,

(i) $[\eta_\omega(x), \eta_\omega(y)]_\omega = i\{\langle\eta_\omega(x), \eta_\omega(y)\rangle_\omega^- - \langle\eta_\omega(x), \eta_\omega(y)\rangle_\omega^+\}$. *That is,* $[X, Y]_\omega = i\{\langle X, Y\rangle_\omega^- - \langle X, Y\rangle_\omega^+\}$ *on* $\mathcal{H}_\omega$.

(ii) $\langle\eta_\omega(x), \eta_\omega(y)\rangle_\omega^+ = \langle\eta_\omega(x), \eta_\omega(y)\rangle_\omega + \frac{i}{2}[\eta_\omega(x), \eta_\omega(y)]_\omega$. *That is* $\langle X, Y\rangle_\omega^+ = \langle X, Y\rangle_\omega + \frac{i}{2}[X, Y]_\omega$ *on* $\mathcal{H}_\omega$.

(iii) $\langle\eta_\omega(x), \eta_\omega(y)\rangle_\omega^- = \langle\eta_\omega(x), \eta_\omega(y)\rangle_\omega - \frac{i}{2}[\eta_\omega(x), \eta_\omega(y)]_\omega$. *That is,* $\langle X, Y\rangle_\omega^- = \langle X, Y\rangle_\omega - \frac{i}{2}[X, Y]_\omega$ *on* $\mathcal{H}_\omega$.

4. Main results

In this section, we formulate the inequalities of the generalized uncertainty relation on *-algebras. Also we show some inequalties associated with the uncertainty relation.

Theorem 5. *We have the following on the complex Hilbert space* $(\mathcal{H}_\omega, \langle , \rangle)$ *and the real Hilbert space* $(\mathcal{H}_\omega^h, \langle , \rangle)$.

(1) $\langle X, X\rangle_\omega \geq \frac{1}{2}\langle X, X\rangle_\omega^+, \ \ X \in \mathcal{H}_\omega$.

(2) $\langle X, X\rangle_\omega \geq \frac{1}{2}\langle X, X\rangle_\omega^-, \ \ X \in \mathcal{H}_\omega$.

(3) $\langle X_1, X_1\rangle_\omega + \langle X_2, X_2\rangle_\omega \geq [X_1, X_2]_\omega, \ \ X_1, X_2 \in \mathcal{H}_\omega^h$.

(4) $\langle X_1, X_1\rangle_\omega \langle X_2, X_2\rangle_\omega \geq \frac{1}{4}[X_1, X_2]_\omega^2, \ \ X_1, X_2 \in \mathcal{H}_\omega$.

Proof.

(1)(2) For any $x \in \mathcal{H}_\omega$,

$$\langle X, X\rangle_\omega^\pm = \langle X, X\rangle_\omega \pm \frac{i}{2}[X, X]_\omega.$$

So $\langle X, X\rangle_\omega^+ + \langle X, X\rangle_\omega^- = 2\langle X, X\rangle_\omega$. Here by $\langle X, X\rangle_\omega^\pm \geq 0$, we have

$$\langle X, X\rangle_\omega^\pm \leq 2\langle X, X\rangle_\omega.$$

(3) We suppose $X_1, X_2 \in \mathcal{H}_\omega^h$ and $X \equiv X_1 + iX_2 \in \mathcal{H}_\omega$. Then we have

$$\begin{aligned} 0 &\leq \langle X, X\rangle_\omega^- \\ &= \langle X_1 + iX_2, X_1 + iX_2\rangle_\omega^- \\ &= \langle X_1, X_1\rangle_\omega^- + \langle X_2, X_2\rangle_\omega^- + i\langle X_2, X_1\rangle_\omega^- - i\langle X_1, X_2\rangle_\omega^- \end{aligned}$$

For any $X_1, X_2 \in \mathcal{H}_\omega^h$, since we have

$$\begin{aligned} \langle X_1, X_1\rangle_\omega^- &= \langle X_1, X_1\rangle_\omega^+ = \langle X_1, X_1\rangle_\omega \\ \langle X_1, X_2\rangle_\omega^- &= \overline{\langle X_1, X_2\rangle_\omega^+} \end{aligned}$$

we obtain

$$\begin{aligned} & i\langle X_2, X_1\rangle_\omega^- - i\langle X_1, X_2\rangle_\omega^- \\ &= -i\left\{\langle X_1, X_2\rangle_\omega^- - \langle X_2, X_2\rangle_\omega^-\right\} \\ &= -i\left\{\langle X_1, X_2\rangle_\omega^- - \overline{\langle X_1, X_2\rangle_\omega^-}\right\} \\ &= -i\left\{\langle X_1, X_2\rangle_\omega^- - \langle X_1, X_2\rangle_\omega^+\right\} \\ &= -[X_1, X_2]_\omega \end{aligned}$$

Since we also have $\langle X_1, X_1\rangle_\omega^- = \langle X_1, X_1\rangle_\omega$, $\langle X_2, X_2\rangle_\omega^- = \langle X_2, X_2\rangle_\omega$, we obtain

$$0 \leq \langle X_1, X_1\rangle_\omega + \langle X_2, X_2\rangle_\omega - [X_1, X_2]_\omega.$$

(4) For any $x, y \in A$, we have

$$\begin{aligned}
|[x,y]_\omega| &= |i(\omega(xy^*) - \omega(y^*x))| \\
&\leq |\omega(xy^*)| + |\omega(y^*x)| \\
&\leq \omega(xx^*)^{\frac{1}{2}}\omega(yy^*)^{\frac{1}{2}} + \omega(x^*x)^{\frac{1}{2}}\omega(y^*y)^{\frac{1}{2}} \\
&= (\langle x,x\rangle_\omega^-)^{\frac{1}{2}}(\langle y,y\rangle_\omega^-)^{\frac{1}{2}} + (\langle x,x\rangle_\omega^+)^{\frac{1}{2}}(\langle y,y\rangle_\omega^+)^{\frac{1}{2}} \\
&\leq (\langle x,x\rangle_\omega^- + \langle x,x\rangle_\omega^+)^{\frac{1}{2}}(\langle y,y\rangle_\omega^- + \langle y,y\rangle_\omega^+)^{\frac{1}{2}} \\
&= (2\langle x,x\rangle_\omega)^{\frac{1}{2}}(2\langle y,y\rangle_\omega)^{\frac{1}{2}} \\
&= 2(\langle x,x\rangle_\omega)^{\frac{1}{2}}(\langle y,y\rangle_\omega)^{\frac{1}{2}}
\end{aligned}$$

Thus we have

$$|[\eta_\omega(x), \eta_\omega(y)]_\omega| \leq 2(\langle \eta_\omega(x), \eta_\omega(x)\rangle_\omega)^{\frac{1}{2}}(\langle \eta_\omega(y), \eta_\omega(y)\rangle_\omega)^{\frac{1}{2}}$$

Since A/N_ω is dence in $\mathcal{H}_\omega$, we therefore obtain

$$|[X_1, X_2]_\omega| \leq 2(\langle X_1, X_1\rangle_\omega)^{\frac{1}{2}}(\langle X_2, X_2\rangle_\omega)^{\frac{1}{2}}, \quad {}^\forall X_1, X_2 \in \mathcal{H}_\omega.$$

■

From the inequality (4) of the Theorem 4.1, the generalized uncertainty relation can be formulated in our general situation. We hope to report on the application to the entropic uncertainty relations (Chap.16 in [6]) in the future.

References

[1] A.S.Holevo, *Commutation superoperator of a state and its applications to the noncommutative statistics*, Rep. Math. Phys., Vol.12,pp.251-271(1977).

[2] A.S.Holevo, *Probablistics and Statistical Aspects of Quantum Theory*, North-Holland, 1982.

[3] A.Inoue, *Tomita-Takesaki theory in algebras of unbounded operators*, Lecture Notes in Mathematics No.1699, Springer-Verlag, 1988.

[4] M.Takesaki, *Theory of operator algebras I*, Springer-Verlag, 1979.

[5] O.Bratteli and W.Robinson, *Operator Algebras and Quantum Statistical Mechanics 1*, Springer-Verlag, 1987.

[6] M.Ohya and D.Petz, *Quantum entropy and its use*, Springer-Verlag, 1993.

21

THE HARTOGS EXTENSION THEOREM FOR THE MULTI-REGULAR FUNCTIONS TAKING VALUES IN A CLIFFORD ALGEBRA

Le Hung Son
Hanoi University of Technology
E-mal: lehung@netnam.org.vn

Nguyen Thanh Van
Hanoi University of Natural Sciences
E-mail: vanhuong6996@hn.vnn.vn

Abstract: This paper deals with the inhomogeneous systems of the form $D_{x(h)}g = f_h$ when f_h have compact supports. It is proved that the Hartogs extension theorem is valid for the multi-regular functions.

AMS-classification: 30G35, 35N05

1. Introduction

The theory of regular functions taking values in a Clifford algebra was studied in [1]. It is a natural generalization to higher dimensions of the theory of holomorphic functions in one complex variable. It is proved that many important properties of holomorphic functions in one complex variable may be extended to the regular functions which are solutions of the generalized Cauchy-Riemann system and take a very essential place in theoretical physics and in other mathematical applications.

In Section 2, some basic notations and definitions of Clifford algebra, used frequently in the paper, are introduced. Also, we present the definition of the multi-regular functions used in Section 5. That is the generalization of the regular functions to higher dimensions.

Section 3 focusses on some properties of the functions taking values in a Clifford algebra of the class C^1. The integral formula is proved for those functions and in the special case an integral formula of functions $f \in C_0^1(\Omega, \mathcal{A})$, which have the compact supports, is obtained.

In Section 4, we show the necessary condition for the solvability of the systems $D_{x^{(h)}} g = f_h$. Applying this condition and the properties, which are considered in Section 3, we are able to solve the inhomogeneous systems $D_{x^{(h)}} g = f_h$ when $f_h \in C_0^k(\mathbb{R}^{(m+1)n}, \mathcal{A})$. At the same time, we can show the necessary and sufficient condition such that the mentioned systems have solutions with compact supports.

Section 5 deals with the Hartogs extension theorem for the multi-regular functions by using the results which are studied both in Section 3 and Section 4.

2. Preliminaries

Let $\mathbb{R}^m$ be a m-dimensional Euclidean space and $e_1, ..., e_m$ be an orthonormal basis of $\mathbb{R}^m$. Then a basis of $\mathcal{A}$ is given by

$$\{e_A : A \subset \mathcal{N};\ \mathcal{N} = \{1, 2, ..., m\}\}$$

where $e_i = e_{\{i\}}, i = 1, .., m, e_0 = e_\emptyset$ *and* $e_A = e_{\alpha_1} \cdots e_{\alpha_h}$, $A = \{\alpha_1, ..., \alpha_h\}$ with $1 \leq \alpha_1 < \cdots < \alpha_h \leq m$. The product in $\mathcal{A}$ is determined by the relations

$$e_i e_j + e_j e_i = -2\delta_{ij} e_0 \quad \text{for} \quad i, j = 1, ..., m$$

where $e_0 = 1$ is the identity of $\mathcal{A}$. We define an involution in $\mathcal{A}$ as follows. Let a be an element of $\mathcal{A}$ which can be represented as

$$a = \sum_A a_A e_A, \quad a_A \in \mathbb{R}.$$

Put $\bar{a} = \sum_A a_A \bar{e}_A$ where $\bar{e}_A = \bar{e}_{\alpha_h} \bar{e}_{\alpha_{h-1}} \cdots \bar{e}_{\alpha_1}$,

$$\bar{e}_j = -e_j \quad \text{for} \quad j = 1, ..., m; \quad \bar{e}_0 = e_0.$$

For $n \leq m$, the space $\mathbb{R}^{n+1}$ can be embedded in $\mathcal{A}$ as follows.

Hence $x = (x_0, x_1, ..., x_n)$ will be identified with

$$x_0 + \vec{x} = x_0 e_0 + \sum_{j=1}^{n} x_j e_j.$$

For other definitions, we refer the readers to [1].

Let Ω be an open set of the Euclidean space

$$\underbrace{\mathbb{R}^{m+1}(x^{(1)}) \times \mathbb{R}^{m+1}(x^{(2)}) \times \cdots \times \mathbb{R}^{m+1}(x^{(n)})}_{n} = \mathbb{R}^{(m+1)n}.$$

We consider the functions f, defined in Ω and taking values in the Clifford algebra $\mathcal{A}$. Then

$$f = \sum_{A} f_A(x^{(1)}, ..., x^{(n)}) e_A \quad \text{where} \quad x^{(j)} = (x_0^{(j)}, ..., x_m^{(j)}) \in \mathbb{R}^{m+1}(x^{(j)})$$

for $j = 1, ..., n$ and f_A are real-valued functions.

Further, we introduce the generalized Cauchy-Riemann operators

$$D_{x^{(j)}} = \sum_{i=0}^{m} e_i \frac{\partial}{\partial x_i^{(j)}}, \quad j = 1, ..., n.$$

Definition 2.1. A function $f : \Omega \rightarrow \mathcal{A}$ is called multi-regular in Ω if $f \in C^1(\Omega, \mathcal{A})$ and satisfies the systems

$$D_{x^{(j)}} f = 0 \quad \text{for} \quad j = 1, ..., n. \tag{2.1}$$

Note that, in case $n = 1$, the multi-regular functions reduce to the regular functions (in sense of R. Delanghe (see [1])). Thus the multi-regular functions are the natural generalizations to higher dimensions of the regular functions.

In case $n = 2$ the multi-regular functions are quite different from the biregular functions, which were studied by F. Brackx and W. Pincket (see [2]).

Remark 2.1. Let $f \in C^2(\Omega, \mathcal{A})$ be a multi-regular function, then

$$\Delta_{x^{(j)}} f = \overline{D}_{x^{(j)}} D_{x^{(j)}} f = D_{x^{(j)}} \overline{D}_{x^{(j)}} f = 0 \quad \text{for} \quad j = 1, ..., n.$$

Hence

$$\Delta f = \sum_{j=1}^{n} \Delta_{x^{(j)}} f = 0.$$

This means that f is harmonic in Ω. Then f is real analytic and the Uniqueness Theorem is valid for the multi-regular functions f.

3. Some properties of functions taking values in the Clifford algebra $\mathcal{A}$

Lemma 3.1. *Let Ω be an open set of $\mathbb{R}^{m+1}$ and $f, g \in C^1(\Omega, \mathcal{A})$. Then we have*

$$D_x(fg) = (D_x f)g + \sum_{i=0}^{m} e_i f \frac{\partial g}{\partial x_i}$$

where

$$D_x = \sum_{i=0}^{m} e_i \frac{\partial}{\partial x_i}, \quad x = (x_0, ..., x_m) \in \Omega. \tag{3.1}$$

Proof. Suppose that $f = \sum_A f_A e_A, \quad g = \sum_B g_B e_B$ then

$$fg = \sum_{A,B} f_A g_B e_A e_B \,. \tag{3.2}$$

We have

$$D_x(fg) = \sum_{i=0}^{m} e_i \frac{\partial (fg)}{\partial x_i} \,. \tag{3.3}$$

It is clear that

$$\frac{\partial (fg)}{\partial x_i} = \frac{\partial}{\partial x_i}\Big(\sum_{A,B} f_A g_B e_A e_B\Big) = \sum_{A,B} \frac{\partial (f_A g_B)}{\partial x_i} e_A e_B$$

$$= \sum_{A,B} \Big(\frac{\partial f_A}{\partial x_i} g_B + \frac{\partial g_B}{\partial x_i} f_A\Big) e_A e_B = \sum_{A,B} \Big(\frac{\partial f_A}{\partial x_i} g_B e_A e_B + f_A \frac{\partial g_B}{\partial x_i} e_A e_B\Big)$$

$$= \sum_{A,B} \Big(\frac{\partial f_A}{\partial x_i} e_A (g_B e_B) + f_A e_A \frac{\partial g_B}{\partial x_i} e_B\Big)$$

$$= \Big(\sum_A \frac{\partial f_A}{\partial x_i} e_A\Big)\Big(\sum_B g_B e_B\Big) + \Big(\sum_A f_A e_A\Big)\Big(\sum_B \frac{\partial g_B}{\partial x_i} e_B\Big)$$

$$= \frac{\partial f}{\partial x_i} g + f \frac{\partial g}{\partial x_i} \quad \text{for} \quad i = 0, 1, ..., m. \tag{3.4}$$

It follows from (3.3), (3.4) that

$$D_x(fg) = \sum_{i=0}^{m} e_i \Big(\frac{\partial f}{\partial x_i} g + f \frac{\partial g}{\partial x_i}\Big)$$

$$= \Big(\sum_{i=0}^{m} e_i \frac{\partial f}{\partial x_i}\Big) g + \sum_{i=0}^{m} e_i f \frac{\partial g}{\partial x_i} = (D_x f) g + \sum_{i=0}^{m} e_i f \frac{\partial g}{\partial x_i}.$$

Lemma 3.2. *Let A be an open set of $\mathbb{R}^{m+1}$ and D be a domain with smooth boundary, such that $\overline{D} \Subset A$, $0 \notin \partial D$. If $g \in C^1(A, \mathcal{A})$, then*

$$\int_D e_i E(y) \cdot D_y g(y) dy - \int_D D_y\big(E(y) e_i g(y)\big) dy =$$

$$= \int_{\partial D} \big(e_i E(y) d\sigma_y\, g(y) - d\sigma_y\, E(y) e_i g(y)\big)$$

for $i = 0, ..., m,$ *where* $dy = dy_0 \wedge dy_1 \wedge \ldots \wedge dy_m.$
Proof.

a) Assume that $i = 0$ then $e_i = 1$. We will prove that

$$\int_D E.Dg dy - \int_D D(Eg) dy = \int_{\partial D} (E d\sigma_y\, g - d\sigma_y\, Eg). \tag{3.5}$$

If $0 \notin \overline{D}$, then according to Lemma 9.2 (see [1]), we have

$$\int_{\partial D} E d\sigma_y\, g = \int_D (ED \cdots g + E \cdots Dg) dy = \int_D E.Dg dy \tag{3.6}$$

because $ED = 0$ (see [1]) and

$$\int_{\partial D} d\sigma_y\, Eg = \int_D D(Eg) dy. \tag{3.7}$$

From (3.6), (3.7) it follows (3.5).

If $0 \in D$, then we choose $R > 0$, such that $\overline{B}(0, R) \subset D$ where $B(0, R)$ is the ball in $\mathbb{R}^{m+1}$. Using (3.5) for the domain $(D \setminus B)$, we obtain

$$\int_{D\setminus B} E \cdot Dg dy - \int_{D\setminus B} D(Eg) dy = \int_{\partial(D\setminus B)} (E d\sigma_y\, g - d\sigma_y\, Eg). \tag{3.8}$$

Clearly, if $R \to 0^+$, then the left hand side of (3.8) tends to

$$\int_D E \cdot Dg dy - \int_D D(Eg) dy, \text{ (see [1], Theorem 9.5).} \tag{3.9}$$

The right hand side of (3.8) has the form

$$\int_{\partial D} (E d\sigma_y\, g - d\sigma_y\, Eg) - \int_{\partial B(0,R)} (E d\sigma_y\, g - d\sigma_y\, Eg). \tag{3.10}$$

Applying the method used in the proof of Theorem 9.5 (see [1]), we have

$$\lim_{R\to 0^+} \int_{\partial B(0,R)} E d\sigma_y\, g = g(0). \tag{3.11}$$

Similarly, we can prove that

$$\lim_{R\to 0^+} \int_{\partial B(0,R)} d\sigma_y\, E g = g(0). \tag{3.12}$$

Hence we get

$$\lim_{R\to 0^+} \int_{\partial B(0,R)} (E d\sigma_y\, g - d\sigma_y\, E g) = g(0) - g(0) = 0. \tag{3.13}$$

By (3.8), (3.9), (3.10) and (3.13), we obtain (3.5).

b) Using a similar method as in the proof a), we are able to prove the Lemma for $i = 2, ..., m$.

Thus Lemma 3.2 is proved.

Lemma 3.3. *Let D be a bounded domain in $\mathbb{R}^{m+1}$ with smooth boundary, $f \in C^1(D, \mathcal{A})$, $x \in D$, then*

$$f(x) = \int_{\partial D} d\sigma_y\, E(y-x) f(y) - \int_D D_y\big(E(y-x) f(y)\big) dy. \tag{3.14}$$

Proof. According to Theorem 9.5 (see [1]), we have

$$f(x) = \int_{\partial D} E(y-x) d\sigma_y\, f(y) - \int_D E(y-x) \cdot D_y f(y) dy. \tag{3.15}$$

Put $y - x = \eta$, $g(\eta) = f(x+\eta)$, then the right hand side of (3.15) reduces to

$$\int_{\partial D'} E(\eta) d\sigma_\eta\, g(\eta) - \int_{D'} E(\eta) \cdot D_\eta g(\eta) d\eta. \tag{3.16}$$

By Lemma 3.2, (3.16) is of the form

$$\int_{\partial D'} d\sigma_\eta\, E(\eta) g(\eta) - \int_{D'} D_\eta\big(E(\eta) g(\eta)\big) d\eta$$

$$= \int_{\partial D} d\sigma_y\, E(y-x) f(y) - \int_D D_y\big(E(y-x) f(y)\big) dy.$$

So we have (3.14).

Corollary 3.1. *Let D be a domain as in Lemma 3.3. If $f \in C_0^1(D, \mathcal{A})$, then*

$$f(x) = -\int_D D_y\big(E(y-x)f(y)\big)dy \quad \text{for} \quad x \in D. \tag{3.17}$$

4. The inhomogeneous generalized Cauchy-Riemann systems

In this section, we consider the system

$$D_{x^{(h)}} g = f_h \quad \text{for} \quad h = 1, ..., n; \quad n \geq 2, \tag{4.1}$$

where f_h are defined in an open set $\Omega \subset \mathbb{R}^{(m+1)n}$ and take values in the Clifford algebra $\mathcal{A}$.

Note that in case $n = 1$ the system (4.1) was studied in [1].

Lemma 4.1. *Suppose that $f_h \in C^1(\Omega, \mathcal{A})$. If $g \in C^2(\Omega, \mathcal{A})$ is a solution of (4.1), then*

$$\sum_{i=0}^{m} e_i E(\eta) \frac{\partial f_h}{\partial x_i^{(1)}} = \sum_{i=0}^{m} D_\eta\Big(E(\eta) e_i \frac{\partial g}{\partial x_i^{(h)}}\Big), \tag{4.2}$$

$$\sum_{i=0}^{m} e_i E(\eta) \frac{\partial f_1}{\partial x_i^{(h)}} = \sum_{i=0}^{m} e_i E(\eta) . D_\eta \frac{\partial g}{\partial x_i^{(h)}} \tag{4.3}$$

for $h = 1, ..., n;$ *where*

$$\begin{aligned} \frac{\partial f_h}{\partial x_i^{(1)}} &= \frac{\partial f_h}{\partial x_i^{(1)}}(x^{(1)} + \eta, x^{(2)}, ..., x^{(n)}), \\ \frac{\partial f_1}{\partial x_i^{(h)}} &= \frac{\partial f_1}{\partial x_i^{(h)}}(x^{(1)} + \eta, x^{(2)}, ..., x^{(n)}), \\ \frac{\partial g}{\partial x_i^{(h)}} &= \frac{\partial g}{\partial x_i^{(h)}}(x^{(1)} + \eta, x^{(2)}, ..., x^{(n)}). \end{aligned}$$

Proof. The right hand side of (4.2) has the form

$$\sum_{i=0}^{m} D_\eta\Big(E(\eta) e_i \frac{\partial g}{\partial x_i^{(h)}}\Big) = D_\eta\Big(E(\eta) \sum_{i=0}^{m} e_i \frac{\partial g}{\partial x_i^{(h)}}\Big)$$

$$= D_\eta(E(\eta) D_{x^{(h)}} g) = D_\eta(E(\eta) f_h) \quad \text{because} \quad D_{x^{(h)}} g = f_h. \tag{4.4}$$

By Lemma 3.1, (4.4) reduces to

$$
\begin{aligned}
D_\eta(E(\eta)f_h) &= (D_\eta E(\eta))f_h + \sum_{i=0}^{m} e_i E(\eta)\frac{\partial f_h}{\partial \eta_i} \quad \text{where} \quad \eta = (\eta_0, ..., \eta_m) \\
&= 0 + \sum_{i=0}^{m} e_i E(\eta)\frac{\partial f_h}{\partial x_i^{(1)}} \\
&= \sum_{i=0}^{m} e_i E(\eta)\frac{\partial f_h}{\partial x_i^{(1)}} \quad \text{since} \quad E(\eta) \quad \text{is regular (see [1]).}
\end{aligned}
$$

Thus (4.2) is proved.

On the other hand, we get

$$
\begin{aligned}
D_\eta \frac{\partial g}{\partial x_i^{(h)}} &= D_{x^{(1)}} \frac{\partial g}{\partial x_i^{(h)}} = \sum_{j=0}^{m} e_j \frac{\partial}{\partial x_j^{(1)}}\Big(\frac{\partial g}{\partial x_i^{(h)}}\Big) \\
&= \sum_{j=0}^{m} e_j \frac{\partial}{\partial x_j^{(1)}}\Big(\sum_B e_B \frac{\partial g_B}{\partial x_i^{(h)}}\Big) \\
&= \sum_{j=0}^{m}\sum_B e_j e_B \frac{\partial^2 g_B}{\partial x_j^{(1)}\partial x_i^{(h)}} = \sum_{j=0}^{m}\sum_B e_j e_B \frac{\partial}{\partial x_i^{(h)}}\Big(\frac{\partial g_B}{\partial x_j^{(1)}}\Big) \\
&= \frac{\partial}{\partial x_i^{(h)}}\Big(\sum_{j=0}^{m} e_j \frac{\partial}{\partial x_j^{(1)}}\big(\sum_B e_B g_B\big)\Big) = \frac{\partial}{\partial x_i^{(h)}}\Big(\sum_{j=0}^{m} e_j \frac{\partial g}{\partial x_j^{(1)}}\Big) \\
&= \frac{\partial}{\partial x_i^{(h)}}(D_{x^{(1)}} g) = \frac{\partial f_1}{\partial x_i^{(h)}} \,. \qquad (4.5)
\end{aligned}
$$

It follows from (4.5) that

$$
e_i E(\eta)\frac{\partial f_1}{\partial x_i^{(h)}} = e_i E(\eta)\cdot D_\eta \frac{\partial g}{\partial x_i^{(h)}} \quad \text{for} \quad i = 0, ..., m. \qquad (4.6)
$$

Hence we obtain (4.3).

Theorem 4.1. *Suppose that* $f_1, ..., f_n \in C_0^k(\mathbb{R}^{(m+1)n}, \mathcal{A})$, $k \geq 2$, *then the system (4.1) has at least a solution* $g \in C_0^k(\mathbb{R}^{(m+1)n}, \mathcal{A})$ *if and only if the following condition is satisfied*

$$
\int_{\mathbb{R}^{m+1}} \sum_{i=0}^{m} e_i E(\eta)\frac{\partial f_h}{\partial x_i^{(1)}} d\eta = \int_{\mathbb{R}^{m+1}} \sum_{i=0}^{m} e_i E(\eta)\frac{\partial f_1}{\partial x_i^{(h)}} d\eta \qquad (4.7)
$$

for $h = 2, ..., n$ *and for all* $(x^{(1)}, ..., x^{(n)}) \in \mathbb{R}^{(m+1)n}$ *where* $\dfrac{\partial f_h}{\partial x_i^{(1)}}$ *and* $\dfrac{\partial f_1}{\partial x_i^{(h)}}$ *are defined as in Lemma 4.1.*

When the conditions of the theorem are satisfied, then the solution $g \in C_0^k(\mathbb{R}^{(m+1)n}, \mathcal{A})$ is uniquely determined.

1. The necessary condition

Suppose that the system (4.1) has a solution $g \in C_0^k(\mathbb{R}^{(m+1)n}, \mathcal{A})$, we will prove that (4.7) is valid.

According to Lemma 4.1, we get

$$\sum_{i=0}^{m} e_i E(\eta) \frac{\partial f_h}{\partial x_i^{(1)}} = \sum_{i=0}^{m} D_\eta \Big(E(\eta) e_i \frac{\partial g}{\partial x_i^{(h)}} \Big), \tag{4.8}$$

$$\sum_{i=0}^{m} e_i E(\eta) \frac{\partial f_1}{\partial x_i^{(h)}} = \sum_{i=0}^{m} e_i E(\eta). D_\eta \frac{\partial g}{\partial x_i^{(h)}}. \tag{4.9}$$

Since $g \in C_0^k(\mathbb{R}^{(m+1)n}, \mathcal{A})$ then $\dfrac{\partial g}{\partial x_i^{(h)}}$ also have such a property.

It follows from Lemma 3.2 that

$$\int_{\mathbb{R}^{m+1}} \sum_{i=0}^{m} D_\eta \Big(E(\eta) e_i \frac{\partial g}{\partial x_i^{(h)}} \Big) d\eta = \int_{\mathbb{R}^{m+1}} \sum_{i=0}^{m} e_i E(\eta). D_\eta \frac{\partial g}{\partial x_i^{(h)}} d\eta. \tag{4.10}$$

By (4.8), (4.9) and (4.10) we have (4.7).

2. The sufficient condition

Assume that (4.7) is satisfied. Set

$$\begin{aligned} g(x^{(1)}, ..., x^{(n)}) &= - \int_{\mathbb{R}^{m+1}} E(y - x^{(1)}) f_1(y, x^{(2)}, ..., x^{(n)}) dy \\ &= - \int_{\mathbb{R}^{m+1}} E(\eta) f_1(x^{(1)} + \eta, x^{(2)}, ..., x^{(n)}) d\eta, \end{aligned} \tag{4.11}$$

then

$$D_{x^{(h)}} g(x^{(1)}, ..., x^{(n)}) = - \int_{\mathbb{R}^{m+1}} D_{x^{(h)}} (E(\eta) f_1(x^{(1)} + \eta, x^{(2)}, ..., x^{(n)})) d\eta. \tag{4.12}$$

Using Lemma 3.1 the right hand side of (4.12) has the form

$$-\int_{\mathbb{R}^{m+1}} \Big[\big(D_{x^{(h)}}E(\eta)\big)f_1(x^{(1)}+\eta, x^{(2)},...,x^{(n)})$$

$$+\sum_{i=0}^{m} e_i E(\eta)\frac{\partial f_1}{\partial x_i^{(h)}}(x^{(1)}+\eta, x^{(2)},...,x^{(n)})\Big]d\eta. \tag{4.13}$$

i) If $h=1$, then $D_{x^{(h)}}E(\eta)=0$ because $E(\eta)$ is regular.

ii) If $h\geq 2$, then $D_{x^{(h)}}E(\eta)=0$ because $E(\eta)$ is independent on $x^{(h)}$.

Hence (4.13) reduces to

$$-\int_{\mathbb{R}^{m+1}} \sum_{i=0}^{m} e_i E(\eta)\frac{\partial f_1}{\partial x_i^{(h)}}(x^{(1)}+\eta, x^{(2)},...,x^{(n)})d\eta. \tag{4.14}$$

From (4.7) and Corollary 3.1, (4.14) becomes

$$-\int_{\mathbb{R}^{m+1}} \sum_{i=0}^{m} e_i E(\eta)\frac{\partial f_h}{\partial x_i^{(1)}}(x^{(1)}+\eta, x^{(2)},...,x^{(n)})d\eta$$

$$=-\int_{\mathbb{R}^{m+1}} D_y\big(E(y-x^{(1)})f_h(y,x^{(2)},...,x^{(n)})\big)dy = f_h(x^{(1)},...,x^{(n)}) \tag{4.15}$$

for $h=1,...,n$.

According to (4.12)-(4.15), then g, which is defined by (4.11), is a solution of (4.1).

At last we will show that g has the compact support. It is easy to see that $g(x^{(1)},...,x^{(n)})=0$ for all $x^{(1)}\in\mathbb{R}^{m+1}$ if $\sqrt{\sum_{i=2}^{n}|x^{(i)}|^2}$ is large enough. So $g=0$ in the unbounded connected component of $\mathbb{R}^{(m+1)n}$. On the other hand, f_h have compact supports, then g is multi-regular outside of a compact dics C of $\mathbb{R}^{(m+1)n}$.

From the Uniqueness Theorem for the multi-regular functions (see Remark 2.1), it follows that $g=0$ outside C. Hence g has a compact support.

We are able to prove that, the solution $g\in C_0^k(\mathbb{R}^{(m+1)n},\mathcal{A})$ of system (4.1), which is defined by (4.11), is uniquely determined.

In fact indeed, suppose that g and g_1 are solutions with compact support of (4.1). Then we have

$$D_{x^{(h)}}(g-g_1)=f_h-f_h=0 \text{ in } \mathbb{R}^{(m+1)n} \text{ for all } h=1,\cdot,n.$$

Hence $\varphi = g - g_1$ is a multi-regular function in $\mathbb{R}^{(m+1)n}$. Since g and g_1 have compact support, then φ has also compact support. According to the Uniqueness Theorem for multi-regular functions, it follows that $\varphi = 0$ in the whole $\mathbb{R}^{(m+1)n}$. Thus $g = g_1$ in $\mathbb{R}^{(m+1)n}$.

5. The Hartogs extension theorem for multi-regular functions

In this section we prove the following theorem

Theorem 5.1. *Let A be an open set of $\mathbb{R}^{(m+1)n}$, $n \geq 2$ and K be a compact set of A, such that $A \setminus K$ is connected. Suppose that $f \in C^k, k \geq 2$ is multi-regular in $A \setminus K$, then there exists an unique multi-regular function F in A such that $F = f$ in $A \setminus K$.*

Proof. Since f is multi-regular in $A \setminus K$ then f is harmonic in $A \setminus K$. Take $\varphi \in C_0^\infty(A, \mathcal{A})$ with real values such that $\varphi = 1$ in an open neighbourhood of K.

Put $f_0 = (1 - \varphi)f$, then $f_0 \in C^k(A, \mathcal{A})$ and f_0 is multi-regular in $A \setminus \{\operatorname{supp} \varphi\}$.

We define

$$h_j = \begin{cases} D_{x^{(j)}} f_0 & \text{in } A \\ 0 & \text{in } \mathbb{R}^{(m+1)n} \setminus A \end{cases} \quad \text{for} \quad j = 1, ..., n.$$

Obviously, $h_j = 0$ in $\mathbb{R}^{(m+1)n} \setminus \{\operatorname{supp} \varphi\}$.

We will prove that h_j satisfies (4.7). This means that

$$M_s(x^{(1)}, ..., x^{(n)}) = \int_{\mathbb{R}^{m+1}} \sum_{i=0}^{m} e_i E(\eta) \times$$

$$\times \left(\frac{\partial h_s}{\partial x_i^{(1)}} - \frac{\partial h_1}{\partial x_i^{(s)}} \right) (x^{(1)} + \eta, x^{(2)}, ..., x^{(n)}) d\eta = 0 \quad \text{for} \quad 2 \leq s \leq n.$$

For each fixed $(x^{(2)}, ..., x^{(n)}) \in \mathbb{R}^{(m+1)}(x^{(2)}) \times \cdots \times \mathbb{R}^{(m+1)}(x^{(n)})$, set

$$\begin{aligned} A^{(1)} &= \{x \in \mathbb{R}^{m+1} : (x, x^{(2)}, ..., x^{(n)}) \in A\}, \\ B &= \{x \in \mathbb{R}^{m+1} : (x, x^{(2)}, ..., x^{(n)}) \in \{\operatorname{supp} \varphi\}\}. \end{aligned}$$

Obviously, $B \subseteq A^{(1)}$.

Let $D^{(1)}$ be an open bounded set of $\mathbb{R}^{m+1}$ with smooth boundary, such that

$$B \subseteq D^{(1)}, \ \overline{D^{(1)}} \subseteq A^{(1)}.$$

In addition, we set

$$D_*^{(1)} = \{\eta \in \mathbb{R}^{m+1} : \eta + x^{(1)} \in D^{(1)}\}.$$

Then we have

$$\begin{aligned} M_s(x^{(1)}, ..., x^{(n)}) &= \\ &= \int_{D^{(1)}} \sum_{i=0}^{m} e_i E(y - x^{(1)}) \Big(\frac{\partial h_s}{\partial x_i^{(1)}} - \frac{\partial h_1}{\partial x_i^{(s)}} \Big) (y, x^{(2)}, ..., x^{(n)}) dy \\ &= \int_{D_*^{(1)}} \sum_{i=0}^{m} e_i E(\eta) \Big(\frac{\partial h_s}{\partial x_i^{(1)}} - \frac{\partial h_1}{\partial x_i^{(s)}} \Big) (x^{(1)} + \eta, x^{(2)}, ..., x^{(n)}) d\eta. \end{aligned}$$

From $\eta \in D_*^{(1)}$, it follows that

$$h_s(x^{(1)} + \eta, x^{(2)}, ..., x^{(n)}) = D_{x^{(s)}} f_0(x^{(1)} + \eta, x^{(2)}, ..., x^{(n)}).$$

According to Lemma 4.1, we obtain

$$\sum_{i=0}^{m} e_i E(\eta) \frac{\partial h_s}{\partial x_i^{(1)}} = \sum_{i=0}^{m} D_\eta \Big(E(\eta) e_i \frac{\partial f_0}{\partial x_i^{(s)}} \Big), \tag{5.1}$$

$$\sum_{i=0}^{m} e_i E(\eta) \frac{\partial h_1}{\partial x_i^{(s)}} = \sum_{i=0}^{m} e_i E(\eta).D_\eta \frac{\partial f_0}{\partial x_i^{(s)}} \tag{5.2}$$

for all $\eta \in D_*^{(1)}$; $\quad s = 1, ..., n$.

Using (5.1), (5.2) and Lemma 3.2, we get

$$\begin{aligned} M_s(x^{(1)}, ..., x^{(n)}) &= \int_{\partial D_*^{(1)}} \sum_{i=0}^{m} \Big(d\sigma_\eta \, E(\eta) e_i \frac{\partial f_0}{\partial x_i^{(s)}} - e_i E(\eta) d\sigma_\eta \, \frac{\partial f_0}{\partial x_i^{(s)}} \Big) \\ &= \int_{\partial D_*^{(1)}} d\sigma_\eta \, E(\eta) \cdots D_{x^{(s)}} f_0 - \int_{\partial D_*^{(1)}} \sum_{i=0}^{m} e_i E(\eta) d\sigma_\eta \, \frac{\partial f_0}{\partial x_i^{(s)}}. \end{aligned}$$

Because $f_0 = f$ in $A \backslash \{supp \varphi\}$ and f is multi-regular in $A \setminus K$ then

$$M_s(x^{(1)}, ..., x^{(n)}) = - \int_{\partial D_*^{(1)}} \sum_{i=0}^{m} e_i E(\eta) d\sigma_\eta \, \frac{\partial f}{\partial x_i^{(s)}} (x^{(1)} + \eta, x^{(2)}, ..., x^{(n)}).$$

We have

$$\Delta M_s(x^{(1)},...,x^{(n)}) = -\int\limits_{\partial D_*^{(1)}} \sum_{i=0}^{m} e_i E(\eta) d\sigma_\eta \times$$

$$\times \frac{\partial}{\partial x_i^{(s)}}(\Delta f)(x^{(1)}+\eta, x^{(2)},...,x^{(n)}) = 0 \quad \text{since } f \text{ is harmonic.}$$

Hence M_s is harmonic in $\mathbb{R}^{(m+1)n}$.

On the other hand, $h_j = 0$ in $\mathbb{R}^{(m+1)n} \setminus \{\operatorname{supp}\varphi\}$, then $M_s = 0$ in a non-empty open set $\sigma \in \mathbb{R}^{(m+1)n}$, thus it follows from the Uniqueness Theorem for the harmonic functions that

$$M_s = 0 \quad \text{in} \quad \mathbb{R}^{(m+1)n}.$$

According to Theorem 4.1, there exists $g \in C_0^k(\mathbb{R}^{(m+1)n}, \mathcal{A})$ such that

$$D_{x^{(j)}} g = h_j \quad \text{for} \quad j = 1, 2, ..., n.$$

Because $h_j = 0$ in $\mathbb{R}^{(m+1)n} \setminus \{\operatorname{supp}\varphi\}$ then g is multi-regular in $\mathbb{R}^{(m+1)n} \setminus \{\operatorname{supp}\varphi\}$.

Since g has compact support then $g = 0$ for $x = (x^{(1)},...,x^{(n)})$ with $\sqrt{\sum_{i=1}^{n} |x^{(i)}|^2}$ large enough. From the Uniqueness Theorem for multi-regular function (see Remark 2.1), it follows that $g \equiv 0$ in the unbounded connected component C of $\mathbb{R}^{(m+1)n} \setminus \{\operatorname{supp}\varphi\}$.

Put $F = f_0 - g$, then F is multi-regular in A and $F = f$ in $C \cap A$. The fact is that $A \setminus K$ is connected then using the Uniqueness Theorem for the multi-regular functions it follows that $F = f$ in $A \setminus K$ and F is unique.

Acknowledgements. We would like to express our special thanks to Dr. Nguyen Canh Luong, Department of Applied Mathematics, Hanoi University of Technology for his precious assistance and encouragement during completion of this paper.

References

[1] F. Brackx, R. Delanghe and F. Sommen, *Clifford Analysis*, Pitman Research Notes in Math. **76** (1982).

[2] F. Brackx and W. Pincket, *A Bochner-Martinelli Formula for the Biregular Functions of Clifford Analysis.* Complex Variables, **4** (1984).

[3] F. Brackx and W. Pincket, *Two Hartogs Theorems for Null Solutions of Overdetermined System in Euclidean Space.* Complex Variables, **5** (1985).

[4] R.P. Gilbert and J.L. Buchanan, *First Order Elliptic Systems, A Functions Theoretic Approach.* Math. in Sc. and Engineering, **163**. Academic Press, New York 1983.

[5] B. Goldschmidt, *A Theorem About the Representation of Linear Combinations in Clifford Algebra.* Beiträge Zur Algebra und Geometrie **13** (1982), 21-24.

[6] M. Hervé, *Analytic and Plurisubharmonic Functions in Finite and Infinite Dimensional Space.* Lecture Notes in Mathematics **198**, Springer Verlag, Berlin 1971.

[7] Le Hung Son, *Extension Problem for Functions With Values in a Clifford Algebra*, Archiv fuer Mathematik, **55** (1990), 146-150.

[8] Le Hung Son, *Some New Results of Clifford Analysis in Higher Dimensions.* Finite or Infinite Dimensional Complex Analysis. Lecture Notes in Pure and Applied Mathematics, **214**, 245-265, Marcel Dekker, Inc. New York. Basel.

[9] D. Pertici, *Funzioni Regolari di più Variabili Quaternioniche*, Ann. Math. Pura e Appl. Serie IV **151** (1988), 39-65.

22

THE EXTENSION OF ENTIRE FUNCTIONS OF NUCLEAR TYPE ON LOCALLY CONVEX SPACES

Masaru Nishihara
Department of computer science and engineering
Fukuoka Institute of Technology
Fukuoka 811-0295 Japan

1. Introduction

Throughout this paper all locally convex spaces are assumed to be complex and Hausdorff. Let E and F are be locally convex spaces. We assume that E is a closed subspace of F. Let f be an entire function on E. Then we consider the problem to ask whether f can be extended to F holomorphically, that is, the problem to ask if Hahn-Banach theorem for entire functions on locally convex spaces is valid. It is known that in general Hahn-Banach theorem for entire functions is not valid. For example, by Aron-Berner [1] an entire function f on c_0 can be extended holomorphically to l^∞ if and only if f is bounded on every bounded subsets of E. On the other hand, in case E is a nuclear space, Meise-Vogt [6] showed that f can be extended holomorphically to every locally convex space F which is of Hilbert type if and only if f is uniformly bounded type on E. Nishihara [8] showed that the condition that F is of Hilbert can be removed. An entire function on a nuclear space is of nuclear type if it is of uniformly bounded type (cf. Nishihara [9]). Thus it is natural to be concerned for the holomorphic extension for entire functions of nuclear type on a locally convex space as a generalization of the above results. In this paper we will study the holomorphic extension for entire functions of nuclear type on a locally convex space E to F.

2. Notations and preliminaries

In this section we collect some notation, definitions and basic properties of locally convex spaces. Let E be a locally convex space. We denote by $cs(E)$ the set of all continuous seminorms on E. We denote by $P_a({}^nE)$ (resp. $P({}^nE)$) the space of all n-homogeneous polynomials(resp. all continuous n-homogeneous polynomials) from E into $\boldsymbol{C}$. We denote by $\otimes_{n,s}E$ the n-fold symmetric tensor product of E. We denote by δ_n the canonical mapping of E into $\otimes_{n,s}E$ defined by

$$\delta_n(x) = x \otimes \cdots \otimes x \in \otimes_{n,s}E$$

for every $x \in E$. If $p \in P_a({}^nE)$, then there exists a unique linear functional $j_n^*(p)$ on $\otimes_{n,s}E$ such that $p = j_n^*(p) \circ \delta_n$. Moreover the mapping

$$j_n^* : P_a({}^nE) \to (\otimes_{n,s}E)^*$$

is bijective.

For a subset S of the locally convex space E, we denote by S° the polar set of S, that is, $S^\circ = \{\varphi \in E' ; |\varphi(x)| \leq 1 \quad \text{for every} \quad x \in S\}$. For a complex valued function f on S, we set

$$|f|_S = \sup\{|f(x)| ; \quad \text{for every} \quad x \in S\}.$$

For every $\alpha \in cs(E)$ we set

$$B_\alpha(r) = \{x \in E ; \alpha(x) < r\}.$$

We define seminorms $\pi^\alpha_{n,E}$ and $\epsilon^\alpha_{n,E}$ on $\otimes_{n,s}E$ by

$$\pi^\alpha_{n,E}(\theta) = \inf\{\sum_{i=1}^{l} \alpha(x_i)^n ; \theta = \sum_{i=1}^{l} x_i \otimes \cdots \otimes x_i\} \tag{1}$$

and

$$\epsilon^\alpha_{n,E}(\theta) = \sup\{|\sum_{i=1}^{l} \varphi(x_i)^n| ; \varphi \in B_\alpha(1)^\circ\}. \tag{2}$$

The π or projective topology(resp. ϵ or injective topology) on $\otimes_{n,s}E$ is the locally convex topology generated by the set $\{\pi^\alpha_{n,E} ; \alpha \in cs(E)\}$ (resp. $\{\epsilon^\alpha_{n,E} ; \alpha \in cs(E)\}$) of seminorms defined by (1)(resp. (2)). We denote by $\otimes_{n,s,\pi}E$ (resp. $\otimes_{n,s,\epsilon}E$) the locally convex space $\otimes_{n,s}E$ endowed with π-topology (resp. ϵ-topology). The mapping $j_n^* : P({}^nE) \to (\otimes_{n,s,\pi}E)'$ is bijective. However even if E is a topological linear subspace of a locally convex space F, $\otimes_{n,s,\pi}E$ is not always a topological linear subspace of the locally convex space $\otimes_{n,s,\pi}F$. It seems that this is one of the reasons why the problem of extending continuous polynomials on E to F is in general difficult.

Lemma 1. *Let F be a locally convex space and let E be a topological linear subspace of F. We assume that $\alpha \in cs(E)$ and $\beta \in cs(F)$ satisfy $\beta|_E = \alpha$. Then the restriction mapping:*

$$B_\beta(1)^\circ \to B_\alpha(1)^\circ$$

is surjective.

Proof. Let $\varphi \in B_\alpha(1)^\circ$. Then we have

$$|\varphi(x)| \leq \alpha(x) = \beta(x)$$

for every $x \in E$. By Hahn-Banach theorem, there exists $\psi \in E'$ such that $\psi|_E = \varphi$ and $|\psi(x)| \leq \beta(x)$ for every $x \in F$. This implies that $\psi \in B_\beta(1)^\circ$ and $\psi|_E = \varphi$. This completes the proof.

Lemma 2. *With the notation of lemma 2.1,*

$$\epsilon^{\alpha}_{n,\,E}(\theta) = \epsilon^{\beta}_{n,\,F}(\theta) \quad \text{for every} \quad \theta \in \otimes_{n,s,\epsilon}E. \tag{3}$$

Thus $\otimes_{n,s,\epsilon}E$ is a topological subspace of $\otimes_{n,s,\epsilon}F$

Proof. Let $\theta = \sum_{i=1}^{l} x_i \otimes \cdots \otimes x_i \in \otimes_{n,s}E$. By lemma 2.1 we have

$$\begin{aligned}
\epsilon^{\alpha}_{n,\,E}(\theta) &= \sup\{|\sum_{i=1}^{l} \varphi(x_i)^n|\,;\, \varphi \in B_\alpha(1)^\circ\} \\
&= \sup\{|\sum_{i=1}^{l} \psi(x_i)^n|\,;\, \psi \in B_\beta(1)^\circ\} \\
&= \epsilon^{\beta}_{n,\,F}(\theta)
\end{aligned}$$

This completes the proof.

We denote by $H(E)$ the vector space of all entire functions on E. An entire function f on E is said to be of *uniformly bounded type* if there exists a neighborhood V of 0 in E such that $|f|_{rV} < +\infty$ for every positive number r. We denote by $H_{ub}(E)$ the space of all entire functions on E which are of uniformly bounded type.

Lemma 3. *Let f be an entire function on E. Then the following statements are equivalent.*

(a) $f \in H_{ub}(E)$.

(b) *Let* $f(x) = \sum_{n=0}^{\infty} p_n(x)$ *be the Taylor expansion of f at 0. Then there exists a balanced convex open neighborhood V of 0 in E such that*

$$\sum_{n=0}^{\infty} |p_n|_V r^n < \infty \tag{4}$$

for every $r > 0$.

Proof. We show that (a) implies (b). Since $f \in H_{ub}(E)$, there exists a balanced convex open neighborhood V of 0 in E such that

$$|f|_{rV} < +\infty \quad \text{for every} \quad r > 0.$$

We set $M(r) = |f|_{2rV}$ for every $r > 0$. Then we have $|p_n|_{2rV} \leq M(r)$, and so $|p_n|_{rV} \leq \frac{1}{2^n} M(r)$. Therefore we have

$$\sum_{n=0}^{\infty} |p_n|_{rV} \leq 2M(r) < +\infty.$$

Thus

$$\sum_{n=0}^{\infty} |p_n|_V r^n = \sum_{n=0}^{\infty} |p_n|_{rV} \leq 2M(r) < +\infty$$

and we have proved that (a) implies (b).

It follows from the following inequality that (b) implies (a)

$$|f|_{rV} \leq \sum_{n=0}^{\infty} |p_n|_{rV} = \sum_{n=0}^{\infty} |p_n|_V r^n < +\infty$$

for every $r > 0$. This completes the proof.

3. Entire functions of nuclear type

An n-homogeneous polynomial $p \in P(^nE)$ is said to be of *nuclear type* if there exist a convex balanced open neighborhood U of 0 in E, $(\phi_j)_{j=1}^{\infty} \subset U^\circ$ and $(\lambda_j) \in l^1$ such that

$$p(x) = \sum_{j=1}^{\infty} \lambda_j \left(\phi_j(x)\right)^n$$

for all $x \in E$. We denote by $P_N(^nE)$ the space of all nuclear n-homogeneous polynomials on E. We set

$$\pi_A(p) = |p|_{N,A} = \inf\left\{\sum_{j=1}^{\infty} \lambda_j |\phi_j(x)|^n \,;\, p = \sum_{j=1}^{\infty} \lambda_j\,(\phi_j)^n\right\}$$

for $p \in P(^nE)$ and a subset A of E. Let f be an entire function on E. We denote by $f = \sum_{n=0}^{\infty} \frac{\hat{d}^n f(0)}{n!}$ the Taylor series expansion of f at 0 in E let

$$\pi_V(f) = \sum_{n=0}^{\infty} \pi_V\left(\frac{\hat{d}^n f(0)}{n!}\right).$$

Let V be a convex balanced subset of a locally convex space E. We set

$$H_N^{\infty}(V) = \left\{f \in H(V)\,;\, \hat{d}^n f(0) \in P_N(^nE) \quad \text{for all } n \text{ and } \quad \pi_V(f) < \infty\right\}.$$

If $\mathcal{V} = (V_n)_{n=1}^{\infty}$ is an increasing countable balanced open cover of E, let

$$H_{N,\mathcal{V}}(E) = \{f \in H(E)\,;\, f|V_n \in H_N^{\infty}(V_n) \quad \text{for all } n\}.$$

We set

$$H_N(E) = \bigcup_{\mathcal{V}} H_{N,\mathcal{V}}(E)$$

where $\mathcal{V}$ ranges over all increasing countable convex balanced open covers of E. An entire function f on E is said to be of *nuclear type* if $f \in H_N(E)$. For every convex balanced open neighborhood U, let

$$\mathcal{V}(U) = \{nU\,;\, n \geq 1\}.$$

We set

$$H_{N,ub}(E) = \bigcup_{U} H_{N,\mathcal{V}(U)}(E)$$

where U ranges over all convex balanced open neighborhood of 0 in E. $f \in H(E)$ is said to be of *uniformly bounded nuclear type* if $f \in H_{N,ub}(E)$. We have the following relation:

$$H_{N,ub}(E) \subset H_N(E) \quad \text{and} \quad H_{N,ub}(E) \subset H_{ub}(E).$$

Proposition 1. *Let $p \in P_N(^nE)$ and let U be a convex balanced open neighborhood of 0 such that there exists $(\phi_j) \subset U^\circ$ and $(\lambda_j) \in l^1$ with $p = \sum_{j=1}^{\infty} \lambda_j \phi_j^n$. Then we have*

$$|j_n^*(p)(\theta) \leq \pi_U(p)\epsilon_{n,E}^{\alpha_U}(\theta)$$

for every $\theta \in \otimes_{n,s} E$ where α_U is the Minkovski functional of U. Thus p is of ϵ type.

Proof. Let $\theta = \sum_{k=1}^{l} x_k \otimes \cdots \otimes x_k \in \otimes_{n,s} E$. Then we have

$$\begin{aligned}
|j_n^*(p)(\theta)| &= |\sum_{k=1}^{l}\sum_{j=1}^{\infty} \lambda_j \phi_j(x_k)^n| \\
&= |\sum_{j=1}^{\infty}\sum_{k=1}^{l} \lambda_j \phi_j(x_k)^n| \\
&\leq \sum_{j=1}^{\infty}|\sum_{k=1}^{l} \lambda_j \phi_j(x_k)^n| \\
&= \sum_{j=1}^{\infty}\left|\sum_{k=1}^{l} \lambda_j (|\phi_j|_U)^n \left(\frac{\phi_j}{(|\phi_j|_U)}(x_k)\right)^n\right| \\
&= \sum_{j=1}^{\infty} \lambda_j (|\phi_j|_U)^n \left|\sum_{k=1}^{l} \left(\frac{\phi_j}{(|\phi_j|_U)}(x_k)\right)^n\right| \\
&\leq \sum_{j=1}^{\infty} \lambda_j (|\phi_j|_U)^n \epsilon_{n,E}^{\alpha_U}(\theta).
\end{aligned}$$

Thus we have

$$|j_n^*(p)(\theta)| \leq \pi_U(p)\epsilon_{n,E}^{\alpha_U}(\theta)$$

for every $\theta \in \otimes_{n,s} E$. This completes the proof.

4. The main theorems

In this section we will show that every entire function of uniformly bounded nuclear type on a locally convex space E can be extended holomorphically to any locally convex space F so that E is a topological linear subspace of F.

Theorem 1. *Let E and F be locally convex spaces, and let E be a topological linear subspace of F. Let f be an entire function on E which is of uniformly bounded nuclear type. Then there exists $\tilde{f} \in H_{ub}(F)$ such that $\tilde{f}|_E = f$.*

Proof. Let $f(x) = \sum_{n=0}^{\infty} \frac{\hat{d}^n f(0)}{n!}(x)$ be the Taylor expansion of f at 0. We set $p_n = \frac{\hat{d}^n f(0)}{n!}$. Since $f \in H_{N,ub}(E)$, there exists a balanced convex open neighborhood V of 0 in E and $\beta \in cs(F)$ such that $\beta|E = \alpha_V$ and

$$\sum_{n=0}^{\infty} \pi_{mV}(p_n) < \infty$$

for every $m \geq 1$ where α_V is the Minkovski functional of V. By Proposition 4 we have

$$|j_n^*(p_n)(\theta)| \leq \pi_V(p_n)\epsilon_{n,E}^{\alpha_V}(\theta)$$

for every $\theta \in \otimes_{n,s}E$. Then by Lemma 2 and Hahn-Banach theorem we have

$$\epsilon_{n,F}^{\beta}| \otimes_{n,s} E = \epsilon_{n,s}^{\alpha_U}.$$

By Proposition 4 there exists $A_n \in (\otimes_{n,s,\epsilon}F)'$ such that

$$A_n| \otimes_{n,s} E = j_n^*$$

and

$$|A_n(\theta)| \leq \pi_V(p_n)\epsilon_{n,F}^{\beta}(\theta)$$

for every $\theta \in \otimes_{n,s}F$. We set

$$\tilde{p_n}(x) = A_n(x \otimes \cdots \otimes x)$$

for every $x \in F$. Then we have $\tilde{p_n}|E = p_n$ and $|\tilde{p_n}(x)| \leq \pi_V(p_n)\beta(x)^n$ for every $x \in F$ and every $n \leq 1$. For every $r > 0$, we have

$$\begin{aligned} \sum_{n=0}^{\infty} |\tilde{p_n}|_{rB_\beta(1)} &\leq \sum_{n=0}^{\infty} \pi_V(p_n)r^n \\ &= \sum_{n=0}^{\infty} \pi_{rV}(p_n) < \infty. \end{aligned}$$

Therefore we can define $\tilde{f} \in H(F)$ by letting

$$\tilde{f}(x) = \sum_{n=0}^{\infty} \tilde{p_n}(x)$$

for every $x \in F$. Then we have $\tilde{f}|E = f$ and $\tilde{f} \in H_{ub}(F)$. This completes the proof.

References

[1] R.M Aron and P.D. Berner, *A Hahn-Banach extension theorem for analytic mappings*, Bull. Soc. Math. France, 106(1978), 3–24.

[2] S. Dineen, *Complex Analysis in Locally Convex Spaces*, North-Holland Math. Studies, Amsterdam 57(1981).

[3] S. Dineen, *Complex Analysis on Infinite Dimensional Spaces*, Springer Monographs in Mathematics,(1999).

[4] G. Köthe, *Topological Vector Spaces I*, Springer Verlag, (1969).

[5] G. Köthe, *Topological Vector Spaces II*, Springer Verlag, (1979).

[6] R. Meise and D. Vogt, *Extension of entire functions on nuclear locally convex spaces*, Proc. Amer. Math. Soc., 92-4(1984), 495–500.

[7] J. Mujica, *Complex Analysis in Banach Spaces*, North-Holland Math. Studies, Amsterdam, 120(1986).

[8] M. Nishihara, *The extension of holomorphic functions on a nuclear space*, Finite or Infinite Dimensinal Complex Analysis, Ed. J. Kajiwara, Zhong Li and K. H. Shon, Lecture notes in pure and applied mathematics, Vol. 214, pp. 389–394, 2000, Marcel Dekker. INC.

[9] M. Nishihara, *A Hahn-Banach extension theorem for entire functions of nuclear type* , Preprint.

[10] D. Carando and I. Zalduendo, *A Hahn-Banach theorem for integrable polynomials*, Proc. Amer. Math. Soc., 127(1999), 241–250.

23

SOME PROPERTIES OF THE INHOMOGENEOUS GENERALIZED CAUCHY-RIEMANN SYSTEMS IN QUATERNIONIC ALGEBRA AND ITS APPLICATIONS

Nguyen Thanh Van
Department of Mathematics
Hanoi University of Science
Nguyen Trai Road 334, 10000 Hanoi Vietnam
E-mail: vanhuong6996@hn.vnn.vn

Abstract: This paper deals with the way of solving the inhomogeneous generalized Cauchy-Riemann systems in quaternionic algebra. The Cousin problem for regular functions with parameter is considered in the paper, too.

AMS-classification: 30G35, 35N05

1. Introduction

The theory of functions over the field of quarternions $\mathbb{H}$ was discovered by Hamilton (see [5]) at the end of the 19th century. By means of an anlogue of the Cauchy-Riemann equations, Fueter (see [3,4]) identified a special class of quaternionic functions which play the role of holomorphic functions

The functions of this class are called *regular functions* which have many important properties of holomorphic functions of a complex variable.

Latter, Subbery (see [12]) added a certain number of new results in the theory of regular function, defined by Fueter (see [3]).

Recently, Pertici (see [8]) has studied the theory of regular functions of several quaternionic variables and generalized some theorems from complex analysis of several variables as an analogue of the Bochner-Martinelli formula, the Hartog's theorems...

In Section 2, some basic notations and definitions of quaternionic algebra, used frequently in the paper, are introduced.

A Runge-type theorem for regular functions with parameter is proved in Section 3.

The inhomogeneous generalized Cauchy-Riemann systems in quaternionic algebra are considered in Section 4.

Section 5 deals with the Cousin problem for regular functions with parameter and some applications are given in Section 6.

2. Preliminaries

The field $\mathbb{H}$ of quaternions

$$q = x_0 + ix_1 + jx_x + kx_3, \quad x_0, x_1, x_2, x_3 \in \mathbb{R} \tag{2.1}$$

is a four-dimensional non-commutative $\mathbb{R}$-field generated by four base elements $1, i, j$ and k with the following non-commutative multiplication rule

$$i^2 = j^2 = k^2 = -1, \; ij = -ji = k, \; jk = -kj = i = i, \; ki = -ik = j. \tag{2.2}$$

We define by

$$\overline{q} = x_0 - ix_1 - jx_2 - kx_3 \quad \text{the conjugate of } q. \tag{2.3}$$

The norm of a quaternionic number q is defined by

$$|q| = \sqrt{x_0^2 + x_1^2 + x_2^2 + x_3^2}. \tag{2.4}$$

For other definitions, we refer the readers to [8,12].

Let Ω be an open set of $\mathbb{H}$ and $f : \Omega \to \mathbb{H}$ be a functions. Then f has the form

$$f = f_0 + if_1 + jf_2 + kf_3 \tag{2.5}$$

where $f_\alpha = f_\alpha(x_0, x_1, x_2, x_3)$ are real-valued functions for $\alpha = 0, 1, 2, 3$.

The Cauchy-Fueter operator is defined by

$$\frac{\partial}{\partial \overline{q}} = \frac{\partial}{\partial x_0} + i\frac{\partial}{\partial x_1} + j\frac{\partial}{\partial x_2} + k\frac{\partial}{\partial x_3}. \tag{2.6}$$

Then we have

Definition 2.1 (see [8]). The function f is said to be a regular function in Ω if $f \in C^1(\Omega, \mathbb{H})$ and f satisfies

$$\frac{\partial f}{\partial \overline{q}} = 0 \quad \text{in} \quad \Omega. \tag{2.7}$$

3. The Runge theorem

In this section, we prove a Runge-type theorem for regular functions with parameter.

Let Ω_1 be an open set of $\mathbb{H}$ with smooth boundary and Ω_2 be an open set of $\mathbb{R}^n$. Let K_i be compact subsets of Ω_i, such that $\Omega_i \setminus K_i$ have no component which are relatively compact in Ω_i, $i = 1, 2$.

Put $\Omega = \Omega_1 \times \Omega_2$, $K = K_1 \times K_2$, then K is a compact subset of Ω.

Definition 3.1. A function $f : \Omega \to \mathbb{H}$ is said to be regular in Ω_1 and harmonic in Ω_2 if $f \in C^1(\Omega_1 \times t)$ and $f \in C^2(q \times \Omega_2)$ for each fixed $t \in \Omega_2, q \in \Omega_1$ and satisfies the sytem

$$\begin{cases} \dfrac{\partial f(q,t)}{\partial \overline{q}} = 0 \\ \Delta_t f(q,t) = 0, \end{cases} \quad \text{in} \quad \Omega.$$

where Δ_t is Laplace operator.

Denoting by $\mathcal{R}_{\mathcal{H}}(\mathbb{H} \times \mathbb{R}^n, \mathbb{H})$ the set of all functions which are regular in $q \in \mathbb{H}$ and harmonic in $t \in \mathbb{R}^n$. Then we have

Theorem 3.1. *If $f : \Omega \to \mathbb{H}$ is regular in Ω_1 and harmonic in Ω_2, then f is uniformly approximable in K by functions in $\mathcal{R}_{\mathcal{H}}(\mathbb{H} \times \mathbb{R}^n, \mathbb{H})$.*

Proof. Let $\varepsilon > 0$, we will show that there exists a function $R(q,t) \in \mathcal{R}_{\mathcal{H}}(\mathbb{H} \times \mathbb{R}^n, \mathbb{H})$, such that

$$\left\| f(q,t) - R(q,t) \right\|_K < \varepsilon. \tag{3.1}$$

Since the function f is regular in Ω_1 for each fixed $t \in \Omega_2$, then using the Cauchy-Fueter integral formula (see [8]) for $q \in \Omega_1$ and $t \in \Omega_2$, we have

$$f(q,t) = \frac{1}{2\pi^2} \int_{\partial\Omega_1} \frac{\overline{u} - \overline{q}}{|u-q|^4} Du f(u,t) \tag{3.2}$$

where $u = u_0 + iu_1 + ju_2 + k\mu_3$

$Du = du_1 \wedge du_2 \wedge du_3 - idu_0 \wedge du_2 \wedge du_3 + jdu_0 \wedge jdu_1 \wedge du_3 - kdu_0 \wedge du_1 \wedge du_2.$

Put

$$d\widehat{u}_0 = du_1 \wedge du_2 \wedge du_3, \; d\widehat{u}_1 = du_0 \wedge du_2 \wedge du_3$$
$$d\widehat{u}_2 = du_0 \wedge du_1 \wedge du_3, \; d\widehat{u}_3 = du_0 \wedge du_1 \wedge du_2,$$

and denote $1, i, j, k$ by e_α, $\alpha = 0, 1, 2, 3$.

From (3.2), we have

$$f(q,t) = \frac{1}{2\pi^2} \int_{\partial\Omega_1} \frac{\overline{u} - \overline{q}}{|u-q|^4} \Big(\sum_{\alpha=0}^{3} (-1)^\alpha e_\alpha f(u,t) d\widehat{u}_\alpha \Big). \tag{3.3}$$

Since the Riemann-Sums which approximate the integral on the right hand side of (3.3) do so uniformly on K, then there exists a partition of $\Gamma = \partial\Omega_1$, such that

$$\Big\| f(q,t) - \frac{1}{2\pi^2} \sum_{l=1}^{p} \frac{\overline{u} - \overline{q}}{|u-q|^4} \Big(\sum_{\alpha=0}^{3} (-1)^\alpha e_\alpha f(u,t) \Delta\Gamma^{(l)} \Big) \Big\|_K < \frac{\varepsilon}{2}, \tag{3.4}$$

where $\Delta\Gamma^{(l)}$ is the measure of $\Gamma^{(l)}$, $l = 1, ..., p$, $\Gamma = \bigcup_{l=1}^{p} \Gamma^{(l)}$. Denoting

$f_l(t) = \frac{1}{2\pi^2} \sum_{\alpha=0}^{3} (-1)^\alpha e_\alpha f(u_l,t) \Delta\Gamma^{(l)}$, then

$$\Big\| f(q,t) - \sum_{l=1}^{p} \frac{\overline{u}_l - \overline{q}}{|u_l - q|^4} f_l(t) \Big\|_K < \frac{\varepsilon}{2}. \tag{3.5}$$

For each fixed $u_l \in \partial\Omega_1$, the function $\frac{\overline{u}_l - \overline{q}}{|u_l - q|^4}$ is regular in Ω_1 (see [8]),

According to Theorem 4 (see [8]), there exists a regular function $P_l(q) \in \mathcal{R}(\mathbb{H}, \mathbb{H})$, such that

$$\Big\| \frac{\overline{u}_l - \overline{q}}{|u_l - q|^4} - P_l(q) \Big\|_{K_1} < \frac{\varepsilon}{2^{l+2} \|f_l(t)\|_{K_2}}, \tag{3.6}$$

where $\mathcal{R}(\mathbb{H}, \mathbb{H})$ is the set of all functions which are regular in $\mathbb{H}$.

Similarly, for each fixed l, the function $f_l(t)$ is harmonic on $t \in \Omega_2$. This means that

$$f_l(t) = f_0^{(l)}(t) + i f_1^{(l)}(t) + j f_2^{(l)}(t) + k f_3^{(l)}(t)$$

where $f_\alpha^{(l)}$ are real-valued and harmonic functions. Using Theorem 3.2.31 (see [7]), there exists functions $Q_\alpha^{(l)}(t) \in \mathcal{H}(\mathbb{R}^n, \mathbb{R})$, which is the set of all real-valued and harmonic functions in $\mathbb{R}^n$, such that

$$\|f_\alpha^{(l)}(t) - Q_\alpha^{(l)}(t)\|_{K_2} < \frac{1}{4} \cdot \frac{\varepsilon}{2^{l+2}\|P_l(q)\|_{K_1}} \quad \text{for} \quad \alpha = 0, 1, 2, 3. \tag{3.7}$$

Put $Q_l(t) = Q_0^{(l)}(t) + iQ_1^{(l)}(t) + jQ_2^{(l)}(t) + kQ_3^{(l)}(t)$, then by (3.7), we get

$$\|f_l(t) - Q_l(t)\|_{K_2} \leq \sum_{\alpha=0}^{3} \|f_\alpha^{(l)}(t) - Q_\alpha^{(l)}(t)\|_{K_2} < \frac{\varepsilon}{2^{l+2}\|P_l(q)\|_{K_1}}. \tag{3.8}$$

From (3.6) and (3.8), it follows that

$$\begin{aligned} &\left\|\frac{\overline{u}_l - \overline{q}}{|u_l - q|^4} f_l(t) - P_l(q)Q_l(t)\right\|_K \\ \leq &\left\|\frac{\overline{u}_l - \overline{q}}{|u_l - q|^4} - P_l(q)\right\|_{K_1} \|f_l(t)\|_{K_2} + \|P_l(q)\|_{K_1} \|f_l(t) - Q_l(t)\|_{K_2} \\ < &\frac{\varepsilon}{2^{l+2}} + \frac{\varepsilon}{2^{l+2}} = \frac{\varepsilon}{2^{l+1}} \quad \text{for every} \quad l = 1, ..., p. \end{aligned} \tag{3.9}$$

Denote

$$R(q,t) = \sum_{l=1}^{p} P_l(q)Q_l(t). \tag{3.10}$$

Since $P_l(q) \in \mathcal{R}(\mathbb{H}, \mathbb{H})$, $Q_l(t) \in \mathcal{H}(\mathbb{R}^n, \mathbb{H})$, then

$$\frac{\partial R(q,t)}{\partial \overline{q}} = \sum_{l=1}^{p} \frac{\partial P_l(q)}{\partial \overline{q}} Q_l(t) = 0, \tag{3.11}$$

$$\Delta_t R(q,t) = \sum_{l=1}^{p} P_l(q)(\Delta_t Q_l(t)) = 0. \tag{3.12}$$

By (3.11), (3.12) we get $R(q,t) \in \mathcal{R}_{\mathcal{H}}(\mathbb{H} \times \mathbb{R}^n, \mathbb{H})$. Using (3.9), (3.10), we obtain

$$\begin{aligned} &\left\|\sum_{l=1}^{p} \frac{\overline{u}_l - \overline{q}}{|u_l - q|^4} f_l(t) - R(q,t)\right\|_K \leq \sum_{l=1}^{p} \left\|\frac{\overline{u}_l - \overline{q}}{|u_l - q|^4} f_l(t) - P_l(q)Q_l(t)\right\|_K \\ < &\sum_{l=1}^{p} \frac{\varepsilon}{2^{l+1}} < \frac{\varepsilon}{2}. \end{aligned} \tag{3.13}$$

According to (3.5) and (3.13), we obtain (3.1), which completes the proof.

Remark 3.1. Let $\mathcal{A}$ be a 2^m-dimensional real Clifford algebra. Denote by $\mathcal{R}_{\mathcal{H}}(\mathbb{R}^{m+1} \times \mathbb{R}^n, \mathcal{A})$ the set of all functions which are regular in $x = (x_0, ..., x_m) \in \mathbb{R}^{m+1}$ and harmonic in $t = (t_1, ..., t_n) \in \mathbb{R}^n$.

Let Ω_1 be an open set with smooth boundary of $\mathbb{R}^{m+1}$ and Ω_2 be an open set of $\mathbb{R}^n$. Let K_i be compact subsets of Ω_i, such that $\Omega_i \setminus K_i$ have no component which is relatively compact in Ω_i, $i = 1, 2$.

Put $\Omega = \Omega_1 \times \Omega_2$, $K = K_1 \times K_2$, then K is compact subset of Ω. Using a similar method as in the proof of Theorem 3.1 and Runge's theorem for regular functions taking values in the Clifford $\mathcal{A}$ (see [2]), we can prove the following theorem

Theorem 3.2. *If $f : \Omega \to \mathcal{A}$ is regular in Ω_1 and harmonic in Ω_2, then f is uniformly approximable in K by functions in $\mathcal{R}_{\mathcal{H}}(\mathbb{R}^{m+1} \times \mathbb{R}^n, \mathcal{A})$.*

4. The inhomogeneous systems

Let Ω_1 and Ω_2 be simply connected domains of $\mathbb{H}$ and $\mathbb{R}^n$, respectively. Suppose that $\partial\Omega_1$ is smooth.

We consider the system

$$\begin{cases} \partial f(q,t) = g(q,t) & (4.1) \\ \Delta_t f(q,t) = 0 & (4.2) \end{cases}$$

where $g(q,t) \in C^\infty(\Omega_1, \mathbb{H})$ for each fixed $t \in \Omega_2$ and $g(q,t)$ is harmonic in $t \in \Omega_2$ for each fixed $q \in \Omega_1$.

Theorem 4.1. *The system (4.1), (4.2) has at least a solution $f(q,t)$, such that $f(q,t) \in C^\infty(\Omega_1, \mathbb{H})$ for each fixed $t \in \Omega_2$.*

Proof. (a) We assume that $g \in C_0^\infty(\Omega_1, \mathbb{H})$.

Define the function

$$f(q,t) = -\frac{1}{2\pi^2} \int_{\mathbb{H}} \frac{\overline{u} - \overline{q}}{|u-q|^4} g(u,t)\theta$$

where $u = u_0 + iu_1 + ju_2 + ku_3$, $\theta = du_0 \wedge du_1 \wedge du_2 \wedge du_3$. It is easy to show that $f(q,t) \in C^\infty(\mathbb{H}, \mathbb{H})$ for each fixed $t \in \Omega_2$. According to Green's theorem II (see [8]), it follows that f is a solution of Equation (4.1).

Since $\Delta_t g(q,t) = 0$, then

$$\Delta_t f(q,t) = -\frac{1}{2\pi^2} \int_{\mathbb{H}} \frac{\overline{u} - \overline{q}}{|u-q|^4} (\Delta_t g(u,t))\theta = 0$$

in Ω_2 for each fixed $q \in \Omega_1$.

This means that f satisfies (4.2). Hence f is a solution of system (4.1), (4.2).

(b) Let $g \in C^\infty(\Omega_1, \mathbb{H})$ and $\{G_\nu\}$ be an exhaustion of G with $G_\nu \Subset G_{\nu+1} \Subset \cdots$ where G_ν are simply connected domains, such that $G_\nu = G_\nu^{(1)} \times G_\nu^{(2)}$, $G_\nu^{(1)} \Subset G_{\nu+1}^{(1)} \Subset \cdots$, $G_\nu^{(2)} \Subset G_{\nu+1}^{(2)} \Subset \cdots$, and that $G_\nu^{(1)}, G_\nu^{(2)}$ are simply connected domains of Ω_1 and Ω_2, respectively.

It is clear that $\{G_\nu^{(1)}\}$, $\{G_\nu^{(2)}\}$ are open exhaustions of Ω_1 and Ω_2, respectively.

We define the function $g_\nu \in C^\infty(\Omega_1, \mathbb{H})$, such that

$$g_\nu = \begin{cases} g & \text{in} \quad G_\nu^{(1)} \times \Omega \\ 0 & \text{in} \quad (\Omega_1 \setminus G_\nu^{(1)}) \times \Omega_2, \end{cases}$$

and consider the system

$$\begin{cases} \dfrac{\partial f(q,t)}{\partial \overline{q}} = g_\nu(q,t) \\ \Delta_t f(q,t) = 0. \end{cases} \tag{4.3}$$

From step (a) of the proof, it follows that system (4.3) has at least a solution $f(q,t)$, such that $f(q,t) \in C^\infty(\Omega_1, \mathbb{H})$ for each fixed $t \in \Omega_2$.

It is shown that we can find solutions f_ν of system (4.3), such that

$$\|f_{\nu+1} - f_\nu\|_{G_\nu} < \frac{1}{2^\nu} \qquad \text{for} \quad \nu = 1, 2, \ldots \tag{4.4}$$

We choose

$$f_1 = -\frac{1}{2\pi^2} \int_{\partial\Omega_1} \frac{\overline{u} - \overline{q}}{|u-q|^4} g_1(u,t)\theta,$$
$$\widetilde{f}_2 = -\frac{1}{2\pi^2} \int_{\partial\Omega_1} \frac{\overline{u} - \overline{q}}{|u-q|^4} g_2(u,t)\theta.$$

It is clear that

$$\frac{\partial(\widetilde{f}_2 - f_1)}{\partial \overline{q}} = \frac{\partial \widetilde{f}_2}{\partial \overline{q}} - \frac{\partial f_1}{\partial \overline{q}} = g_2 - g_1 \quad \text{in} \quad G_1^{(1)},$$
$$\Delta_t(\widetilde{f}_2 - f_1) = \Delta_t \widetilde{f}_2 - \Delta_t f_1 = 0 \quad \text{in} \quad \Omega_2.$$

According to Theorem 3.1, there exists a function $R_1(q,t) \in \mathcal{R}_\mathcal{H}(\mathbb{H} \times \mathbb{R}^n, \mathbb{H})$, such that

$$\|\widetilde{f}_2 - f_1 - R_1\|_{G_1} < \frac{1}{2}.$$

Denoting $f_2 = \widetilde{f}_2 - R_1$, we get

$$\frac{\partial f_2}{\partial \overline{q}} = \frac{\partial \widetilde{f}_2}{\partial \overline{q}} - \frac{\partial R_1}{\partial \overline{q}} = g_2 - 0 = g_2,$$
$$\Delta_t f_2 = \Delta_f \widetilde{f}_2 - \Delta_t R_1 = 0.$$

This means that (4.4) is valid for $\nu = 1$.

Using a similar method, we can construct all f_ν, $\nu = 2, 3, \ldots$ for which the condition (4.4) is valid.

The sequence $\{f_\nu\}$ converges uniformly in every compact subset $K_1 \subset \Omega_1$ to the function

$$f = f_1 + \sum_{\nu=1}^{\infty} (f_{\nu+1} - f_\nu).$$

In every $G_\mu^{(1)}$, the function f is the sum of a finite number of functions from C^∞ and of a series of regular functions $(f_{\nu+1} - f_\nu)$ which converges uniformly in every compact subset of Ω_1, then $f \in C^\infty(\Omega_1, \mathbb{H})$.

On the other hand,

$$\frac{\partial f_{\nu+1}}{\partial \overline{q}} = \frac{\partial f_\nu}{\partial \overline{q}} = g \quad \text{in} \quad G_\mu^{(1)} \quad \text{for} \quad \nu \geq \mu.$$

Hence it follows that $\dfrac{\partial f}{\partial \overline{q}} = \lim\limits_{\nu \to \infty} \dfrac{\partial f_\nu}{\partial \overline{q}} = g$ in Ω_1 for each fixed $t \in \Omega_2$.

By (4.4), we get

$$\left\| f_{\nu+1} - f_\nu \right\|_{G_\nu^{(2)}} < \frac{1}{2^\nu} \quad \text{for each fixed} \quad q \in G_\nu^{(1)}.$$

This means that $f(q,t)$ is the sum of a series of harmonic functions $(f_{\nu+1} - f_\nu)$ in t, which converges uniformly in every compact subset $K_2 \subset \Omega_2$. It follows that $f(q,t)$ is harmonic in Ω_2 for each fixed $q \in \Omega_1$ (see [1]).

5. The Cousin problem

Problem I. Let $\Omega_1 \subset \mathbb{H}$ be a simply connected domain with smooth boundary and $\Omega_2 \subset \mathbb{R}^n$ be a simply connected domain. Let $\{U_\alpha\}$, $(\alpha \in I)$ be an open covering of Ω_1.

Put $V_\alpha = U_\alpha \times \Omega_2$, then $\{V_\alpha\}$ is an open covering of $\Omega = \Omega_1 \times \Omega_2$ and $V_\alpha \cap V_\beta = (U_\alpha \cap U_\beta) \times \Omega_2$.

In every $V_\alpha \cap V_\beta$, let $h_{\alpha\beta} \in C^\infty(U_\alpha \cap U_\beta, \mathbb{H})$, such that

1. $\frac{\partial h_{\alpha\beta}}{\partial \overline{q}} = 0$ in $U_\alpha \cap U_\beta$
2. $\Delta_t h_{\alpha\beta} = 0$ in Ω_2
3. $h_{\alpha\beta} + h_{\beta\alpha} = 0$ in $V_\alpha \cap V_\beta$
4. $h_{\alpha\beta} + h_{\beta\gamma} + h_{\gamma\alpha} = 0$ in $V_\alpha \cap V_\beta \cap V_\gamma$, $\alpha, \beta, \gamma \in I$.

Find functions $h_\alpha \in \mathcal{R}_{\mathcal{H}}(V_\alpha, \mathbb{H})$ such that

$$h_\beta - h_\alpha = h_{\alpha\beta} \quad \text{in} \quad V_\alpha \cap V_\beta,$$

where $\mathcal{R}_{\mathcal{H}}(V_\alpha, \mathbb{H})$ is the set of all functions which are regular in $q \in U_\alpha$ and harmonic in $t \in \Omega_2$.

Theorem 5.1. *There exists functions $h_\alpha \in \mathcal{R}_{\mathcal{H}}(V_\alpha, \mathbb{H})$, such that $h_\beta - h_\alpha = h_{\alpha\beta}$ in $V_\alpha \cap V_\beta$.*

Proof. Let $\{\phi_j\}$ be a subordinate partition of unity for the covering $\{U_\alpha\}$ with $\phi_j \in C_0^\infty(U_{\alpha_j}, \mathbb{R})$.

For every fixed β, we define the function

$$g_\beta = \sum_j h^*_{\alpha_j\beta}, \text{ where } h^*_{\alpha_j\beta} = \begin{cases} \phi_j h_{\alpha_j\beta} & \text{in} \quad U_{\alpha_j} \cap U_\beta \\ 0 & \text{in} \quad U_\beta \setminus U_{\alpha_j}. \end{cases} \tag{5.2}$$

It is clear that $g_\beta \in C^\infty(U_\beta, \mathbb{H})$, $\Delta_t g_\beta = 0$ and

$$\begin{aligned} g_\beta - g_\alpha &= \sum_j h^*_{\alpha_j\beta} - \sum_j h^*_{\alpha_j\alpha} = \sum_j \phi_j h^*_{\alpha_j\beta} - \sum_j \phi_j h_{\alpha_j\alpha} \\ &= \sum_j \phi_j (h_{\alpha\alpha_j} - h_{\alpha_j\beta}) = \sum_j \phi_j h_{\alpha\beta} = h_{\alpha\beta}. \end{aligned} \tag{5.2}$$

Hence $\dfrac{\partial g_\beta}{\partial \overline{q}} - \dfrac{\partial g_\alpha}{\partial \overline{q}} = \dfrac{\partial (g_\beta - g_\alpha)}{\partial \overline{q}} = \dfrac{\partial h_{\alpha\beta}}{\partial \overline{q}} = 0$ and $\Delta_t(g_\beta - g_\alpha) = \Delta_t h_{\alpha\beta} = 0$ because $h_{\alpha\beta} \in \mathcal{R}_{\mathcal{H}}(V_\alpha \cap V_\beta, \mathbb{H})$. This means that $\left\{\dfrac{\partial g_\alpha}{\partial \overline{q}}\right\}$ defines a function $g(q,t)$ in Ω, such that $g(q,t) \in C^\infty(\Omega_1, \mathbb{H})$ for each fixed $t \in \Omega_2$.

From (5.1), from condition (1) and because of the fact that ϕ_j are real-valued functions, we obtain (see [8])

$$\begin{aligned} \frac{\partial g}{\partial \overline{q}} &= \frac{\partial}{\partial \overline{q}} \Big(\sum_j \phi_j h_{\alpha_j\alpha} \Big) = \sum_j \frac{\partial \phi_j h_{\alpha_j\alpha}}{\partial \overline{q}} \\ &= \sum_j \left(\frac{\partial \phi_j}{\partial \overline{q}} h_{\alpha_j\beta} + \phi_j \frac{\partial h_{\alpha_j\alpha}}{\partial \overline{q}} \right) = \sum_j \frac{\partial \phi_j}{\partial \overline{q}} h_{\alpha_j\alpha}. \end{aligned} \tag{5.3}$$

By (5.3), by condition (2) and because of the fact that ϕ_j are independent on t, it follows that

$$\Delta_t\Big(\frac{\partial g_\alpha}{\partial \overline{q}}\Big) = \sum_j \frac{\partial \phi_j}{\partial \overline{q}}(\Delta_t h_{\alpha_j \alpha}) = 0.$$

Thus, $g(q,t)$ is harmonic in $t \in \Omega_2$ for each fixed $q \in \Omega_1$.

Now, we consider the system

$$\begin{cases} \dfrac{\partial h}{\partial \overline{q}} & = -g \\ \Delta_t h & = 0 \end{cases} \tag{5.4}$$

According to Theorem 4.1, the system (5.4) has a solution $h(q,t)$, such that $h(q,t) \in C^\infty(\Omega_1, \mathbb{H})$ for each fixed $t \in \Omega_2$, and $h(q,t)$ is harmonic in $t \in \Omega_2$ for each fixed $q \in \Omega_1$.

Define

$$h_\alpha = h + g_\alpha \quad \text{in every} \quad V_\alpha, \tag{5.5}$$

then

$$\frac{\partial h_\alpha}{\partial \overline{q}} = \frac{\partial h}{\partial \overline{q}} + \frac{\partial g_\alpha}{\partial \overline{q}} = -g + g = 0$$

and $\Delta_t h_\alpha = \Delta_t h + \Delta_t g_\alpha = 0$ in $V_\alpha = U_\alpha \times \Omega_2$, $\alpha \in I$. Thus, $h_\alpha \in \mathcal{R}_{\mathcal{H}}(V_\alpha, \mathbb{H})$.

It follows from (5.2), (5.5) that

$$h_\beta - h_\alpha = (h + g_\beta) - (h + g_\alpha) = g_\beta - g_\alpha = h_{\alpha\beta}.$$

So h_α which are defined by (5.5) satisfy Theorem 5.1.

The Theorem is proved.

Problem II. Suppose that the sets Ω_1, Ω_2 are defined as in Problem I. Let $\{U_\alpha\}$, $(\alpha \in I)$ be an open covering of $\Omega = \Omega_1 \times \Omega_2$, such that $U_\alpha = U_\alpha^{(1)} \times U_\alpha^{(2)}$ and that $\{U_\alpha^{(1)}\}$, $\{U_\alpha^{(2)}\}$ are the open coverings of Ω_1 and Ω_2, respectively.

In every $U_\alpha \cap U_\beta$, let $h_{\alpha\beta}(q,t) \in C^\infty(U_\alpha^{(1)} \cap U_\beta^{(1)}, \mathbb{H})$ satisfy

1. $\dfrac{\partial h_{\alpha\beta}}{\partial \overline{q}} = 0$ in $U_\alpha^{(1)} \cap U_\beta^{(1)}$
2. $\Delta_t h_{\alpha\beta} = 0$ in $U_\alpha^{(2)} \cap U_\beta^{(2)}$
3. $h_{\alpha\beta} + h_\beta h_{\beta\alpha} = 0$ in $U_\alpha \cap U_\beta$
4. $h_{\alpha\beta} + h_{\beta\gamma} + h_{\gamma\alpha} = 0$ in $U_\alpha \cap U_\beta \cap U_\gamma, \quad \alpha, \beta, \gamma \in I.$

Find $h_\alpha \in \mathcal{R}_{\mathcal{H}}(U_\alpha, \mathbb{H})$, such that

$$h_\beta - h_\alpha = h_{\alpha\beta} \quad \text{in} \quad U_\alpha \cap U_\beta,$$

where $\mathcal{R}_{\mathcal{H}}(U_\alpha, \mathbb{H})$ is the set of all functions which are regular in $q \in U_\alpha^{(1)}$ and harmonic in $t \in U_\alpha^{(2)}$.

Note that there is difference between the two these problems. In Problem I, the given functions $h_{\alpha\beta}$ are harmonic in the whole Ω_2 while in Problem II, $h_{\alpha\beta}$ has such a property in $(U_\alpha^{(2)} \cap U_\beta^{(2)}) \subset \Omega_2$. Besides, $h_{\alpha\beta}$, in the first case must be harmonic in Ω_2. In contrast, h_α, in the second case need only to be harmonic in $U_\alpha^{(2)}$.

Using a similar method as in the proof of Theorem 5.1, we are able to prove the following theorem

Theorem 5.2. *There exists functions $h_\alpha \in \mathcal{R}_{\mathcal{H}}(U_\alpha, \mathbb{H})$, such that*

$$h_\beta - h_\alpha = h_{\alpha\beta} \quad in \quad U_\alpha \cap U_\beta, \quad \alpha, \beta \in I.$$

6. Applications

Example 6.1. Suppose that the sets Ω_1, Ω_2 and $\{V_\alpha\}$ are defined as in **Problem I**.

We denote by $\mathcal{R}_{\mathcal{H}}(\Omega_1 \times \Omega_2, \mathbb{H})$ the set of all functions which are regular in $q \in \Omega$ and harmonic in $t \in \Omega_2$, and by $\widetilde{\mathcal{R}}_{\mathcal{H}}(\Omega_1 \times \Omega_2, \mathbb{H})$ the set of all functions $f(q,t)$ which possess the following properties:

1. f is almost everywhere regular on $q \in \Omega_1$ except the set P_t for each fixed $t \in \Omega_2$ (P_t is regarded as the set of singularities of the function f)
2. f can not be extended as a regular function at any point of P_t.

In every V_α, let $f_\alpha \in \widetilde{\mathcal{R}}_{\mathcal{H}}(V_\alpha, \mathbb{H})$ be the functions satisfying the following condition

$$f_\alpha - f_\beta \in \mathcal{R}_{\mathcal{H}}(V_\alpha \cap V_\beta, \mathbb{H}), \quad \alpha, \beta \in I. \tag{6.1}$$

Then we have

Theorem 6.1. *There exists a function $f \in \widetilde{\mathcal{R}}_{\mathcal{H}}(\Omega_1 \times \Omega_2, \mathbb{H})$, such that*

$$f - f_\alpha \subset \mathcal{R}_{\mathcal{H}}(V_\alpha, \mathbb{H}) \quad for\ all \quad \alpha \in I. \tag{6.2}$$

Proof. Set $f_\alpha - f_\beta = h_{\alpha\beta}$, then the $h_{\alpha\beta}$ satisfy the assumption of Problem I.

According to Theorem 5.1, there exists $h_\alpha \in \mathcal{R}_{\mathcal{H}}(V_\alpha, \mathbb{H})$, such that

$$h_\beta - h_\alpha = h_{\alpha\beta}. \tag{6.3}$$

From (6.2), (6.3), we have

$$h_\beta - h_\alpha = f_\alpha - f_\beta.$$

Hence $h_\alpha + f_\alpha = h_\beta + f_\beta$ in $V_\alpha \cap V_\beta$, $\alpha, \beta \in I$.

This means that $\{h_\alpha + f_\alpha\}$ defines a function $f \in \widetilde{\mathcal{R}}_{\mathcal{H}}(\Omega_1 \times \Omega_2, \mathbb{H})$ such that $f - f_\alpha = h_\alpha \in \mathcal{R}_{\mathcal{H}}(V_\alpha, \mathbb{H})$ for all $\alpha \in I$.

The theorem is proved.

Example 6.2. Suppose that the sets Ω_1, Ω_2 and $\{U_\alpha\}$ are defined as in Problem II. Let $\mathcal{R}_{\mathcal{H}}(\Omega_1 \times \Omega_2, \mathbb{H})$ and $\widetilde{\mathcal{R}}_{\mathcal{H}}(\Omega_1 \times \Omega_2, \mathbb{H})$ be defined as in Ecample 6.1.

In every $U_\alpha = U_\alpha^{(1)} \times U_\alpha^{(2)}$, let $f_\alpha \in \widetilde{\mathcal{R}}_{\mathcal{H}}(U_\alpha, \mathbb{H})$ be functions satisfying the following condition

$$f_\alpha - f_\beta \in \mathcal{R}_{\mathcal{H}}(U_\alpha \cap U_\beta, \mathbb{H}), \quad \alpha, \beta \in I. \tag{6.4}$$

According to a similar method as in the proof of Theorem 6.1 and by Theorem 5.2, we obtain

Theorem 6.2. *There exists a function $f \in \mathcal{R}_{\mathcal{H}}(\Omega_1 \times \Omega_2, \mathbb{H})$, such that*

$$f - f_\alpha \in \mathcal{R}_{\mathcal{H}}(U_\alpha, \mathbb{H}) \quad \textit{for all} \quad \alpha \in I.$$

Acknowledgements

My special thanks go to Professor Dr. Le Hung Son and Dr. Nguyen Canh Luong, Department of Applied Mathematics, Hanoi University of Technology for their precious remarks and correction of this paper.

I also acknowledge my gratitude to Professor Dr. Tutschke, Department of Mathematics D, Graz University of Technology for his active support. The paper could not have been complited without his assistance.

References

[1] S. Bochner and W. T. Martin, *Several Complex Variables*, Princeton University Press, 1948.

[2] F. Brackx, R. Delanghe and F. Sommen, *Clifford Analysis*, Research Notes in Mathematics **76**, Pitman Books Ltd., London 1982.

[3] R. Fueter, *Ueber die Analytische Darstellung der Regulaeren Functionen Einer Quaternionen Variable*, Comment. Math. Helv. **8** (1947), 371-378.

[4] R. Fueter, *Die Funktionentheorie der Differentialgleichungen* $\Delta u = 0$ *und* $\Delta\Delta u = 0$ *mit Vier Rellen Variablen*, Comment. Math. Helv. **7** (1935), 307-330.

[5] W. R. Hamilton, *Elements of Quaternions*, Longmans Greene, London, 1866.

[6] L. Hoermander, *An Introduction to Complex Analysis in Several Variables*, North Holland Publishing Co., Amsterdam 1973.

[7] L. Hoermander, *Notions of Convexity*, Birkhäuser, Boston-Basel-Berlin, 1994

[8] D. Pertici, *Funzioni Regolari di piu Variabili Quaternioniche*, Ann. Math. Pura e Appl. Serie IV **151** (1988), 39-65.

[9] B. V. Shabat, *Introduction to Complex Analysis*, Nauka, Moscow, 1969 (in Russian).

[10] Le Hung Son, *Cousin Problem for Biregular Functions with Values in a Clifford Algebra*, Complex Variables: Theory and Applications, **20** (1992), 255-263.

[11] Le Hung Son, *Some Remarks on Functions of Several Quaternionic Variables*, Proceedings of the Fifth Vietnamese Mathematical Conference 169-177.

[12] A. Sudbery, *Quaternionic Analysis*, Math. Proc. Camb. Phil. Soc. **85** (1979), 199-225.

24

THE FUNDAMENTAL SOLUTION OF THE HYPERBOLIC DIRAC-EQUATION

D. Eelbode*
Ghent University, Building S-22, Galglaan 2
B-9000 Ghent, Belgium
E-mail: deef@cage.rug.ac.be

F. Sommen†
Ghent University, Building S-22, Galglaan 2
B-9000 Ghent, Belgium
E-mail: fs@cage.rug.ac.be

Abstract: In this paper the hyperbolic unit ball is identified with the manifold of rays within the future null cone. By means of the induced Clifford algebra structure there, a definition of Dirac operators on sections of homogeneous line bundles is obtained. We then construct the fundamental solution for this Dirac-equation in hyperbolic space, in case of an odd dimension. This construction uses a geometrical picture, which enables us to transform the original equation to an equation expressed in terms of coordinates on a parabolic surface. Eventually, this will lead to a solution of the hyperbolic equation that can be interpreted as a modulated version of the fundamental solution of the Euclidian Dirac-equation.

1. Introduction

Let $e_1, \ldots, e_m$ be an orthonormal basis for the Euclidian space $\mathbb{R}^m$. The Clifford algebra $\mathbb{R}_{p,q}$ is then defined as the universal real algebra

*senior research assistant supported by N.F.W.O. Belgium.
†research assistant supported by N.F.W.O. Belgium

generated by the following relations:

$$\begin{array}{ll} e_i^2 = 1 & i = 1 \dots p \\ e_i^2 = -1 & i = p+1 \dots p+q = m \\ e_i e_j + e_j e_i = 0 & i \neq j \,. \end{array}$$

Within these settings, the Dirac operator $\sum_{i=1}^{m} \sigma_i e_i \partial_{x_i}$ - with $\sigma_i = \pm 1$, depending on the fact whether $e_i^2 = \pm 1$ - arises as a natural Clifford-valued first-order operator.

In Clifford analysis, one deals with the corresponding function theory. For a first study, we refer to [2]. The study of the invariance group of null solutions of the Euclidian Dirac operator, which coincides with the group of Moebius transformations, is another topic of major importance. One possible approach is to make a projective identification between the points in the Euclidian space $\mathbb{R}^m$ and the rays in the null cone in $\mathbb{R}^{m+1,1}$ (see e.g.[16]); hence the Moebius group corresponds to the group $Spin(m+1,1)$. This method has been the main theme in several works on Clifford analysis (see [6], [13], [17]). The study of the Clifford analysis on hyperbolic spaces is also related with this approach, due to the fact that the subgroup $Spin(m,1)$ of Moebius transformations leaving the unit sphere invariant is the isometry group of these non-Euclidian geometrics. For the hyperbolic unit ball or the hyperbolic half plane we refer to the work of Heinz Leutwiler and his students (see [15], [9], [12]) and to the one of Cnops (see [5], [6]).

Where the above attempts generalize the Dirac operator for Spin1 fields, in this paper we work with a generalization of the Dirac operator for Spin$\frac{1}{2}$ fields, as seen in [3]. Again, this uses a homogeneous version of the hyperbolic ball whereby points are identified with rays in the future null cone and the isometry group corresponds to the Lorentz group Spin(m,1). Monogenic functions on the hyperbolic space are then defined as homogeneous solutions of the space-time Dirac operator.

Finally, we also mention the Q_p-spaces. The scale of Q_p-spaces for the complex plane was introduced and studied by several authors (see [1] e.g.), and in [11] a similar scale was introduced for quaternion-valued functions. Later on, this scale was then further generalized by Delanghe and Cnops to the case of the Dirac operator in $\mathbb{R}^n$ (see [7]). Recently, based on the fact that Q_p-spaces depend strongly on the hyperbolic geometry, Cerejeiras, Kähler and Sommen gave a definition for Q_p-spaces in the projective model (see [4]).

Our current purpose is to introduce on $\mathbb{R}^{m,1}$ the Clifford algebra $\mathbb{R}_{1,m}$ with generators $(e_1, \ldots, e_m, \epsilon)$ obeying the following relations:

$$
\begin{aligned}
&e_i^2 = -1 && i = 1 \ldots m \\
&\epsilon^2 = 1 && \\
&e_i e_j + e_j e_i = 0 && i \neq j \\
&e_i \epsilon + \epsilon e_i = 0 && i = 1 \ldots m ,
\end{aligned}
$$

in order to describe Clifford-analysis on a projective hyperbolic space. We may speak of a *projective* hyperbolic space, because with each point in the hyperbolic unit ball $B(1)$ we will associate a ray in the future cone. Essentially, this is what was done in [4], where the authors considered the following conformal embedding of $\mathbb{R}^m$ into $\mathbb{R}^{m+2}$,

$$
\begin{aligned}
\mathbf{x} \in \mathbb{R}^m \longrightarrow \boldsymbol{X} &= (x_1, \ldots, x_m, \frac{1-r^2}{2}, \frac{1+r^2}{2}) \\
&= (X_1, \ldots, X_m, X_{m+1}, X_{m+2}) \in \mathbb{R}^{m+2} .
\end{aligned}
$$

Next, they considered the positive half ray inside $NC_+ = NC \cap \{\boldsymbol{X} | X_{m+2} > 0\}$:

$$
ray(\boldsymbol{X}) = \{\boldsymbol{Y} = \lambda \boldsymbol{X} | \lambda > 0\} ,
$$

and regarded the mapping $\mathbf{x} \rightarrow ray(\boldsymbol{X})$ as the embedding of $\mathbb{R}^m$ into the manifold of rays, $ray(NC_+)$. This mapping is injective and covers all of NC_+, except for the ray $ray(0, -1, 1)$, which leads to the definition of a single point at infinity $\{\infty\}$ and to the one point compactification $\mathbb{R}^m \cup \{\infty\} = S^m$. Here, S^m may be topologically identified with a sphere on the FC, as $S^m \cong NC_+ \cap \{X | X_{m+2} = 1\}$.

Note that the map from $\mathbb{R}^m \rightarrow S^m \backslash \{\infty\}$ given by $\mathbf{x} \rightarrow \boldsymbol{X} \rightarrow ray(X) \rightarrow \boldsymbol{Y} \in S^m \backslash \{\infty\}$ is nothing but the stereographic projection:

$$
\begin{aligned}
Y_j &= \frac{2x_j}{1+r^2} , \quad j = 1, \ldots, m \\
Y_{m+1} &= \frac{1-r^2}{1+r^2} ,
\end{aligned}
$$

which is angle preserving.

In [4], it was also proven that the Lorentz group $SO(m+1,1)$, mapping rays into rays and preserving NC_+, leads to an angle preserving mapping on $ray(NC_+) = S^m$. This means that Lorentz transformations can be identified with Moebius transformations.

As for the geometry of the hyperbolic unit ball $B(1) \subseteq \mathbb{R}^m$, it suffices to say that $B(1)$ corresponds to the set of $ray(X)$ inside the half space $X_{m+1} > 0$. If one then determines the projective hyperbolic geometry or Lobatchevskii space by stating that "spheres of dimension d" correspond to sets of the form $S \cap B(1)$ - whereby S intersects S^{m-1} under a right angle, one has that $SO(m,1)$ is the automorphism group of hyperbolic projective geometry.

This is true because the group $SO(m,1)$ leaving the coordinate X_{m+1} invariant, is the subspace of $SO(m+1,1)$ leaving the unit sphere S^{m-1} invariant. Hence, as it corresponds to the Moebius transformations leaving $B(1)$ invariant, it maps any sphere S (of any dimension $\leq m-1$) intersecting S^{m-1} under a right angle into a dito sphere. Notice that $SO(m,1)$ also preserves the hyperbolic metric on $B(1)$, given by $ds_H^2 = dX_{m+2}^2 - \sum_{j=1}^{m+1} X_j^2$, which however doesn't correspond to the metric on the parabolic surface we will use further in this paper.

Finally, the projection from $\mathbb{R}^{m+1,1}$ to $\mathbb{R}^{m,1}$,

$$(X_1, \ldots, X_m, X_{m+1}, X_{m+2}) \to (X_1, \ldots, X_m, X_{m+2}) \; ,$$

preserves all the subsets of the projective geometry, but we will come to this point in the next section.

2. The Geometrical Picture

Let us start with the introduction of coordinates on $\mathbb{R}^{m,1} : (X_1, \ldots, X_{m+1})$. Denoting $X_{m+1} = T$, there is a simple identification between $\mathbb{R}^{m,1}$ and $\mathbb{R}_{1,m}^{(1)}$:

$$(X_1 \ldots X_m, T) \to \boldsymbol{X} + \epsilon T, \quad \boldsymbol{X} = \sum_{j=1}^{m} X_j e_j \, . \tag{1}$$

Here $\mathbb{R}^{m,1}$ is in fact nothing but the hyperbolic space - as it stands for an $(m+1)$-dimensional real space, endowed with an inproduct derived from Minkowski's metric - and $\mathbb{R}_{1,m}^{(1)}$ is nothing but the set of vectors in the real Clifford algebra $\mathbb{R}_{1,m}$. The nullcone in hyperbolic space is now given by those vectors in $\mathbb{R}_{1,m}^{(1)}$ for which $(\boldsymbol{X}+\epsilon T)^2 = T^2 - \sum_{j=1}^{m} X_j^2 = 0$, while the future cone (FC) corresponds to the set of vectors $\boldsymbol{X}+\epsilon T$ with $(\boldsymbol{X}+\epsilon T)^2 > 0$ and $T > 0$.

Our purpose is to find a function $E(\mathbf{X}, T)$, defined for all vectors $\boldsymbol{X}+\epsilon T$

in the future cone, which is the homogeneous fundamental solution of the hyperbolic Dirac-equation:

$$(-\partial_{\boldsymbol{X}} + \epsilon\partial_T)E(\boldsymbol{X},T) = 0 \; . \tag{2}$$

Let us start from a $\mathbb{R}_{1,m}$-valued function $E(\boldsymbol{x})$ on $B(1) \subseteq \mathbb{R}^m$, which may be identified with a function $E(\boldsymbol{X}+\epsilon T)$ - $\boldsymbol{X}+\epsilon T \in P$ - simply by putting

$$E(\boldsymbol{X}+\epsilon T) = E(\boldsymbol{x}), \;\; X_j = x_j, \;\; T = \frac{1+x^2}{2} \; . \tag{3}$$

As we are looking for a function defined on the whole FC, we need to extend this function $E(\boldsymbol{X}+\epsilon T)$ in a suitable unique way such that it corresponds to an invariant object on the manifold of rays,

$$ray(FC) \;\; = \;\; \{\, \{\lambda(\boldsymbol{X}+\epsilon T) \mid \lambda > 0\} \mid (\boldsymbol{X}+\epsilon T) \in FC\} \; , \tag{4}$$

which is the true hyperbolic space, as explained in the previous section. This may be done by considering the α-homogeneous "Clifford-line" bundle E^α over $ray(FC)$ given by the equivalence relation:

$$(\boldsymbol{X}+\epsilon T, c) \propto (\lambda(\boldsymbol{X}+\epsilon T), \lambda^\alpha c) \qquad c \in \mathbb{R}_{1,m} \; , \tag{5}$$

the sections of which have the form

$$(\boldsymbol{X}+\epsilon T, F(\boldsymbol{X}+\epsilon T)) \propto (\lambda(\boldsymbol{X}+\epsilon T), \lambda^\alpha F((\boldsymbol{X}+\epsilon T)) \; . \tag{6}$$

This means that $E(\boldsymbol{X}+\epsilon T)$ has to satisfy the following homogeneity condition:

$$E(\lambda(\boldsymbol{X}+\epsilon T)) = \lambda^\alpha E(\boldsymbol{X}+\epsilon T) \; . \tag{7}$$

This homogeneity condition can be interpreted as follows: once we have choosen a surface Σ in the future cone FC, such that each ray intersects this surface in a unique point, it suffices to know the value $F(\boldsymbol{X}+\epsilon T)$ only for the points lying on this surface. The value of $F(\boldsymbol{X}'+\epsilon T')$ for points $(\boldsymbol{X}'+\epsilon T') \notin \Sigma$ can then be regarded as a certain homogeneity factor λ^α, multiplied with $F(\boldsymbol{X}+\epsilon T)$, with $\boldsymbol{X}+\epsilon T$ the unique point of intersection of Σ and the ray connecting $\boldsymbol{X}'+\epsilon T'$ with the origin.

As for Σ, we take the parabolic surface P with equation $T = \frac{1+|\boldsymbol{X}|^2}{2}$. Given $(\boldsymbol{X}+\epsilon T) \in FC$, consider the ray through the origin and $(\boldsymbol{X}+\epsilon T)$. This ray doesn't intersect P in a unique point, unless one focuses on the point of intersection characterized by a coordinate $T < 1$. This point of intersection has the following form:

$$\mathbf{x} + \epsilon\frac{1+r^2}{2}, \;\; \mathbf{x} \in B(1) \; . \tag{8}$$

This means that for each $(\boldsymbol{X} + \epsilon T) \in FC$ we can determine $\lambda > 0$ in such a way that we have:

$$\boldsymbol{X} + \epsilon T = \lambda(\mathbf{x} + \epsilon\frac{1+r^2}{2}), \quad \mathbf{x} \in B(1) \tag{9}$$

One can easily verify that λ is a solution of the following equation:

$$\lambda^2 - 2\lambda T + |\boldsymbol{X}| = 0 \ . \tag{10}$$

This means that

$$\lambda = T \pm \sqrt{T^2 - |\boldsymbol{X}|^2} \ . \tag{11}$$

In order to determine which solution to discard, consider an arbitrary element of $\mathbb{R}_{1,m}$ lying on the parabola P - say $\mathbf{x} + \epsilon\frac{1+|x|^2}{2}$ with $\mathbf{x} \in B(1)$ (so that $T < 1$). Because $\mathbf{x} + \epsilon\frac{1+|x|^2}{2} \in P$, λ should be 1. But only if we consider the solution with the positive sign, we will find that

$$\lambda = T + \sqrt{T^2 - |\boldsymbol{X}|^2} = \frac{1+|x|^2}{2} + \frac{1-|x|^2}{2} = 1 \ .$$

We can now consider a new coordinate-system in the future cone. Instead of working with $\boldsymbol{X}$ and T, we will rewrite (2) in terms of λ and $\mathbf{x} \in B(1)$. Here, λ needs to be interpreted as a sort of stretching factor, used to transform the vector $\boldsymbol{X} + \epsilon T$ into a point on the parabola along the ray that joins the origin and $\boldsymbol{X} + \epsilon T$. This goes as follows:

$$\partial_{\boldsymbol{X}} = \sum_{i=1}^{m} e_i \partial_{X_i} \tag{12}$$

with

$$\partial_{X_i} = \sum_{j=1}^{m} \frac{\partial x_j}{\partial X_i}\partial_{x_j} + \frac{\partial \lambda}{\partial X_i}\partial_\lambda. \tag{13}$$

Because

$$x_j = \frac{X_j}{\lambda} \tag{14}$$

and

$$\lambda = T + \sqrt{T^2 - |\boldsymbol{X}|^2}, \tag{15}$$

we will eventually find:

$$\partial_{\boldsymbol{X}} = \frac{(1-|\mathbf{x}|^2)\partial_{\mathbf{x}} + 2\mathbf{x}\mathbb{E}_{\mathbf{x}} - 2\mathbf{x}\mathbb{E}_\lambda}{\lambda(1-|\mathbf{x}|^2)}. \tag{16}$$

Analogue calculations can be made for ∂_T, leading to:

$$\partial_T = \frac{-2\mathbb{E}_{\mathbf{x}}}{\lambda(1-|\mathbf{x}|^2)} + \frac{2\mathbb{E}_\lambda}{\lambda(1-|\mathbf{x}|^2)} \,. \tag{17}$$

Combining (16) and (17), one finds the hyperbolic Dirac-operator, rewritten in terms of the coordinates $\mathbf{x} \in B(1)$ and λ:

$$-\partial_{\boldsymbol{X}} + \epsilon\partial_T = \frac{\mathbf{x}+\epsilon}{\lambda(1-|\mathbf{x}|^2)}(-(\mathbf{x}+\epsilon)\partial_{\mathbf{x}} - 2\mathbb{E}_{\mathbf{x}} + 2\mathbb{E}_\lambda) \,. \tag{18}$$

But as $-\mathbf{x}\partial_{\mathbf{x}} = \partial_{\mathbf{x}}\mathbf{x} + 2\mathbb{E}_{\mathbf{x}} + m\mathbb{I}$ and $-\epsilon\partial_{\mathbf{x}} = \partial_{\mathbf{x}}\epsilon$, we also have that:

$$-\partial_{\boldsymbol{X}} + \epsilon\partial_T = \frac{\mathbf{x}+\epsilon}{\lambda(1-|\mathbf{x}|^2)}(\partial_{\mathbf{x}}(\mathbf{x}+\epsilon) + m\mathbb{I} + 2\mathbb{E}_\lambda) \,. \tag{19}$$

Now using the homogeneity condition (7), we propose the following definition for $E(\boldsymbol{X},T)$:

$$E(\boldsymbol{X},T) := \lambda^\alpha E(\mathbf{x}), \quad \boldsymbol{X} + \epsilon T = \lambda(\mathbf{x} + \epsilon\frac{1+|\mathbf{x}|^2}{2}) \,. \tag{20}$$

This is equivalent with stating that

$$E(\boldsymbol{X},T) := \left(T + \sqrt{T^2 - |\boldsymbol{X}|^2}\right)^\alpha E(\mathbf{x}), \quad \mathbf{x} = \frac{\boldsymbol{X}}{T + \sqrt{T^2 - |\boldsymbol{X}|^2}} \in B(1). \tag{21}$$

If one now uses (19) and (20) to express that $(-\partial_{\boldsymbol{X}} + \epsilon\partial_T)E(\boldsymbol{X},T) = 0$, it is easily seen that solving (2) is equivalent with finding a function $E(\mathbf{x})$, defined for $\mathbf{x} \in B(1)$ such that

$$(\partial_{\mathbf{x}}(\mathbf{x}+\epsilon) + (m+2\alpha))E(\mathbf{x}) = 0 \tag{22}$$

Our task now is to solve this equation, and for reasons that will become clear later, we restrict ourselves to the case of an odd dimension.

The geometrical picture described troughout this section is very similar to the one that was used in [4], but there the authors chose to work with a parabolic surface in $\mathbb{R}^{m+2}$. That is why they needed the projection from $\mathbb{R}^{m+1,1}$ to $\mathbb{R}^{m,1}$, of which we spoke at the end of the introduction. Eventually, this means that with each $\mathbf{x} \in B(1)$ one can associate a point on the parabolic surface P:

$$\mathbf{x} \in B(1) \longrightarrow (x_1, \ldots, x_m, \frac{1+r^2}{2}) \in P\ . \tag{23}$$

As mentioned above, this doesn't affect the subsets of projective hyperbolic geometry, in the sense that hyperbolic subspaces of $B(1)$ will now correspond to the intersections of P with planes through the origin. However, the above mapping is no isometry, and the parabolic surface P is not $SO(m,1)$-invariant. This is because $SO(m,1)$ corresponds to the hyperbolic isometry group, and acts transitively on the hyperboloid $H \leftrightarrow T_{m+2}^2 - \sum_m^{j=1} X_j^2 = 1$. In other words, it is the following mapping that can be considered as an isometric embedding of the hyperbolic space:

$$\mathbf{x} \in B(1) \rightarrow \boldsymbol{X} \in P \rightarrow \frac{2}{1-r^2}\boldsymbol{X} \in H\ . \tag{24}$$

However, troughout this paper we will not work with this weight-factor $w(\mathbf{x})$.

3. Construction of the Fundamental Solution

In order to solve (22) we will work in the even subalgebra $\mathbb{R}_{1,m}^{(+)}$, generated by the bivectors $e_\imath \epsilon = \epsilon_\imath$. This means that instead of working with $\partial_{\mathbf{x}}$ and $\mathbf{x}$, we will work with:

$$\partial_{\mathbf{x}}^{(\epsilon)} := \sum_{\imath=1}^{m} \epsilon_\imath \partial_{X_\imath}$$

and

$$\mathbf{x}^{(\epsilon)} := \sum_{\imath=1}^{m} \epsilon_\imath x_\imath\ .$$

Reason for this is the following simplification:

$$\begin{aligned} & (\partial_{\mathbf{x}}(\mathbf{x}+\epsilon) + (m+2\alpha))E(\mathbf{x}) = 0 \\ \Leftrightarrow\ & (\partial_{\mathbf{x}}\epsilon^2(\mathbf{x}+\epsilon) + (m+2\alpha))E(\mathbf{x}) = 0 \\ \Leftrightarrow\ & (\partial_{\mathbf{x}}^{(\epsilon)}(-\mathbf{x}^{(\epsilon)}+1) + (m+2\alpha))E(\mathbf{x}) = 0 \\ \Leftrightarrow\ & \partial_{\mathbf{x}}^{(\epsilon)}(\mathbf{x}^{(\epsilon)}-1)E(\mathbf{x}) = (m+2\alpha)E(\mathbf{x}) \end{aligned} \tag{25}$$

Solving (25) is again equivalent with solving the original Dirac-equation (2). To solve (25) we will use Frobenius' method, but to avoid an overloaded notation we will suppress the (ϵ)-indices. Formally we will thus be working over $\mathbb{R}_{m,0} \cong \mathbb{R}_{1,m}^{(+)}$. As a solution for (25) we propose the following function, which is the product of the Cauchy-kernel and an analytic function:

$$\begin{aligned} E(\mathbf{x}) &= \frac{\mathbf{x}}{A_m|\mathbf{x}|^m}\sum_{k=0}^{\infty} c_k\mathbf{x}^k && (26)\\ &= \frac{\mathbf{x}}{A_m|\mathbf{x}|^m}\left[\sum_{k=0}^{\infty} c_{2k}r^{2k} + \mathbf{x}\sum_{k=0}^{\infty} c_{2k+1}r^{2k}\right] && (27)\\ &= \frac{1}{A_m}\left[\mathbf{x}\sum_{k=0}^{\infty} c_{2k}r^{2k-m} + \sum_{k=0}^{\infty} c_{2k+1}r^{2+2k-m}\right] && (28)\end{aligned}$$

To determine $c_k(\forall k \in \mathbb{N})$, we plug the previous expression into equation (25) and try to solve the resulting equations. First, let us calculate $\partial_{\mathbf{x}}\mathbf{x}E(\mathbf{x})$:

$$\partial_{\mathbf{x}}\mathbf{x}E(\mathbf{x}) = \frac{1}{A_m}\partial_{\mathbf{x}}\sum_{k=0}^{\infty} c_{2k}r^{2k+2-m} + \frac{1}{A_m}\partial_{\mathbf{x}}\mathbf{x}\sum_{k=0}^{\infty} c_{2k+1}r^{2k+2-m}\ . \qquad (29)$$

As we are formally working in $\mathbb{R}_{m,0}$, we have that

$$\begin{aligned} \partial_{\mathbf{x}}\mathbf{x} &= m\mathbb{I} + \mathbb{E}_{\mathbf{x}} - \Gamma_{\mathbf{x}} && (30)\\ \partial_{\mathbf{x}} &= \omega\left(\partial_r + \frac{1}{r}\Gamma_{\mathbf{x}}\right)\ , && (31)\end{aligned}$$

but as we are considering radial functions, it is allowed to drop the angular operator $\Gamma_{\mathbf{x}}$. This means that

$$\begin{aligned} \partial_{\mathbf{x}}\mathbf{x}E(\mathbf{x}) &= \frac{1}{A_m}\omega\partial_r\sum_{k=0}^{\infty} c_{2k}r^{2k+2-m} + \frac{1}{A_m}(m+\mathbb{E})\sum_{k=0}^{\infty} c_{2k+1}r^{2k+2-m}\\ &= \frac{1}{A_m}\mathbf{x}\sum_{k=0}^{\infty}(2k+2-m)c_{2k}r^{2k-m} +\\ &\quad +\frac{1}{A_m}\sum_{k=0}^{\infty}(2k+2)c_{2k+1}r^{2k+2-m}\ . && (32)\end{aligned}$$

Analogue calculations can be made to find $-\partial_{\mathbf{x}}E(\mathbf{x})$:

$$
\begin{aligned}
-\partial_{\mathbf{x}}E(\mathbf{x}) &= \\
&= -\frac{1}{A_m}\partial_{\mathbf{x}}\mathbf{x}\sum_{k=0}^{\infty} c_{2k}r^{2k-m} - \frac{1}{A_m}\partial_{\mathbf{x}}\sum_{k=0}^{\infty} c_{2k+1}r^{2k+2-m} \\
&= -\frac{1}{A_m}(m+\mathbb{E}_{\mathbf{x}})\sum_{k=0}^{\infty} c_{2k}r^{2k-m} - \frac{1}{A_m}\omega\partial_r\sum_{k=0}^{\infty} c_{2k+1}r^{2k+2-m} \\
&= -\frac{1}{A_m}\sum_{k=0}^{\infty} 2kc_{2k}r^{2k-m} - \frac{1}{A_m}\mathbf{x}\sum_{k=0}^{\infty}(2+2k-m)c_{2k+1}r^{2k-m} \qquad (33)
\end{aligned}
$$

Using (28), (32) and (33) to express the fact that $\partial_{\mathbf{x}}(\mathbf{x}-1)E(\mathbf{x}) = (m+2\alpha)E(\mathbf{x})$, we find a system consisting of the following two equations:

$$
\begin{aligned}
&\sum_{k=0}^{\infty}(2+2k)c_{2k+1}r^{2+2k-m} - \sum_{k=0}^{\infty}(2k)c_{2k}r^{2k-m} = \qquad (34)\\
&= \sum_{k=0}^{\infty}(m+2\alpha)c_{2k+1}r^{2+2k-m}
\end{aligned}
$$

$$
\begin{aligned}
&\mathbf{x}\left[\sum_{k=0}^{\infty}(2+2k-m)c_{2k}r^{2k-m} - \sum_{k=0}^{\infty}(2+2k-m)c_{2k+1}r^{2k-m}\right] = \\
&= \sum_{k=0}^{\infty}(m+2\alpha)c_{2k}r^{2k-m}\,. \qquad (35)
\end{aligned}
$$

This means that

$$
\left\{
\begin{array}{lll}
(k-\alpha-\frac{m}{2})c_{2k-1} &= kc_{2k} & (\forall k \in \mathbb{N}_0) \\
(1+k-\alpha-\frac{m}{2})c_{2k} &= (1+k-\frac{m}{2})c_{2k+1} & (\forall k \in \mathbb{N})
\end{array}
\right. \qquad (36)
$$

Eliminating c_{2k} leads to a recursive relation for the odd coefficients,

$$
c_{2k+1} = \frac{1+k-m-\alpha}{1+k-\frac{m}{2}}\frac{k-\alpha-\frac{m}{2}}{k}\,c_{2k-1} \qquad (\forall k \in \mathbb{N}_0) \qquad (37)
$$

while eliminating c_{2k+1} (after replacing k for $k+1$ in the first equation of (36)) leads to a recursive relation for the even coefficients:

$$
c_{2k+2} = \frac{1+k-m-\alpha}{1+k-\frac{m}{2}}\frac{1+k-\alpha-\frac{m}{2}}{1+k}\,c_{2k} \qquad (\forall k \in \mathbb{N}) \qquad (38)
$$

Choosing $c_0 = 1$, we have that $c_1 = \frac{1-m-\alpha}{1-\frac{m}{2}}$, and plugging this respectively into (38) and (37) we find:

$$c_{2k+1} = \frac{1+k-m-\alpha}{1+k-\frac{m}{2}} \cdots \frac{1-m-\alpha}{1-m} \cdot \frac{k-\alpha-\frac{m}{2}}{k} \cdots \frac{1-\alpha-\frac{m}{2}}{1} \tag{39}$$

$$= \left(\frac{1-m-\alpha}{1-\frac{m}{2}}\right) \frac{(2-m-\alpha)\dots(1+k-m-\alpha)}{(2-\frac{m}{2})\dots(k+1-\frac{m}{2})} \cdot \times$$
$$\times \frac{(1-\alpha-\frac{m}{2})\dots(k-\alpha-\frac{m}{2})}{k!} \tag{40}$$

and

$$c_{2k} = \frac{k-m-\alpha}{k-\frac{m}{2}} \cdots \frac{1-m-\alpha}{1-m} \cdot \frac{k-\alpha-\frac{m}{2}}{k} \cdots \frac{1-\alpha-\frac{m}{2}}{1} \tag{41}$$

$$= \frac{(1-m-\alpha)\dots(k-m-\alpha)}{(1-\frac{m}{2})\dots(k-\frac{m}{2})} \cdot \frac{(1-\alpha-\frac{m}{2})\dots(k-\alpha-\frac{m}{2})}{k!} \tag{42}$$

Introducing Appell's symbol, defined as

$$\begin{aligned} (a)_n &= a(a+1)\dots(a+n-1) \\ (a)_0 &= 1\,, \end{aligned}$$

expressions (40) and (42) can be simplified:

$$c_{2k+1} = \left(\frac{1-m-\alpha}{1-\frac{m}{2}}\right) \frac{(1-\alpha-\frac{m}{2})_k(2-m-\alpha)_k}{k!(2-\frac{m}{2})_k} \tag{43}$$

$$c_{2k} = \frac{(1-\alpha-\frac{m}{2})_k(1-m-\alpha)_k}{k!(1-\frac{m}{2})_k} \tag{44}$$

Using the definition of the hypergeometric function $F(a,b,c;t)$, defined as

$$F(a,b,c;t) = \sum_{k=0}^{\infty} \frac{(a)_k(b)_k}{(c)_k k!} t^k,$$

we see that $E(\mathbf{x})$ can be written as follows:

$$E(\mathbf{x}) = \frac{\mathbf{x}}{A_m|\mathbf{x}|^m}\left[\sum_{k=0}^{\infty} c_{2k}r^{2k} + \mathbf{x}\sum_{k=0}^{\infty} c_{2k+1}r^{2k}\right] \tag{45}$$

$$= \frac{\mathbf{x}}{A_m|\mathbf{x}|^m} M(\alpha, 1-m, \mathbf{x}) \tag{46}$$

with

$$M(\alpha,1-m,\mathbf{x}) = M_1(\alpha,1-m,|\mathbf{x}|^2) + \mathbf{x}\left(\frac{1-m-\alpha}{1-\frac{m}{2}}\right)\times$$
$$\times M_2(\alpha,1-m,|\mathbf{x}|^2) \quad (47)$$

and

$$\begin{cases} M_1(\alpha,1-m,|\mathbf{x}|^2) = F(1-\alpha-\frac{m}{2},1-\alpha-m,1-\frac{m}{2};|\mathbf{x}|^2) \\ M_2(\alpha,1-m,|\mathbf{x}|^2) = F(1-\alpha-\frac{m}{2},2-\alpha-m,2-\frac{m}{2};|\mathbf{x}|^2) \end{cases} \quad (48)$$

Actually, this formulae illustrate that the fundamental solution of the hyperbolic Dirac-equation is nothing but a modulated version of the classical fundamental solution (i.e. $E(\mathbf{x})$ satisfying $\partial_{\mathbf{x}}E(\mathbf{x}) = 0$).

The time has now come to make a few remarks:

1 Though Frobenius' method leads to a solution for equation (25), this doesn't necessarily mean that this solution is well-defined. Everything depends on the presence of the two hypergeometric functions in the modulation factor $M(\alpha,1-m,\mathbf{x})$. In general, a hypergeometric function can be ill-defined if one of the parameters in the denominator is a negative integer. In our case, this means that expression (46) may be ill-defined if

$$1-\frac{m}{2} \in -\mathbb{N} \ \vee \ 2-\frac{m}{2} \in -\mathbb{N}.$$

As the latter condition is incorprated in the first one, this means that the fundamental solution $E(\mathbf{x})$ may be ill-defined if $m \in 2\mathbb{N}_0$. This is the reason why we restricted ourselves to the case of an odd dimension.

2 Now that we have found a fundamental solution for equation (25), it is easy to give the solution for our original problem, solving equation (2). First of all, a solution for equation (22) is constructed, using the fact that

$$\partial_{\mathbf{x}}(\mathbf{x}+\epsilon) + (m+2\alpha))E(\mathbf{x}) = 0 \Leftrightarrow$$
$$\partial_{\mathbf{x}}^{(\epsilon)}(\mathbf{x}^{(\epsilon)}-1)E(\mathbf{x}^{(\epsilon)}) = (m+2\alpha)E(\mathbf{x}^{(\epsilon)}) \ .$$

Indeed, reintroducing ϵ in expression (46) gives us the fundamental solution for equation (22):

$$E(\mathbf{x}) = \frac{\mathbf{x}\epsilon}{A_m|\mathbf{x}|^m} M(\alpha, 1-m, \mathbf{x}\epsilon) \tag{49}$$

$$= \frac{\mathbf{x}\epsilon}{A_m|\mathbf{x}|^m} \left[M_1(\alpha, 1-m, |\mathbf{x}|^2) + \mathbf{x}\epsilon \left(\frac{1-m-\alpha}{1-\frac{m}{2}} \right) \times \right.$$

$$\left. \times M_2(\alpha, 1-m, |\mathbf{x}|^2) \right], \tag{50}$$

with $M_1(\alpha, 1-m, |\mathbf{x}|^2)$ and $M_2(\alpha, 1-m, |\mathbf{x}|^2)$ as in (50):

$$\begin{cases} M_1(\alpha, 1-m, |\mathbf{x}|^2) = F(1-\alpha-\frac{m}{2}, 1-\alpha-m, 1-\frac{m}{2}; |\mathbf{x}|^2) \\ M_2(\alpha, 1-m, |\mathbf{x}|^2) = F(1-\alpha-\frac{m}{2}, 2-\alpha-m, 2-\frac{m}{2}; |\mathbf{x}|^2) \end{cases}$$

Using expression (20) as our definition for $E(\boldsymbol{X}, T)$:

$$E(\boldsymbol{X}, T) := \lambda^\alpha E(\mathbf{x}), \quad \boldsymbol{X} + \epsilon T = \lambda(\mathbf{x} + \epsilon\frac{1+|\mathbf{x}|^2}{2}),$$

we have now constructed the solution for the hyperbolic Dirac-equation:

$$E(\boldsymbol{X}, T) := \left(T + \sqrt{T^2 - |\boldsymbol{X}|^2}\right)^\alpha E(\mathbf{x}), \quad \mathbf{x} = \frac{\boldsymbol{X}}{T + \sqrt{T^2 - |\boldsymbol{X}|^2}} \tag{51}$$

with $E(\mathbf{x})$ as in (49).

3 Now that we have found the fundamental solution $E(\boldsymbol{X}, T)$ of the hyperbolic Dirac-equation, we can look whether it converges. Because $E(\boldsymbol{X}, T)$ is nothing but the product of the classical fundamental solution and a modulation factor, and the latter is a sum of two hypergeometric functions, it is worthwhile to investigate the convergence of a hypergeometric function. In general, a hypergeometric function $F(a, b, c; z)$ converges in the unit disc $|z| < 1$. This means that both $M_1(\alpha, 1-m, |\mathbf{x}|^2)$ and $M_2(\alpha, 1-m, |\mathbf{x}|^2)$ will converge for all $\mathbf{x}$ in the unit ball $B(1)$. But as

$$\mathbf{x} = \frac{\boldsymbol{X}}{\lambda} = \frac{\boldsymbol{X}}{T + \sqrt{T^2 - |\boldsymbol{X}|^2}},$$

$E(\boldsymbol{X}, T)$ converges if

$$\frac{|\boldsymbol{X}|}{T + \sqrt{T^2 - |\boldsymbol{X}|^2}} < 1 \quad \Leftrightarrow \quad |\boldsymbol{X}| < T.$$

In other words, as long as one is considering points $\boldsymbol{X} + \epsilon T$ inside the future cone, $E(\boldsymbol{X}, T)$ converges. This is of course something we expected, because for every $\boldsymbol{X} + \epsilon T$ inside the future cone there is a unique point $\mathbf{x} + \epsilon\frac{1+r^2}{2}$ with $\mathbf{x} \in B(1)$ such that $\boldsymbol{X} + \epsilon T = \lambda(\mathbf{x} + \epsilon\frac{1+r^2}{2})$.

References

[1] Aulaskari, R. and Lappan, P. - Criteria for an analytic function to be Bloch and a harmonic or meromorphic function to be normal, *Complex analysis and its applications*, Pitman Research Notes in Mathematics 305, Longman Scientific & Technical (1994), pp. 136-146, Harlow.

[2] Brackx, F., Delanghe, R., Sommen, F. - *Clifford Analysis*, Research Notes in Mathematics No. 76, Pitman, London, 1982.

[3] Cerejeiras, P., Kähler, U., and Sommen, F. - Clifford Analysis on Projective Hyperbolic Space, to appear in: *Journal of Natural Geometry*

[4] Cerejeiras, P., Kähler, U., and Sommen, F. - *Clifford Analysis on Projective Hyperbolic Space II*

[5] Cerejeiras, P. and Cnops, J. - Hodge Dirac Operators for Hyperbolic Spaces, *Complex Variables*, 2000, **41**, pp. 267-278.

[6] Cnops, J. - *Hurwitz pairs and applications of Moebius transformations*, Habilitation Thesis - Ghent, Belgium, 1994.

[7] Cnops, J. and Delanghe, R. - Moebius invariant spaces in the unit ball, *Appl. Anal.* **73**, n.o 1-2, pp. 45-46, 1999.

[8] Delanghe, R., Sommen, F., and Souček, V. - *Clifford algebras and spinorvalued functions*, Kluwer Academic Publishers, 1992.

[9] Eriksson-Bique, S.-L. and Leutwiler, H. - Hypermonogenic Functions, **in:** *Clifford algebras and their applications in mathematical physics - Volume 2: Clifford analysis*, **ed.** Ryan, John et al., Birkhäuser. Prog. Phys. **19**, 287-302 (2000).

[10] Gel'fand, I.M., Minlos, R.A., and Shapiro, Z. Ya. - *Representations of the rotation and Lorentz groups and their applications*, Pergamon Press, 1963.

[11] Gürlebeck, K., Kähler, U., Shapiro, M.V. and Tovar, L.M. - On Q_p-spaces of quaternionic-valued functions, *Complex Variables, Theory Appl.* **39**, n.o 2, pp. 115-135, 1999.

[12] Hempfling, T. - The Dirac operator in $\mathbb{R}_+^{d+1}$ with hyperbolic metric and modified Clifford analysis, **in:** *Dirac operators in analysis*, ed.

Ryan, John et al., Pitman Res. Notes Math. Ser. **394**, pp. 95-108, 1998.

[13] Hestenes, D. and Sobczyk, G. - *Clifford Algebra to Geometric Calculus*, Fundamental Theories in Physics, **vol. 5**, D. Reidel, Dordrecht, 1984.

[14] Hochstadt, H. - *Les Fonctions de la Physique Mathématique*, Masson et Cie Editeurs, 1973.

[15] Leutwiler, H. - Modified quaternionic analysis in $\mathbb{R}^3$, *Complex Variables*, 1991, vol. 20, pp. 19-51.

[16] Loo-Keng Hua - *Starting with the unit circle*, Springer-Verlag, New-York Inc., 1981.

[17] Maass, H. - Automorphe Funktionen von mehreren Veränderlichen und Dirichletsche Reihen, *Abh. Math. Sem. Univ. Hamburg*, **16**, 1949, pp. 72-100.

25

INHOMOGENEOUS ORDINARY DIFFERENTIAL EQUATIONS AND LOCAL COHOMOLOGIES AND RESIDUES

S. Tajima
Dept. of Information Engineering
Faculty of Engineering, Niigata University
8050 Ikarashi Niigata, Niigata 950-2181 Japan
E-mail: tajima@geb.ge.niigata-u.ac.jp

1. Introduction

Let $P = P(z, d/dz)$ be a linear ordinary differential operator with holomorphic coefficients in a neighbourgood X of the origin 0 in $\mathbb{C}$. Let $Pu = f$ be an inhomogeneous differential equation in the space $\mathcal{O}_{\{0\}}$ or in $\hat{\mathcal{O}}_{\{0\}}$, where $\mathcal{O}_{\{0\}}$ is the ring of convergent power series and $\hat{\mathcal{O}}_{\{0\}}$ is the ring of formal power series at the origin 0.

In this paper, we consider inhomogeneous equations in the context of algebraic analysis. We describe some recent results [9], [12] concerning the local solvability condition to inhomogeneous differential equations. In section one, we show that the local solvability condition to an inhomogeneous equation can be expressed in terms of residues. A basic ingredient of the argument is the concept of local cohomologies. In section two, we describe a method for computing the local solvability condition to $Pu = f$ in the space $\hat{\mathcal{O}}_{\{0\}}$. In this method, we exploit certain kind of regular singular system of ordinary differential equations. We also show that the notion of standard basis or Gröbner basis in the ring of ordinary differential operators is effectively applicable to solve the problem. In section three, we consider the local solvability condition from an algorithmic point of views. We give an example of computation for illustration.

2. Local cohomologies and residues

Let $\mathcal{O}_X$ be the sheaf on $X \subset \mathbb{C}$ of germs of holomorphic functions. Let $\mathcal{H}^n_{\{0\}}(\mathcal{O}_X)$ be the sheaf of local cohomologies with support at the origin, $\mathcal{H}^n_{[0]}(\mathcal{O}_X)$ the sheaf of algebraic local cohomologies. The stalk of $\mathcal{H}^1_{\{0\}}(\mathcal{O}_X)$ at the origin (rest. $\mathcal{H}^1_{[0]}(\mathcal{O}_X)$) has the structure of FS-(resp. DFS-)topological vector space. Furthermore, $\mathcal{O}_{\{0\}}$ and $\Omega \otimes_{\mathcal{O}_X} \mathcal{H}^1_{\{0\}}(\mathcal{O}_X)$ (resp. $\hat{\mathcal{O}}_{\{0\}}$ and $\Omega \otimes_{\mathcal{O}_X} \mathcal{H}^1_{[0]}(\mathcal{O}_X)$) are mutually dual vector spaces where Ω_X is the sheaf of holomorphic differential forms.

Let $P^* = P^*(z, d/dz)$ be the formal adjoint of the linear differential operator $P = P(z, d/dz)$. We denote by $\mathrm{Ker}(P^*, \mathcal{H}^1_{\{0\}}(\mathcal{O}_X))$ and by $\mathrm{Ker}(P^*, \mathcal{H}^1_{[0]}(\mathcal{O}_X))$ the kernel spaces of the complex

$$P^* : \mathcal{H}^1_{\{0\}}(\mathcal{O}_X) \to \mathcal{H}^1_{\{0\}}(\mathcal{O}_X)$$

and that of the complex

$$P^* : \mathcal{H}^1_{[0]}(\mathcal{O}_X) \to \mathcal{H}^1_{[0]}(\mathcal{O}_X),$$

respectively.

Let $\mathrm{Res}\langle\, , \rangle$ denote the following residue pairings:

$$\mathrm{Res}\langle\, , \rangle : \mathcal{O}_{\{0\}} \times \mathcal{H}^1_{\{0\}}(\mathcal{O}_X) \to \mathbb{C},$$
$$\mathrm{Res}\langle\, , \rangle : \hat{\mathcal{O}}_{\{0\}} \times \mathcal{H}^1_{[0]}(\mathcal{O}_X) \to \mathbb{C}.$$

In the above pairing, we identify $\Omega_X \otimes \mathcal{H}^1_{\{0\}}(\mathcal{O}_X)$ (resp. $\Omega_X \otimes \mathcal{H}^1_{[0]}(\mathcal{O}_X)$) with $\mathcal{H}^1_{\{0\}}(\mathcal{O}_X)$ (resp. $\mathcal{H}^1_{[0]}(\mathcal{O}_X)$).

Theorem 1.([9])

(1) *Let f be a germ at 0 of holomorphic functions. Then, the inhomogeneous differential equation $Pu = f$ has a solution u in $\mathcal{O}_{\{0\}}$ if and only if $f \in \mathcal{O}_{\{0\}}$ satisfies the following condition:*

$$\mathrm{Res}\langle f, \sigma\rangle = 0, \ \forall \sigma \in \mathrm{Ker}(P^*, \mathcal{H}^1_{\{0\}}(\mathcal{O}_X)).$$

(2) *Let f be a formal power series at the origin. Then, the inhomogeneous equation $Pu = f$ has a formal solution u in $\hat{\mathcal{O}}_{\{0\}}$ if and only if $f \in \hat{\mathcal{O}}_{\{0\}}$ satisfies the following condition:*

$$\mathrm{Res}\langle f, \sigma\rangle = 0, \ \forall \sigma \in \mathrm{Ker}(P^*, \mathcal{H}^1_{[0]}(\mathcal{O}_X)).$$

Sketch of the proof. Consider the two complexes

$$\begin{array}{ccccccc} 0 \to & \mathcal{O}_{\{0\}} & \xrightarrow{P} & \mathcal{O}_{\{0\}} & \to 0 \\ & \updownarrow & & \updownarrow & \\ 0 \leftarrow & \mathcal{H}^1_{\{0\}}(\mathcal{O}_X) & \xleftarrow{P^*} & \mathcal{H}^1_{\{0\}}(\mathcal{O}_X) & \leftarrow 0 \end{array}$$

which are dual to each other. Standard duality argument yields that $\operatorname{Ker}(P^*, \mathcal{H}^1_{\{0\}}(\mathcal{O}_X))$ is the dual space of $\operatorname{Coker}(P, \mathcal{O}_{\{0\}})$. Since the duality between $\mathcal{O}_{\{0\}}$ and $\mathcal{H}^1_{\{0\}}(\mathcal{O}_X)$ is defined by the residue pairing, we have the following non-degenerate pairing ([2], [3])

$$\operatorname{Res}\langle\, ,\, \rangle : \operatorname{Coker}(P, \mathcal{O}_{\{0\}}) \times \operatorname{Ker}(P^*, \mathcal{H}^1_{\{0\}}(\mathcal{O}_X)) \to \mathbb{C}.$$

Let $f \in \mathcal{O}_{\{0\}}$. Then the inhomogeneous equation $Pu = f$ has a solution $u \in \mathcal{O}_{\{0\}}$, i.e., $f = 0$ in $\operatorname{Coker}(P, \mathcal{O}_{\{0\}})$ if and only if f satisfies the following condition:

$$\operatorname{Res}\langle f, \sigma\rangle = 0, \ \forall \sigma \in \operatorname{Ker}(P^*, \mathcal{H}^1_{\{0\}}(\mathcal{O}_X)),$$

which complete the proof of the statement (1). The same argument yields the statement (2).

Example. Put

$$P = (1 + 2z^3 + z^4)z\frac{d}{dz} - 2 + 8z^3 + 6z^4.$$

The origin 0 is a regular singular point of P. The formal adjoint P^* of P is given by

$$P^* = -(1 + 2z^3 + z^4)z\frac{d}{dz} - 3 + z^4.$$

The meromorphic function $g(z) = \dfrac{1}{z^3} + 2 + z$ is a classical solution of the homogeneous equation $P^*g = 0$. We thus have

$$\operatorname{Ker}(P^*, \mathcal{H}^1_{\{0\}}(\mathcal{O}_X)) = \operatorname{Ker}(P^*, \mathcal{H}^1_{[0]}(\mathcal{O}_X)) = \{c[\frac{1}{z^3}] \mid c \in \mathbb{C}\},$$

where $[1/z^3]$ is the algebraic local cohomology class $1/z^3 \mod \mathcal{O}_X (= \dfrac{1}{z^3} + 2 + z \mod \mathcal{O}_X)$. Therefore, in particular, the local solvability conditions in both spaces (i.e., $\mathcal{O}_{\{0\}}$ and $\hat{\mathcal{O}}_{\{0\}}$) coincide. It is consistent with the regular singularity of the differential operator P. Now let $f = c_0 + c_1 z + c_2 z^2 + \cdots$ be a (convergent or formal) power series. Then, according to Theorem 1,

$$\operatorname{Res}\langle f, [\frac{1}{z^3}]\rangle = c_2 = 0$$

is the local solvability condition to the inhomogeneous equation $Pu = f$.

Example. Put $P = 2z^2\frac{d^2}{dz^2} + z\frac{d}{dz} - 6$. The origin is a regular singular point of P. The homogeneous equation $P^*g = 0$ has two linearly independent classical solutions $\frac{1}{z^3}$ and $z^{\frac{1}{2}}$, where $P^* = 2z^3\frac{d^2}{dz^2} + 7z\frac{d}{dz} - 3$ is the formal adjoint of P. Hence we have

$$\mathrm{Ker}(P^*, \mathcal{H}^1_{\{0\}}(\mathcal{O}_X)) = \mathrm{Ker}(P^*, \mathcal{H}^1_{[0]}(\mathcal{O}_X)) = \{c[\frac{1}{z^3}] \mid c \in \mathbb{C}\}.$$

Example. Let $P = (z + 2z^2)z^2\frac{d^2}{dz^2} - (1 + 6z + 2z^2)z\frac{d}{dz} + 2 + 6z$. The formal adjoint is

$$P^* = (z + 2z^2)z^2\frac{d^2}{dz^2} + (1 + 12z + 18z^2)z\frac{d}{dz} + 3 + 24 + 30z^2.$$

We have

$$\mathrm{Ker}(P^*, \mathcal{H}^1_{\{0\}}(\mathcal{O}_X)) = \mathrm{Span}\{\tau_1, \tau_2\},$$
$$\mathrm{Ker}(P^*, \mathcal{H}^1_{[0]}(\mathcal{O}_X)) = \mathrm{Span}\{\tau_1\},$$

where $\tau_1 = \frac{1}{z^3} \bmod \mathcal{O}_X$ and $\tau_2 = \frac{1}{z^5}e^{\frac{1}{z}} \bmod \mathcal{O}_X$.

Let $f = c_0 + c_1z + c_2z^2 + \cdots$ be a convergent power series. Then the inhomogeneous equation $Pu = f$ has a solution u in $\mathcal{O}_{\{0\}}$ if and only if

$$\mathrm{Res}\langle f, \tau_1\rangle = \mathrm{Res}\langle f, \tau_2\rangle = 0.$$

Likewise, the solvability condition to the inhomogeneous equation $Pu = f$ in the space $\hat{\mathcal{O}}_{\{0\}}$ is given by $\mathrm{Res}\langle f, \tau_1\rangle = 0$.

3. A standard basis method

In this section, we consider the local solvability condition of an inhomogeneous equation in the space $\hat{\mathcal{O}}_{\{0\}}$ of formal power series at the origin.

In order to treat explicitly the algebraic local cohomology solutions of the homogeneous adjoint equation, we introduce a regular singular system of linear ordinary differential equations. By exploiting standard basis or Gröbner basis of the regular singular system in the ring of ordinary differential operators, we derive a method for computing the formal solvability condition.

Let P be a linear differential operator with holomorphic coefficients, P^* the formal adjoint of P. Since the space $\mathrm{Ker}(P^*, \mathcal{H}^1_{[0]}(\mathcal{O}_X))$ consisting of all the algebraic local cohomology solutions to the adjoint equation is a finite dimensional vector space, we have the following:

Lemma 2. *There exists a non-negative integer* $\kappa \in \{0, 1, 2, \dots\}$ *s.t.,* $x^\kappa \sigma = 0$, $\forall \sigma \in \mathrm{Ker}(P^*, \mathcal{H}^1_{[0]}(\mathcal{O}_X))$.

Now we define a left ideal I in the sheaf $\mathcal{D}_X$ of ring of ordinary differential operators to be an ideal generated by P^* and x^κ, i.e., $I = \mathcal{D}_X P^* + \mathcal{D}_X x^\kappa$, where κ is an integer which satisfies the conditions in the above lemma.

Remark.

1 The ideal I do not depend on the choice of κ provided that the integer κ satisfies the condition stated in Lemma 2.

2 $I = \mathcal{D}_X$ holds if and only if $\mathrm{Ker}(P^*, \mathcal{H}^1_{[0]}(\mathcal{O}_X)) = \{0\}$.

3 In the case where the origin is a regular singular point of the operator P (and if the inditial equation of P at the origin have non-negative integer roots), we may put
$\kappa = 1+$the largest non-negative integer root of the inditial equation of P.

The left ideal I enjoys the following properties:

Proposition 3([9]) *Let I be the left ideal in $\mathcal{D}_X$ generated by P^* and x^κ, where κ is a non-negative integers satisfying the following condition*

$$x^\kappa \sigma = 0, \ \forall \sigma \in \mathrm{Ker}(P^*, \mathcal{H}^1_{[0]}(\mathcal{O}_X)).$$

Then we have

(1) *The $\mathcal{D}_X$-Module $\mathcal{D}_X/I$ is a regular singular holonomic system supported at the origin.*

(2) $\mathcal{H}om_{\mathcal{D}_X}(\mathcal{D}_X/I, \mathcal{H}^1_{[0]}(\mathcal{O}_X)) = \mathrm{Ker}(P^*, \mathcal{H}^1_{[0]}(\mathcal{O}_X))$.

Example. Put $P = (1 + 2z^3 + z^4)z\frac{d}{dz} - 2 + 8z^3 + 6z^4$. The indicial equation of P at the origin is $\lambda - 2 = 0$. Let $I = \mathcal{D}_X P^* + \mathcal{D}_X z^3$ where P^* is the formal adjoint of P. We have $I = \mathcal{D}_X(z\frac{d}{dz} + 3) + \mathcal{D}_X z^3$.

Example. Put $P = (1+z)z^2\frac{d^2}{dz^2} + (-2+z)\frac{d}{dz} + 2$. The indicial equation of P at the origin is $(\lambda - 1)(\lambda - 2) = 0$. Let $I = \mathcal{D}_X P^* + \mathcal{D}_X z^3$. Then $I = \mathcal{D}_X(z\frac{d}{dz} + 2) + \mathcal{D}_X z^2$. Note that the homogeneous equation $P^*g = 0$ has two linearly independent classical solutions $\frac{1}{z^2}$ and $\frac{1}{z^2}\log z - \frac{1}{z^3}$, whereas

$$\mathcal{H}om_{\mathcal{D}_X}(\mathcal{D}_X/I, \mathcal{H}^1_{[0]}(\mathcal{O}_X)) = \mathrm{Ker}(P^*, \mathcal{H}^1_{[0]}(\mathcal{O}_X)) = \{c[\frac{1}{z^2}] \mid c \in \mathbb{C}\}.$$

For a linear differential operator $R \in \mathcal{D}_X$ of the form

$$R = a_m(z)\frac{d^m}{dz^m} + a_{m-1}(z)\frac{d^{m-1}}{dz^{m-1}} + \cdots + a_0(z),$$

we define the valuation of R, denoted by $v(P)$, to be the valuation at the origin of the coefficient $a_m(z)$. We have the following:

Lemma 4. *Let $\{R_0, R_1, \ldots, R_\ell\}$ be the standard basis of the ideal $I = \mathcal{D}_X P^* + \mathcal{D}_X x^\kappa$ in the ring of $\mathcal{D}_X$ of differential operators with*

$$0 = \mathrm{order} R_0 < \mathrm{order} R_1 < \cdots < \mathrm{order} R_\ell.$$

Then

$$\kappa \geq v(R_0) > v(R_1) > \cdots > v(R_\ell).$$

By applying certain results of Briançon et Maisonobe [1], we get the following result:

Theorem 5.([12]) *Let $\{R_0, R_1, \ldots, R_\ell\}$ be the standard basis of the ideal $I = \mathcal{D}_X P^* + \mathcal{D}_X x^\kappa$. Then we have*

(1) $I = \mathcal{D}_X R_0 + \mathcal{D}_X R_\ell$

(2) $\dim \mathcal{H}om_{\mathcal{D}_X}(\mathcal{D}_X/I, \mathcal{H}^1_{[0]}(\mathcal{O}_X)) = v(R_\ell)$.

Example. Let P be a second order linear ordinary differential operator of the form

$$(3 - 5z^2 - 4z^7 + 2z^9)z^2\frac{d^2}{dz^2} + (-12 - 10z^2 - 68z^7 + 49z^9)z\frac{d}{dz} + 12+$$

$$+30z^2 - 240z^7 + 240z^9.$$

The largest non-negative root of the indicial equation $\lambda^2 - 5\lambda + 4 = (\lambda-1)(\lambda-4) = 0$ is equal to 4. We set $I = \mathcal{D}_X P^* + \mathcal{D}_X z^5$. The standard basis of the ideal I is $\{R_0, R_1, R_2\}$ where

$$R_0 = z^5,\ R_1 = z^3\frac{d}{dz} + 5z^2,\ R_2 = z^2\frac{d^2}{dz^2} + 8z\frac{d}{dz} + 10.$$

We have $\mathcal{H}om_{\mathcal{D}_X}(\mathcal{D}_X/I, \mathcal{H}^1_{[0]}(\mathcal{O}_X)) = \mathrm{Span}\{\tau_1, \tau_2\}$ where $\tau_1 = \dfrac{1}{z^2} \bmod \mathcal{O}_X$ and $\tau_2 = \dfrac{1}{z^5} \bmod \mathcal{O}_X$.

4. An algorithm in the Weyl algebra

In [12], we have constructed a rational algorithm that compute the formal solvability condition to the inhomogeneous differential equation. In this section, we give an outline of the algorithm and show some examples. A key ingredient is the notatin of the Noetherian operator representation of an algebraic local cohomology class (see [11], [13]).

Let D_X denote the Weyl algebra $K[z, d/dz]$ where $K = \mathbb{Q}$. Let $P = P(z, d/dz) \in D_X$ be a linear differential operator with polynomial coefficients of the form

$$P = \sum_{j=0}^{m} p_j(z)\frac{d^j}{dz^j},\ p_j(z) \in K[z],$$

where $p_m(z) = b_1(z)^{\ell_1} b_2(z)^{\ell_2} \cdots b_n(z)^{\ell_n}$, $b_i(z) \in K[z]$ is an irreducible polynomial, $i = 1, 2, \ldots, n$. Let $H^1_{[B_i]}(\mathcal{O}_X)$ be the space of algebraic local cohomology classes supported on $B_i = \{z \in \mathbb{C} \mid b_i(z) = 0\}$. The following algorithm computes the formal solvability condition of the inhomogeneous equation $Pu = f$ at each point $\beta \in B_i$.

Input

- $P = \sum_{j=0}^{m} p_j(z)\frac{d^j}{dz^j} \in D_X$
- $b_i(z) \in K[z]$: an irreducible factor of the coefficient $p_m(z)$.

1 Find a non-negative integer κ_i s.t.,

$$b_i(z)^{\kappa_i}\sigma = 0,\ \forall \sigma \in \mathrm{Ker}(P^*, H^1_{[B_i]}(\mathcal{O}_X)).$$

2 Put $I = D_X P^* + D_X b_i(z)^{\kappa_i}$.

3 Compute the Gröbner basis $\mathrm{Gr}I = \{R_{i,0}, R_{i,1}, \ldots, R_{i,n_i}\}$ of the ideal I with $\mathrm{order}R_{i,j} < \mathrm{order}R_{i,j+1}$, $j = 0, 1, \ldots, n_i - 1$.

4 Put $d_i = v(R_{i,n_i})$.

5 Solve the indicial equation of R_{i,n_i} and determine the order of pole of the linearly independent algebraic local cohomology solution τ_k of the system D_X/I with a condition $\mathrm{order}\tau_1 < \mathrm{order}\tau_2 < \cdots < \mathrm{order}\tau_{d_i}$, where $\mathrm{order}\tau$ denotes the order of the pole of τ.

6 Solve the system of homogeneous equation $R_{i,0}\tau_k = R_{i,n_i}\tau_k = 0$, $k = 1, 2, \ldots, d_i$ for the algebraic local cohomology class τ_k in the form (i.e., the Noetherian operator representation)

$$\tau_k = \sum_{j=0}^{\mathrm{order}\tau_k - 1} (-\frac{d}{dz})^j [c_{i,j,k}(z)\frac{b_i'(z)}{b_i(z)}]$$

where $c_{i,j,k}(z) \in K[z]$ with $\deg c_{i,j,k}(z) < \deg b_i(z)$, $c_{i,\mathrm{order}\tau_k - 1,k}(z) = 1$ and $c_{i,\mathrm{order}\tau_{k'} - 1,k}(z) = 0$ for $k' < k$.

Output

- The Noetherian operator representations of the algebraic local cohomology classes

$\tau_1, \tau_2, \ldots, \tau_{d_i}$.

According to the following result, we make use of the Noetherian operator respresentative of an algebraic local cohomology class in an effective manner:

Proposition 6.([8], [11]) *Let $b(z)$ be an irreducible polynomial in $K[z]$,*

$$B = \{z \in \mathbb{C} \mid b(z) = 0\}.$$

Let $\tau \in H^1_{[0]}(\mathcal{O}_X)$ be an algebraic local cohomology class of the form

$$\tau = \sum_{j=0}^{\mathrm{order}\tau - 1} (-\frac{d}{dz})^j [c_j(z)\frac{b'(z)}{b(z)}]$$

where $c_j(z) \in K[z]$, $\deg c_j(z) < \deg b(z)$. Then at each point $\beta \in B$, we have

$$\mathrm{Res}_\beta\langle f, \tau\rangle = \sum_{j=0}^{\mathrm{order}\tau - 1} c_j(\beta)\frac{d^j f}{dz^j}(\beta).$$

Example. Let

$$\begin{aligned} P &= (10z^8 - 7z^7 - 155z^6 - 289z^5 - 45z^4 + 267x^3 - 905z^2 \\ &\quad -2875z - 2625)\frac{d^2}{dz^2} + (150z^7 - 54z^6 - 1586z^5 - 3730z^4 - \\ &\quad -2902z^3 + 3166z^2 + 3890z - 2150)\frac{d}{dz} \\ &\quad +(480z^6 - 36z^5 - 3036z^4 - 8760z^3 - 8808z^2 + 3084z + 4980), \end{aligned}$$

and $b(z) = z^2 + 2z + 3$. Put $I = D_X P^* + D_X b(z)^2$. The Gröbner basis of the ideal I is given by $\{R_0, R_1\}$ where

$$\begin{aligned} R_0 &= (z^2 + 2z + 3)^2, \\ R_1 &= 6(z^2 + 2z + 3)\frac{d}{dz} + 13z^3 + 29z^2 + 69z + 33. \end{aligned}$$

We have $d = 1$. Put $\tau \in H^1_{[0]}(\mathcal{O}_X)$ in the form

$$\tau = c(z)[\frac{b'(z)}{b(z)}] + (-\frac{d}{dz})[\frac{b'(z)}{b(z)}],$$

where $\deg c(z) \leq 1$. By solving the differential equation $R_1\tau = 0$, we find that the Noetherian operator representative of the algebraic local cohomology solution τ is

$$\tau = -\frac{5}{3}z[\frac{b'(z)}{b(z)}] + (-\frac{d}{dz})[\frac{b'(z)}{b(z)}].$$

Thus the solvability condition to the inhomogeneous differential equation $Pu = f$ at the point $\beta \in B = \{z \mid z^2 + 2z + 3 = 0\}$ is given by

$$-\frac{5}{3}\beta f(\beta) + f'(\beta) = 0.$$

References

[1] J.Briançon et Ph. Maisonobe, *Idéaux de germes d'opérateurs différentiels à une variable,* l'Enseignement Math. **30** (1984), 7–38.

[2] M. Kashiwara, *On the maximally overdetermined system of linear differential equations,I,* Publ. RIMS, Kyoto Univ. **10** (1975), 563–579.

[3] M. Kashiwara and T. Kawai, *On the holonomic system of micro-differential operators III,* Publ. RIMS Kyoto. **17** (1981), 813–979.

[4] H. Komatsu, *On the index of differential operators,* J. Fac. Sc. Univ. Tokyo IA **18** (1971), 379–398.

[5] H. Komatsu and T. Kawai, *Boundary values of hyperfunction solutions of linear partial differential equations*, Publ. RIMS, Kyoto Univ. **7** (1971/1972), 95–104.

[6] B. Malgrange, *Sur les points singuliers des équations differentielles*, l'Enseignement Math. **20** (1974), 147–176.

[7] S. Tajima, Y. Nakamura, *Residue calculus with differential operators*, Kyushu J. Math. **54** (2000), 127–138.

[8] S. Tajima, *The Laurent expansion of an algebraic local cohomology class and L.Ehrenpreis's Noetherian operatrs*, RIMS Kôkyûroku, Kyoto **1138** (2000), 87–95.(in Japanese)

[9] S. Tajima, *On solvability conditions to an inhomogenerous ordinary differential equations I*, RIMS Kôkyûroku, Kyoto **1168** (2000), 66–79.(in Japanese)

[10] S. Tajima and Y. Nakamura, *An algorithm for computing the residue of a rational function via $\mathcal{D}$-modules*, Josai Mathematical Monographs, **2** (2000), 149–158.

[11] S. Tajima, *Algebraic analysis of multivariate Hermite interpolation formulas*, Proceedings of the second ISAAC congress(eds. by H.G.W. Begehr, R.P. Gilbert and J. Kajiwara), Kluwer (2000), 829–838.

[12] S. Tajima, *On solvability conditions to an inhomogenerous ordinary differential equatial equations II*, RIMS Kôkyûroku, Kyoto, to appear. (in Japanese)

[13] S. Tajima, *An algorithm for computing the Noetherian operator representations and its applications to constant coefficients holonomic PDE's.* in Proc. Mathtools'2000, St. Petersbourg, to appear.

[14] N. Takayama, *Kan: A system for computation in algebraic analysis* (1991–), (http://www.math.s.kobe-u.ac.jp).

26

INFINITE DIMENSIONAL CR CATEGORY

Shigeru Takeuchi *
Department of Mathematics, Faculty of Education, Gifu University
Yanagido 1-1, Gifu, Japan

Abstract: The purpose of this paper is to give a few remarks on the infinite dimensional linear CR category; to formulate and construct the complexification and prove the existence and the uniqueness. Some basic morphisms and applications are considered. We refer the readers to [?, ?, ?, ?] for more details of the finite dimensional linear CR category as basic ingredients.

Keywords: CR space, subobject, categorical complexification

1. Complex categories

1.1 Linear complex category

First we recall the basics of the linear complex category. The terminologies surjection and injection are used in the set-theoretic sense throughout the present paper unless otherwise specified; the former means onto map and the latter univalent ones. As usual we have the category of complex vector spaces, denoted $\mathcal{LC}$, whose object is a complex vector space V and whose morphism is a complex linear mapping $f : U \to V$. From the real point of view we can consider the real expression $V = (V, J)$, where V is a real vector space with the real endomorphism $J : V \to V$, such that $J^2 = -1$. Then a morphism $f : U \to V$ is by definition, an $\mathbf{R}$ linear mapping which commutes respective J; $Jf = fJ$.

*Partially supported by Grant in Aid for Scientific Research No.12640167 from Ministry of Education and Science, Japan

1.2 Linear CR category

Definition 1. A triplet (V, R, J) is an (infinite dimensional) CR space, if V is a (possibly infinite dimensional) real vector space, $R = R_V$ is a subspace of V (which may be infinite dimensional) and $J = J_V$ is a complex structure tensor on R. Further if we have $R = V$, V is said to be totally complex. The opposite extreme is the self-evident totally real object $V = (V, 0, J_0)$, i.e., $R = 0$ case. We call $R = R_V$ or more precisely the pair (R, J) the CR structure of $V = (V, R, J)$.

We generalize the definition of CR category in finite dimensional case to that in (finite or) infinite dimensional case. An $\mathbf{R}$ -linear mapping $f : V_1 \to V_2$ of a CR space $V_1 = (V_1, R_1, J_1)$ to another one $V_2 = (V_2, R_2, J_2)$ is a *CR morphism* if f preserves the CR structures R_i, i.e., $f(R_1) \subset R_2$ and commutes the complex structures J_i, i.e., $fJ_1 = J_2 f$. Let U, V and W be CR spaces. Then we obtain the category of (finite or infinite dimensional) CR spaces. Henceforth we omit the adjective CR if there is no fear of confusion.

For the reader's convenience some basics in the category theory are briefly reviewed. Denote the set of morphisms of X to Y by $\mathrm{Hom}(X, Y)$. Then a morphism $f : M \to N$ is *epic* or an *epimorphism* if the induced map $f^* : \mathrm{Hom}(N, L) \to \mathrm{Hom}(M, L)$ is an injection for any L. A (categorical) *quotient* of M is a pair (N, f) of an object N and an epimorphism $f : M \to N$. Similarly a morphism $f : M \to N$ is *monic* or a *monomorphism* if the induced map $f_* : \mathrm{Hom}(L, M) \to \mathrm{Hom}(L, N)$ is an injection for any L. A (categorical) subobject (U, f) of an object V is a pair of an object U and a monomorphism $f : U \to V$.

Set theoretic translations of the above two morphisms are given afterward. Introducing some other basic morphims in the next section we formulate the (categorical) complexification of the CR space V and then compare it with the classical geometric one in some typical cases.

2. Basic morphisms

2.1 Identity and isomorphism

Some basic morphisms are defined as follows. A uniquely defined endomorphism $1 = 1_M : M \hookleftarrow$ of an object M is the *identity* if $1g = g$, $h1 = h$ for any g, h if the compositions are defined. In $\mathcal{LCR}$ we see easily that an endomorphism $f : M \hookleftarrow$ is the identity iff $F(f) : M \hookleftarrow$ is the identity in $\mathcal{CR}$, with respect to the forgetful functor $F : \mathcal{LCR} \to \mathcal{CR}$. A morphism $f : M \to N$ is a *coretraction (or retraction)* if there exists a morphism $g : N \to M$ such that $gf = 1_M$ (or such that $fg = 1_N$). A morphism $f : M \to N$, which is at once a retraction and a coretraction is an *isomorphism.*

2.2 Images and coimages

Definition 2. ((co)image) For a morphism , $f : U \to V$ a subobject $g : I \to V$ of V is an *image* of f and denoted $I = \mathrm{Im}(f)$ or $f(U)$ if

(i) f factors through I; $f = gh$ for some $h : U \to I$,

(ii) (universality condition) it is the smallest in the sense if $g' : I' \to V$ is another one with the property (i), there exists a morphism $g'' : I \to I'$ such that g factors through I'.

The image $I = f(U)$ is said to be *epimorphic* if the canonical morphism $h : U \to I$ in the above is an epimorphism. Dualizing all above we have the notion of coimage of f; for $f : U \to V$ a quotient object $g : U \to Q$ of U is a *coimage* of f and denoted $Q = \mathrm{Coim}(f)$ if

(i') f factors through Q; $f = \pi g$ for some $\pi : Q \to V$,

(ii') it is the largest in the sense if $g' : U \to Q'$ is another one with the property (i'), there exists a morphism $g'' : Q' \to Q$ such that g factors through Q'.

Remark 4. (Uniqueness of (co)image) From the universality condition (ii), or (ii') above we derive the uniqueness of the image I and coimage Q, which we may be justified in denoting Im (f) (or Coim (f)). Hence the uniqueness of h (or π), too.

Then we have

Theorem 1. *The category $\mathcal{LCR}$ has epimorphic images.*

Proof. (1) Let $f : U \to V$ be any morphism. Then as a candidate of the image we take $I = (I, R_I, J_I) := (f(U), f(R_U), J_V|f(R_U))$, where $g : I \to V$ is the set-theoretic canonical inclusion $I \hookrightarrow V$, $h : U \to I$ is to be naturally induced from f itself. We have to show these are well defined object and morphisms of $\mathcal{CR}$ with the required property. First we have the following canonical commutative diagram:

$$\begin{array}{ccccc} 0 & & 0 & & 0 \\ \downarrow & & \downarrow & & \downarrow \\ R_U & \stackrel{h}{\to} & R_I & \stackrel{g}{\to} & R_V \\ \downarrow j_U & & \downarrow j_I & & \downarrow j_V \\ U & \stackrel{h}{\to} & I & \stackrel{g}{\to} & V \end{array}$$

whose columns are exact. The commutativity is shown by the standard arguments. The crucial point is the exactness of the central column, which follows from the commutativity of the diagram below:

$$\begin{array}{ccccc} R_U & \overset{h}{\to} & R_I & \overset{g}{\to} & R_V \\ \downarrow J_U & & \downarrow J_I & & \downarrow J_V \\ R_U & \overset{h}{\to} & R_I & \overset{g}{\to} & R_V \end{array}$$

(2) Now we are to proceed to the next step, i.e. the proof of the universality condition (ii). Let $g' : I' \to V$ be another one with the above property (i). Since we can assume $f(U)$ to be the image of f in $\mathcal{R}$, and g' to be a morphism in $\mathcal{R}$, we have the **R** linear mapping $g'' : I \to I'$ such that the diagram below commutes:

$$\begin{array}{ccc} I & \overset{g''}{\longrightarrow} & I' \\ g \searrow & & \swarrow g' \\ & V & \end{array}$$

There remains to prove g'' defines a morphism , which is a routine work hence omitted.

(3) The last step is to prove $h : U \to I$ is epimorphic, which is immediate from the definition. Q.E.D.

2.3 Monomorphisms and subobjects

Now we can characterize monomorphisms of $\mathcal{LCR}$ in set-theoretic terms. Let $f : U \to V$ be a CR morphism and $I = (f(U), g)$ be the categorical image of f, with the canonical inclusion $g : f(U) \hookrightarrow V$. As was remarked in [1], g is monic and hence

Theorem 2. *In $\mathcal{LCR}$ the categorical image I is a subobject of V.*

Remark 5. It should be noted that a subset V in a CR space $U = (U, R_U, J_U)$ may carry various CR structures (R_V, J_V) with which V is considered a CR subobject $j : V \hookrightarrow U$ of U, where j is the canonical inclusion. Then the set CR(V) of a subobject of U is partially ordered with respect to the canonical inclusion order. Thus the set-theoretic identity 1_V defines various possible morphism $\iota_{1,2} : V_1 = (V, R_1, J_1) \to V_2 = (V, R_2, J_2)$ according to this partial order.

2.4 Epimorphisms

Here we shall characterize the epimorphisms in set-theoretic terms. Let $f : U \to V$ be a CR morphism and $I = (f(U), g)$ be the image of f with the canonical inclusion $g : f(U) \hookrightarrow V$.

Definition 3. (generic surjection [1]) A morphism $f : U \to V$ in the category $\mathcal{LCR}$ is *generically surjective* if the set-theoretic image $f(U)$

of U is generic in V in the sense of [1]; $f(U)$ generates a set-theoretic CR subspace which is equal to the target space $V = (V, R_V, J_V)$ itself, or equivalently V is the smallest CR subspace which contains the set theoretic image $f(U)$. Henceforth we call this operation the *CR closure.*

Then we have the categorical translation of generic surjectivity as follows:

Theorem 3. (c.f. [1]) *In the category $\mathcal{LCR}$ a morphism is epic if it is surjective, and iff it is generically surjective.*

3. CR tensor product

We will consider the tensor product of CR spaces. Let U, V and W be CR spaces. An $\mathbf{R}$-bilinear mapping $f : U \times V \to W$ is a *CR bilinear morphism* if f induces the CR morphisms to W for each fixed element of the respective factors U, V. Before introducing the CR tensor product, we prepare a few notational conventions and a lemma. Considering U, V and W as $\mathbf{R}$-spaces, we have the usual tensor product $T_R = U \otimes_R V$ and the canonical $\mathbf{R}$-bilinear mapping $\iota' : U \times V \to T_R$, such that there exists an $\mathbf{R}$-linear morphism $\phi' : T_R \to W$ for any $\mathbf{R}$-bilinear mapping $g : U \times V \to W$ satisfying $g = \phi' \circ \iota'$. Then take a subspace $I = I(U, V)$:

$$I = \langle (J_U u) \otimes v - u \otimes (J_V v); u \in R_U, v \in R_V \rangle_{\mathbf{R}}$$

of T_R. We put $K_g := \{(u, v) \in U \times V; g(u, v) = 0\}$, the *zero* (locus) of g. Now under these notations we can prove

Lemma 1. $\quad I \subset \langle \iota'(K_g) \rangle = \ker \phi'$.

Theorem 4. ([3]) Let U, V be CR spaces . Then there exist a CR space T and a CR bilinear morphism $\iota : U \times V \to T$ satisfying the following condition:

[T]: For any CR space W and a CR bilinear morphism $g : U \times V \to W$, there exists uniquely a morphism $\phi : T \to W$ such that $\iota \circ \phi = g$.

The pair (ι, T) is determined uniquely up to CR isomorphisms; if there exists another space T' and a bilinear morphism $\iota' : U \times V \to T'$ with the above property, then there exists uniquely an isomorphism $\varphi : T \to T'$ such that $\iota \circ \varphi = \iota'$.

Proof of Theorem. (Existence): Let us take a subspace

$$R_R := \langle R_U \otimes V + U \otimes R_V \rangle_{\mathbf{R}}$$

of T_R which contains I. Then We have quotient spaces $T := T_R / I \supset R_T := R_R / I$. We define a complex structure $J = J_T$ on R_T as follows.

For a semi-primitive element $z = x \otimes v + u \otimes y \in R_T$ define $J(z)$ by

$$J(z) = J_U x \otimes v + u \otimes J_V y, \quad x \in R_U, v \in V, u \in U, y \in R_V.$$

In order to see J is a well defined complex structure on R_T, first for primitive elements $x \otimes v, u \otimes y$ we have, if $v = y, x = u$, the well-definedness of J_T:

$$J_T(x \otimes v) = J_U x \otimes v = u \otimes J_V y = J_T(u \otimes y).$$

Second we have

$$J^2(z) = J(J_U x \otimes v + u \otimes J_V y) = J_U^2 x \otimes v + u \otimes J_V^2 y = -z.$$

Thus we obtain the CR space $T = (T, R_T, J_T)$. Now we shall verify the universality condition (T). Let $W = (W, R_W, J_W)$ be an arbitrary CR space and $g : U \times V \rightarrow W$ be a CR bilinear morphism. For a primitive element $z = u \otimes_{CR} v \in T$ we define a CR morphism $\phi : T \rightarrow W$ by $\phi(z) = g(u, v)$. For an arbitrary $z \in T$ we could define ϕ by bilinearity. So we have only to examine (i) the well definedness of ϕ and the CR-ness of (ii) ϕ and (iii) ι.

(i) First take two representatives $\sum_{i,j} u_i \otimes v_j$, $\sum_{k,l} u'_k \otimes v'_l \in U \otimes_R V$ of a general element $t \in T$. Then we have $\sum_{i,j} u_i \otimes v_j - \sum_{k,l} u'_k \otimes v'_l \in I = I(U, V)$. Then applying the above lemma, we have

$$\begin{aligned}
\sum_{i,j} g(u_i \otimes v_j) - \sum_{k,l} g(u'_k \otimes v'_l) &= \sum \phi' \iota'(u_i, v_j) - \sum \phi' \iota'(u'_k, v'_l) \\
&= \sum \phi'(u_i \otimes v_j) - \sum \phi'(u'_k \otimes v'_l) \\
&= \phi'\left(\sum u_i \otimes v_j - \sum u'_k \otimes v'_l\right) \\
&= 0,
\end{aligned}$$

since we have the commutative diagram as follows:

$$\begin{array}{ccccccc}
K_g & \hookrightarrow & U \times V & \overset{\iota'}{\rightarrow} & T_R & \hookleftarrow & I \\
& \searrow & g \downarrow & \swarrow \iota & \downarrow \pi & \swarrow & \\
& & W & \overset{\phi}{\leftarrow} & T & &
\end{array}$$

(ii) As for the CR-bilinearity of ϕ, first we mention that ϕ is defined so as to satisfy the commutativity relation $\phi' = \phi \circ \pi$, $\pi : T_R \rightarrow T$ denoting the canonical projection. This is justified since we have $\ker \phi' \supset \ker \pi = I$ as the above lemma shows. Then for a primitive element $u \otimes v \in R_T$, we have $u \in R_U$(or $v \in R_V$), $\phi(u \otimes v) = g(u, v) \in g_v(R_U)$ (or $g_u(R_V)$) $\subset R_W$

and $J_W\phi(u\otimes v) = J_W g(u,v) = g(J_U u, v)$ (or $= g(u, J_V v)) = \phi(J_T(u\otimes v))$, respectively.

(iii) To prove the CR-bilinearity of ι, fix $v\in V$. Then we have only to show $\iota_v : U\to T$ is a CR-morphism, since the condition is symmetric w.r.t. U, V. In fact for any $u\in R_U$ we have $\iota_v(u) = u\otimes v\in R_U\otimes_{CR} V\subset R_T$, and $J_T(\iota_v(u)) = (Ju)\otimes v = \iota_v(Ju)$. Since ι is decomposed as $\iota = \pi\circ\iota'$ and the respective factors satisfy the **R**-(bi)linearity, so does ι.

(iv) The uniqueness of the morphism $\phi : T\to W$, follows, as is usual, from the universality condition (T). In fact, if both (ι, T) and (ι', T') satisfy the condition, there exist morphisms $\phi : T\to T'$, and $\psi : T'\to T$ such that $\iota\circ\phi = \iota'$ and $\iota'\circ\psi = \iota$. Then we have $\iota\circ\phi\circ\psi = \iota$ and $\iota'\circ\psi\circ\phi = \iota'$, which imply $\phi\circ\psi = id_{T'}$ and $\psi\circ\phi = id_T$, respectively. Q.E.D.

Under the above situation we give

Definition 4. The pair (T,ι) of T and the CR bilinear morphism ι or simply the CR space T is the *CR tensor product* of U, V and is denoted $U\otimes_{CR} V$.

Taking (finitely) many objects and generalizing the notion of CR-bilinearity in the selfevident manner, we can formulate *CR multilinear morphism* analogously and obtain the following

Theorem 5. *Let U_i be CR spaces. Then there exists a CR space T and a CR multilinear morphism $\iota : \prod_{i=1}^k U_i\to T$ satisfying the following conditions:*

[T]: *For any CR space W and a CR multilinear morphism $g : \prod_{i=1}^k U_i\to W$, there exists uniquely a morphism $\phi : T\to W$ such that $\iota\circ\phi = g$.*
Furthermore the pair (ι, T) is determined uniquely up to CR isomorphisms; if there exists another space T' and a morphism $\iota' : \prod_{i=1}^k U_i\to T'$ with the above property, then there exists uniquely a CR isomorphism $\varphi : T\to T'$ such that $\iota\circ\varphi = \iota'$.

Definition 5. We call the pair (T,ι), or simply the CR space T the *CR tensor product* of $\{U_i; 1\le i\le k\}$ and denote it by $\otimes_{CR}\{U_i; 1\le i\le k\}$ or simply by $\otimes_{i=1}^k U_i$, if there is no fear of confusion.

4. Functorial Properties of CR Tensor Product

In this section we shall list up some of the basic properties of the CR tensor product as a (covariant) functor of the variables and its relations to the Hom_{CR} functor introduced in our previous paper [1]. First we note CR tensor is a natural extension of the usual tensor product in the category of vector spaces:

Remark 6. The embeddings of the categories of real/complex vector spaces into that of CR vector spaces as subcategories are compatible with the respective tensor products. In other words, totally real(or complex) vector spaces yield one and the same (or isomorphic) tensor product irrespective of their categories.

Second we have the *tensor product of CR morphisms*:

Theorem 6. *Let U_i, V_i, $(i = 1,2)$ be CR spaces and let $f_i : U_i \to V_i$, $(i = 1,2)$ be CR morphisms. Then there exists a CR morphism $f_1 \otimes f_2 : U_1 \otimes_{CR} U_2 \to V_1 \otimes_{CR} V_2$, which we may call the CR tensor product of f_1, f_2; we have the following canonical CR morphism β:*

$$\mathrm{Hom}_{cr}(U_1, V_1) \otimes_{CR} \mathrm{Hom}_{cr}(U_2, V_2) \to \mathrm{Hom}_{cr}(U_1 \otimes_{CR} U_2, V_1 \otimes_{CR} V_2),$$

satisfying

$$\beta(f_1 \otimes f_2)(u_1 \otimes u_2) = f_1(u_1) \otimes f_2(u_2).$$

Proof. In fact from the universality condition (T) of the CR tensor product, we have the commutative diagram as follows:

$$\begin{array}{ccccccccc}
U_1 \times U_2 & \overset{f=(f_1,f_2)}{\rightarrow} & (V_1 \times V_2) & \overset{\iota'_V}{\rightarrow} & (V_1 \otimes_R V_2) \\
\iota'_U \downarrow & \searrow \iota_U & \searrow \iota_V \circ f & \downarrow \iota_V & \downarrow \pi_V \\
U_1 \otimes_R U_2 & \overset{\pi_V}{\rightarrow} & U_1 \otimes_{CR} U_2 & \overset{f_1 \otimes f_2}{\rightarrow} & V_1 \otimes_{CR} V_2,
\end{array}$$

where the composed morphism $\iota_V \circ f$ is the CR bilinear morphism involved which is to induce the required CR morphism, denoted $f_1 \otimes f_2$. We deduce easily the CR-bilinearity of β. Q.E.D.

Remark 7. We note this is not a categorical isomorphism in general as the CR structures of the respective spaces differ in general, which is not the case in the classical abelian categories [2].

5. Complexification

Let $U = (U, R, J)$ be a CR space. Then we give the following

Definition 6. (complexification) A pair (V, f) of a totally complex CR space V and a morphism $f : U \to V$ is a (categorical) complexification of U if it satisfies the following condition

[C]: For any complex object W and a morphism $g : U \to W$, there exists uniquely a morphism $h : V \to W$, such that the following diagram commutes:

$$\begin{array}{ccc} U & \xrightarrow{f} & V \\ & g \searrow \quad \swarrow h & \\ & W & \end{array}$$

Remark 8. From the definition it follows immediately that a complexification is, if it exists, unique up to isomorphisms; if there exists another complexification $f' : U \to V'$, then we have a morphism $h : V \to V'$ and a morphism $h' : V' \to V$ such that $hf = f'$ and $h'f' = f$. Then we have $hh'f' = f'$ and $h'hf = f$. From the uniqueness condition above we have $hh' = 1_{V'}$ and $h'h = 1_V$. Thus h, h' are the (categorical) isomorphisms.

Now we are to state the existence of the categorical complexification.

Theorem 7. *There exists the complexification in $\mathcal{LCR}$.*

Proof. Let us construct the complexification using the tensor product operation in $\mathcal{LCR}$ as is introduced in the previous section(cf [3]). Let $U = (U, R, J)$ be a CR space and let $U^C := U \otimes_{\mathbf{R}} \mathbf{C}$ denote the classical complexification of U in $\mathcal{LR}$. Then we can give a description of the *complexification* U' of U in the category $\mathcal{LCR}$ in terms of the CR tensor product as follows:

$$U' = U \otimes_{CR} \mathbf{C},$$

By definition we have $U' = U^C/A$, where $A = \{u + iJ_U u \in U^C;\ u \in R_U\}$ is the complex form of the CR structure R_U of the CR space U. On the other hand we have $U \otimes_{CR} \mathbf{C} = U \otimes_{\mathbf{R}} \mathbf{C}/I(U, \mathbf{C})$, where

$$\begin{aligned} I &= I(U, \mathbf{C}) \\ &= \langle Ju \otimes z - u \otimes iz; u \in R_U,\ z \in \mathbf{C}\rangle \\ &= \langle v \otimes z + Jv \otimes iz; v \in R_U,\ z \in \mathbf{C}\rangle \\ &= \langle v \otimes (x + iy) + Jv \otimes i(x - iy); v \in R_U,\ x, y \in \mathbf{R}\rangle \\ &= \langle xv + iyv + J(ixv - yv); v \in R_U,\ x, y \in \mathbf{R}\rangle \\ &= \langle \{xv - J(yv)\} + iJ\{xv - J(yv)\}; v \in R_U,\ x, y \in \mathbf{R}\rangle. \end{aligned}$$

Thus we have the equality $I = A$, and hence the required formula $U' = U \otimes_{CR} \mathbf{C}$.For the detailed description of the universality condition, the readers are referred to [3].

Then for CR tensor products we have the following basic formulae similar to the usual ones:

Proposition 1. *(1)* $U \otimes_{CR} V \cong V \otimes_{CR} U$ *(commutative law)*

(2) $(U \oplus V) \otimes_{CR} W \cong (U \otimes_{CR} V) \oplus (V \otimes_{CR} W)$ *(distributive law)*

(3) $(U \otimes_{CR} V) \otimes_{CR} W \cong U \otimes_{CR} V \otimes_{CR} W \cong U \otimes_{CR} (V \otimes_{CR} W)$ *(associative law)*

(4) $(U \otimes_{CR} V)' \cong U \otimes_{CR} V' \cong U' \otimes_{CR} V \cong U' \otimes_{CR} V'$ *(absorption law),*

We state, without proof, the canonical relations between the functors Hom_{CR} and $\otimes_{CR}$ as in the finite dimensional case. Let U, V and W be CR spaces of respective CR structures R_U, R_V, R_W. Then as the arguments in the previous section show, for any CR-bilinear morphism $g : U \times V \to W$ there corresponds uniquely a CR morphism $\phi : U \otimes_{CR} V \to W$, denoted $\alpha(g)$. We can endow CR structures with the following spaces ([?, ?]) in such a way as:

Theorem 8. (cf.[1]) *The canonical morphism*

$$\alpha : L(U, V; W) \to \mathrm{Hom}_{CR}(U \otimes_{CR} V, W)$$

is a CR isomorphism.

6. Concluding remarks

We are now developing a non linear version of the theory of (infinite dimensional) CR category, which is now yet under construction. The linear version is the first step towards the more general one, which should play a fundamental role in describing the infinitesimal behavior on the spaces involved.

References

[1] Iwata,K.,Shiga,K. and Takeuchi,S., *Category of CR vector spaces*, Sc. Rep. Fac. Educ., Gifu Univ.**17**(1993),1-18.

[2] Mitchell,B., *Theory of categories*, Academic Press, New York and London, 1965.

[3] Shiga,K.,Takeuchi,S., *CR tensor product*, Sci. Rep. Fac. Educ., Gifu Univ.**21-2**(1997),13-19.

[4] Takeuchi,S., *Some remarks on the category of C/DR spaces*, Sc. Rep. Fac. Educ., Gifu Univ.**23-2**(1999),15-20.

[5] Takeuchi,S., *Categorical images of CR morphisms*, Sc. Rep. Fac. Educ., Gifu Univ.**25-2**(2001),1-5,

MIX
Papier aus verantwortungsvollen Quellen
Paper from responsible sources
FSC® C105338

If you have any concerns about our products,
you can contact us on
ProductSafety@springernature.com

In case Publisher is established outside the EU,
the EU authorized representative is:
Springer Nature Customer Service Center GmbH
Europaplatz 3, 69115 Heidelberg, Germany

Printed by Libri Plureos GmbH
in Hamburg, Germany